A GLOBALLY INTEGRATED

Edited by Steven Bernstein, Jutta Brunnée, David G. Duff, and Andrew J. Green

Canada has been an engaged participant in global climate change negotiations since the late 1980s. Until recently, Canadian policy seemed to be driven in large part by a desire to join in multilateral efforts to address climate change. The current policy, however, is to seek a 'made in Canada' approach to the issue. Recent government-sponsored analytic efforts as well as the government's own stated policies have been focused almost entirely on domestic regulation and incentives, domestic opportunities for technological responses, domestic costs, domestic carbon markets, and the setting of a domestic carbon price at a level that sends the appropriate marketplace signal to produce needed reductions.

A Globally Integrated Climate Policy for Canada is based on the premise that Canada needs an approach that effectively integrates domestic priorities and global policy imperatives. Leading Canadian and international experts explore policy ideas and options from a range of disciplinary perspectives, including science, law, political science, economics, and sociology. Chapters explore the costs of, opportunities for, or imperatives to participating in international diplomatic initiatives and regimes; the opportunities and impacts of regional or global carbon markets; the proper mix of domestic policy tools; the parameters of Canadian energy policy; and the dynamics that propel or hinder the Canadian policy process.

STEVEN BERNSTEIN is an associate professor in the Department of Political Science at the University of Toronto.

JUTTA BRUNNÉE is a professor and holds the Metcalf Chair in Environmental Law in the Faculty of Law at the University of Toronto.

DAVID G. DUFF is a professor in the Faculty of Law at the University of Toronto.

ANDREW J. GREEN is a professor in the Faculty of Law at the University of Toronto.

A Globally Integrated Climate Policy for Canada

Edited by
Steven Bernstein, Jutta Brunnée,
David G. Duff, and Andrew J. Green

UNIVERSITY OF TORONTO PRESS
Toronto Buffalo London

© University of Toronto Press Incorporated 2008
Toronto Buffalo London

www.utppublishing.com

Printed in Canada

ISBN 978-0-8020-9878-8 (cloth)
ISBN 978-0-8020-9596-1 (paper)

Printed on acid-free paper

Library and Archives Canada Cataloguing in Publication

A globally integrated climate policy for Canada / edited by Steven Bernstein ... [et al.].

ISBN 978-0-8020-9878-8 (bound). – ISBN 978-0-8020-9596-1 (pbk.)

1. Climatic changes – Government policy – Canada. 2. Climatic changes – Government policy. I. Bernstein, Steven.

GE190.C3G56 2008   363.738'740971   C2007-906836-7

University of Toronto Press acknowledges the financial assistance to its publishing program of the Canada Council for the Arts and the Ontario Arts Council.

University of Toronto Press acknowledges the financial support for its publishing activities of the Government of Canada through the Book Publishing Industry Development Program (BPIDP).

Contents

Acknowledgments vii

1 Introduction: A Globally Integrated Climate Policy *for* Canada 3
STEVEN BERNSTEIN, JUTTA BRUNNÉE, DAVID G. DUFF,
AND ANDREW J. GREEN

PART ONE: THE NEED FOR ACTION
2 Positive Feedbacks, Dynamic Ice Sheets, and the Recarbonization of the Global Fuel Supply: The New Sense of Urgency about Global Warming 37
THOMAS HOMER-DIXON

PART TWO: CANADA IN THE WORLD
3 Climate Policy beyond Kyoto: The Perspective of the European Union 57
JUTTA BRUNNÉE AND KELLY LEVIN

4 The Future of U.S. Climate Change Policy 79
DAVID B. HUNTER

5 China and India on Climate Change and Development: A Stance That Is Legitimate but Not Sagacious? 104
LAVANYA RAJAMANI

6 Comment – Across the Divide: The Clash of Cultures in Post-Kyoto Negotiations 128
STEVEN BERNSTEIN

PART THREE: GLOBAL REGIME BUILDING – PARAMETERS AND IMPERATIVES FOR CANADA
7 The Global Regime: Current Status of and *Quo Vadis* for Kyoto 137
MATTHEW J. HOFFMANN

8 Grandfathering, Carbon Intensity, Historical Responsibility, or Contract/Converge? 158
J. TIMMONS ROBERTS AND BRADLEY C. PARKS

9 Global Carbon Trading and Climate Change Mitigation in Canada: Options for the Use of the Kyoto Mechanisms 179
MEINHARD DOELLE

PART FOUR: DOMESTIC POLICY TOOLS – THE RIGHT MIX

10 Renewable Energy under the Kyoto Protocol: The Case for Mixing Instruments 203
DAVID M. DRIESEN

11 A Comparative Evaluation of Different Policies to Promote the Generation of Electricity from Renewable Sources 222
DAVID G. DUFF AND ANDREW J. GREEN

12 Bringing Institutions and Individuals into a Climate Policy for Canada 247
ANDREW J. GREEN

PART FIVE: CANADA'S ENERGY POLICY

13 Climate Change and Canadian Energy Policy 261
MARK S. WINFIELD WITH CLARE DEMERSE AND JOHANNE WHITMORE

14 Integrating Climate Policy and Energy Policy 293
IAN H. ROWLANDS

PART SIX: POLICY OBSTACLES AND OPPORTUNITIES

15 A Proposal for a New Climate Change Treaty System 315
SCOTT BARRETT

16 Climate Change and Global Governance: Which Way Ahead? 323
JOHN DREXHAGE

17 Challenges and Opportunities in Canadian Climate Policy 336
KATHRYN HARRISON

Contributors 343

Acknowledgments

This book is based upon a conference that took place at the University of Toronto on 1–2 November 2007. A dual conference and publication project such as this one could not be realized without the support and assistance of many individuals and institutions.

We turn first to individuals. We thank, first of all, the contributors to this volume, who agreed on very short notice to participate in this project, and whose willingness to submit to its tight timelines made the speedy publication of this volume possible. Wynanda Hofstee, of the University of Toronto Faculty of Law, was instrumental in facilitating efficient communication with our authors, and in making sure that nothing got lost in the shuffle of our dual conference and book preparation tracks. We also owe a huge debt of gratitude to the student researchers who assisted us with the editing and preparation of the articles compiled in this book: Wendy Hicks-Casey, Courtney Hood, Nozomi Smith, and Rob Wakulat. Last, but most certainly not least, we are grateful to Virgil Duff, Richard Ratzlaff, and the entire editorial team at the University of Toronto Press for being willing to produce this volume in record speed, and for giving us first-rate editorial support in the process.

Of course, our collective efforts also required institutional and financial support. We are grateful to the following divisions of the University of Toronto: the Faculty of Law, the Centre for International Studies, the Department of Political Science, the School of Public Policy and Governance, the Department of Economics, the Centre for Environment, Hart House, and, in particular, to the University's Roundtable for the Environment of the University of Toronto and the Provost's Office. In addition, we appreciate the financial support provided by the Law Foundation of Ontario.

A GLOBALLY INTEGRATED CLIMATE POLICY FOR CANADA

1 Introduction: A Globally Integrated Climate Policy *for* Canada

STEVEN BERNSTEIN, JUTTA BRUNNÉE,
DAVID G. DUFF, AND ANDREW J. GREEN

1. Introduction

Canada has been an engaged participant in global climate change negotiations since the late 1980s. During much of this time, Canadian policy reflected a desire to participate and take a leading role in these multilateral efforts. More recently, however, the federal government framed its policy initiatives as seeking a 'made in Canada' approach to the issue. The premise of this volume is that neither multilaterally driven policy making nor an inward-looking policy stance will yield effective options for Canada. Neither the expectations of the Canadian public nor Canada's current or future international commitments are likely to be met by such policies. Canada needs an approach that is sensitive to both domestic priorities and global policy imperatives. It needs a globally integrated climate policy that works *for* Canada.

Internationally, Canada is participating in the United Nations process for developing a global plan to follow the expiry in 2012 of the commitment period set out in the *Kyoto Protocol*.[1] It is as yet unclear where this process is headed. For the time being, the United States and major emitters in the developing world, notably China and India, resist binding post-2012 commitments. Nonetheless, the June 2007 G8 meeting in Germany and a September 2007 meeting of the world's 15 major greenhouse gas emitters that was convened by the U.S. administration demonstrated that the major international players are moving toward globally integrated climate-policy making, whether as part of the United Nations process or otherwise.[2]

Canada must be prepared both to participate effectively in the deliberations that shape the future global climate regime and to implement credible measures at home. Unfortunately, the Canadian government, its pledges of global leadership and domestic commitment to meaningful action notwithstanding, has not yet put forth a convincing national approach to climate change. Domestically, in light of a potential election in the near future, the main federal parties are struggling to develop climate change plans. Provincial governments are similarly searching for policies that meet the expectations of Canadian voters.

The imperative to address climate change at the national and global levels simultaneously has never been clearer or more politically salient. This volume aims to provide concrete ideas and advice for Canada's international and domestic policy making at this critical juncture. It brings together analytical work and commentary from multiple disciplinary perspectives, from Canada and abroad, to inform that policy advice. Climate change poses complex and multi-faceted challenges. Effective policy making requires engagement with the full spectrum of these challenges, relying on the analytical tools that are provided by a range of disciplines, including science, economics, ethics, politics, and law. We hope that this volume can facilitate this engagement. Moreover, policy choices must not only be efficient, effective, and legitimate, but also feasible, fair, and sensitive to regional, national, and international political realities. With these parameters in mind, the contributions compiled in this volume offer policy ideas and options that integrate responses to domestic and international imperatives.

In this introductory chapter, we first canvass the Canadian climate policy trajectory from the early stages of international policy making. We then show that, in 2007, climate policy must measure up to a now undeniable fact: potentially dangerous climate change is imminent, and perhaps even underway already. Against this backdrop we explain why we argue for a globally integrated climate policy *for* Canada, and examine what basic parameters such a policy must consider. Finally, we consider a range of policy options – and obstacles – for Canada and conclude that Canada must take action now in order to set itself on a path to make the long-term changes necessary to address climate change.

2. Canada and Climate Policy – Looking Back

In order to understand the policy options for Canada going forward, it is important to know not only where we currently stand on climate issues but also how we got to the present policy juncture. Canada was initially a significant international presence in creating the climate change regime, contributing early on to diplomacy and science. Not only has it recently backed away from this stance, it has also failed – throughout its involvement in global policy making – to back up its international position with effective domestic policies.

2.1. *Hot and Cold: Canada and the Global Regime*

The Canadian government and individual Canadians, especially from the early 1970s to early 1990s, were at the forefront of global environmental concerns and of sustainable development thinking. Government scientists or senior bureaucrats have led transnational or intergovernmental environmental organizations and activities, including the United Nations Environment Programme (Maurice Strong and later Elizabeth Dowdeswell), the World Climate Research Program (Gordon McBean), and the OECD Environment Directorate (Jim MacNeill), and such leadership had been encouraged within the bureaucracy.[3] Canadians have also led major agenda-setting conferences and exercises. Notable examples include Maurice Strong, who was secretary-general of both the 1972 UN Conference on the Human Environment in Stockholm and the 1992 UN Conference on Environment and Development in Rio de Janeiro, and Jim MacNeill, who was secretary-general of the World Commission on Environment and Development. MacNeill worked closely with the chair (Gro Harlem Brundtland) and commissioners (including Strong) to produce the Brundtland Report,[4] known for developing and popularizing the concept of sustainable development.

Early leadership on climate change science and politics was part of this pattern. For example, Jim Bruce, a scientist and assistant deputy minister at Environment Canada, chaired the October 1985 Villach Conference[5] in Austria, which marked the turning point in the development of a sustained transnational scientific research program on the role of greenhouse gases in climate change, aimed at generating consensus and promoting international political attention. Another Canadian – F. Kenneth Hare – chaired the independent Advisory Group on Greenhouse Gases in July 1986, whose work laid much of the basis for current climate change research and whose advice, like that of the Intergovernmental Panel on Climate Change (IPCC) today, influenced governments towards political action.[6] Canada's international influence reached perhaps its zenith with the June 1988 Toronto Conference on The Changing Atmosphere: Implications for Global Security, directed by Howard Ferguson of Environment Canada and then environment minister Tom McMillan, with strong overall support from the Brian Mulroney government. The main conference recommendation, that governments and industry should reduce CO_2 emissions by 20 per cent from 1988 levels by the year 2005 'as an initial global goal,' became a rallying point for proponents of a global convention, which the conference statement also recommended. It also was a major force in creating political momentum toward the negotiation of the 1992 *United Nations Framework Convention on Climate Change* (UNFCCC).[7]

Canada's framing of climate policy and sustainable development – that action to combat climate change must simultaneously enhance competitiveness, promote economic growth, and facilitate innovation – also found resonance in the final outcome of the UNFCCC negotiations. Among the core principles of the UNFCCC is the recognition of 'the need to maintain strong and sustainable economic growth' (Article 4.2).

Despite this early leadership, after 1992 domestic political factors conspired to make Canada a laggard internationally and in its domestic policy response. These factors included internal divisions within the federal Cabinet and between the federal government and provinces, a lack of attention from the prime minister, and resistance from powerful actors in the energy sector and industry more broadly.[8]

As momentum gathered to negotiate a protocol with legally binding commitments, Prime Minister Jean Chretien, previously unengaged in the issue, forcefully intervened, leading to an abrupt policy change on the eve of the Kyoto meeting. While the reasons for the abrupt shift are a matter of some debate,[9] concerns about Canada's international reputation and direct pressure from Canada's allies – including at the Denver G8 Summit in June 1997 and subsequent pressure from EU and American leaders – played a role. In November, Chretien overrode a decision of federal and provincial ministers of environment and energy to merely stabilize emissions at 1990 levels by 2010, and called initially for a three per cent reduction by 2010. Then, in Kyoto, Canada committed to a six per cent reduction of greenhouse gas (GHG) emissions below 1990 levels, one percentage point less than the U.S. commitment at the time. Canada agreed to this cut despite intensive lobbying from industry opposed to a legally binding agreement and a public campaign run by the Canadian Coal Association, which included advertisements warning of 'economic suicide' if Canada signed a deal in Kyoto.[10]

In its public statements, the government played up Canada's position between the United States and the EU and its desire to 'help find common ground in Kyoto.' This stance reflected an attempt to reinvigorate Canada's role as a facilitator of global agreement and compromise. It has obvious parallels to the diplomatic position now asserted by Prime Minister Stephen Harper.

Following the negotiation of the *Kyoto Protocol*, the Chretien government maintained its firm commitment to the agreement. Further negotiations ensued to flesh out the elements of the Kyoto deal, so as to enable states to decide whether to ratify or not. Canada was active in these negotiations and signalled its intention to implement the protocol, despite the U.S. withdrawal and the concerns that this raised about Canadian competitiveness.[11] During this period, notwithstanding ongoing opposition from a number of domestic interests, Canadian public opinion remained generally supportive of ratification.[12]

The path toward ratification in 2002 had remarkable parallels to the run-up to Canada's commitment to emissions reductions in 1997 and the signing of Kyoto.[13] Again, Canadian domestic policy proved woefully inadequate to shift business-as-usual emission trajectories, undercutting Canada's foreign climate policy leverage as it sought to increase its elbow room in meeting the Kyoto commitments. The government had to bargain hard to gain traction on issues such as maximizing allowable activities under the protocol's two 'carbon sink' articles, initially without having the requisite research or diplomatic leverage to convince other parties about the carbon-storage potential of land use, land use change, and forestry activities.

As it turned out, the U.S. withdrawal from Kyoto served to increase Canadian leverage. Now Canada and virtually all of its major negotiating partners in the 'Umbrella Group,'[14] but Japan and Russia in particular, needed to ratify for the protocol to account for 55 per cent of industrialized country emissions, which Kyoto required in order to come into force. Canada thus got its way on forest sinks. Although it did not succeed in getting credits for natural gas exports to the United States,[15] even that proposal received a more serious hearing than it previously had.[16]

As in 1997, however, progress internationally was not matched by significant policy consensus or policy development at home. Thus, when Chretien surprised nearly everyone and signalled at the 2002 World Summit on Sustainable Development that Canada would ratify Kyoto, an even more hotly contested domestic debate erupted than in 1997. An implementation plan that showed signs of haste and lacked specifics was then put forward with strong opposition from the provinces. After Chretien made it clear that the federal plan would prevail despite provincial concerns, 'negotiations between the provinces and the federal government ground to a halt, effectively terminating the joint National Climate Change Process that the federal government had entered into with the provinces after Kyoto.[17]

After the protocol entered into force, Canada again found itself an outlier in negotiations, with its reputation suffering owing to its poor domestic record. For a brief moment it looked as if the tide had turned in both Canadian diplomacy and domestic climate policy. At the first Meeting of the Parties (MOP 1) of the *Kyoto Protocol* in Montreal in 2005, then environment minister Stéphane Dion, acting as president of the conference, helped move discussions forward on a number of contentious issues, including a process to consider future commitments post-2012. Shortly thereafter, however, the defeat of the Paul Martin government and the new Conservative government's decision to shift to a 'made in Canada' climate policy brought yet another policy reversal. It also put then environment minister Rona Ambrose in the unenviable position of being the chair of the UNFCCC Conference of the Parties in 2006, while representing a government that had made clear it had no intention of adhering to its Kyoto commitments or participating in its international mechanisms.[18]

In 2007, Canadian climate policy appears to have come full circle. The Canadian government is again portraying itself as a potential broker between the EU and U.S. positions on climate change. However, as we discuss in section 4, there are some significant differences. Canada is still without a clear vision of how to reconcile contradictory trends in both its foreign and domestic climate policies and, arguably, without much credibility in this role. But it has also taken the unprecedented step of declaring itself not only unable but also unwilling to meet its commitment under the *Kyoto Protocol*, while simultaneously refusing to withdraw from the treaty.

2.2. *The Policy Gap: The Lack of Domestic Action*

Despite its initial international support for climate change action, the Canadian government has done little domestically to reduce greenhouse gas emissions.[19] Again, it is worth taking a closer look at the policy trajectory.

Prior to the Kyoto negotiations, the federal government emphasized mostly information and education programs along with voluntary measures such as a Voluntary Challenge Registry (VCR) for industries. These programs had little apparent effect on emissions, which continued to rise.[20] Immediately following the Kyoto negotiations, the Chretien government launched a National Climate Change Process, a consultation process conducted in conjunction with provinces. As this process was not reaching consensus, the federal government released another plan, again relying on voluntary action and subsidies. In 2002, following Chretien's decision to ratify the *Kyoto Protocol*, it announced another plan – the *Climate Change Plan for Canada*.[21] This plan also relied heavily on voluntary programs, particularly with industry, as well as subsidies. There was very little in the way of actual regulation, no emission limits, and no taxes. There were minor allowances for the purchase of international credits. Importantly, the government also committed to capping abatement costs that industry would have to pay at $15 per tonne.

The next major policy move came from the government of Prime Minister Martin and his environment minister, Stéphane Dion. In 2005, the government proposed Project Green,[22] which relied more heavily on spending, especially internationally. For large emitters, this plan proposed a cap-and-trade system, though it also reduced the share of the emissions reduction burden that these emitters would have to bear and provided that emissions limits could be satisfied by payments to a Technology Investment Fund. To help reach Canada's Kyoto target, Project Green also included a Carbon Fund to finance the purchase of emission credits domestically and internationally. There was also funding of 'partnership' projects with the provinces and territories aimed at reducing emissions. Finally, Project Green contained a number of voluntary measures such as an agreement with the automobile industry to improve fuel efficiency and the One Tonne Challenge encouraging individuals to reduce emissions in their everyday lives.

Project Green was never implemented. The Harper government, which came into power in early 2006, backed away from international purchases of emission credits, declared that Canada could not meet its Kyoto targets, and proposed a 'made in Canada' solution premised on the view that any regulation of greenhouse gases needed to be attentive to the economic and social realities in Canada.[23] The government also clearly stated that it would not introduce a carbon tax. There was continued spending, such as on a tax credit for transit users, as well as continued reliance, at least in the short term, on voluntary agreements (including with the auto sector). On the regulatory side, the government cancelled the cap-and-trade system for large final emitters, instead proposing intensity targets for industry, energy efficiency standards for some products, and future fuel efficiency standards for automobiles.[24] This plan has been criticized on a number of fronts, including that 'the Conservatives' policies for large final emitters tracked previous Liberal efforts, complete with provisions that might more accurately be described as loopholes.'[25]

At best, therefore, the account of domestic policies is one of ineffectiveness. The federal Commissioner of the Environment and Sustainable Development concluded in her 2006 report (prior to the release of the 'made in Canada' plan) that the federal government's response to climate change has been a story of 'inadequate leadership, planning and performance' which 'lacked foresight and direction and has created confusion and uncertainty for those who are trying to deal with it.'[26]

In its 16 October 2007 Speech from the Throne, the Harper government recognized the need for greater action on climate change – including binding national regulations across all major industrial sectors and a 'carbon emissions trading market that will give business the incentive to run cleaner, greener operations.'[27] Whether the throne speech signals a shift toward more effective policies remains to be seen. However, before we take a closer look at Canada's current stance, it is important to recognize that there has been at least one significant change in the context in which Canada must make its policy choices: in 2007 it has become abundantly clear not only that dangerous, human-induced climate change is occurring, but also that the window for global action is rapidly closing.

3. The Need for Action

The recent award of the Nobel Peace Prize to the Intergovernmental Panel on Climate Change (IPCC) is only the latest signal that the scientific debate has entered a new phase, in which political discussion begins by asking *how*, not whether, to respond to the threat of global climate change. The latest IPCC series of reports (its fourth since being created in 1988) concluded that '[w]arming of the climate system is unequivocal, as is now evident from observations of increases in global average air and ocean temperatures, widespread melting of snow and ice, and rising global average sea level.'[28] Moreover, the rate of increase of greenhouse gases (GHGs) in the industrial era is unprecedented in more than 10,000 years, and is 'very likely' (i.e., greater than 90 per cent certainty) the result of human activity, mostly the burning of fossil fuels.

Natural systems throughout the world are already being affected by climate change. By 2100, scientists expect increases in average surface temperatures of approximately 1.1 to 6.4°C, resulting in rising sea levels, increases in heat waves and hot days, more severe storms, and increases in areas affected by drought, among other changes to the earth's climate. Recent data suggests that the developments since the completion of the IPCC report have outpaced even its worst-case scenarios, and that dramatic changes may be underway. In particular, the global climate may be changing at a faster rate than expected, with already manifest effects, because of numerous feedback effects. For example, increasing temperatures have accelerated the melting of the arctic ice cap, leaving more of the Arctic Ocean exposed. As water absorbs more of the sun's radiation than ice, the water warms up, which, in turn, melts the ice even faster.[29] According to the IPCC, these projected changes pose significant consequences for human health, conflict, food security, and movements of displaced populations.[30]

Many experts and some political actors assert that, to avert dangerous climate change, global temperature increases should not exceed 2°C above pre-industrial levels.[31] While it remains contested what constitutes 'dangerous' or 'tolerable' climate change, there is growing consensus that a stabilization target of a 2°C increase would require a 60–80 per cent cut in emissions from industrialized countries by mid-century, with a similar abatement path for developing countries in later years. Emissions must peak somewhere between 2010 and 2020 to achieve this trajectory.

These general findings have set the goalposts for long-term global action. For example, G8 leaders agreed at their 2007 summit to aim for at least 50 per cent reductions by 2050. Equally important, they agreed to do so together, as part of a UN process.[32] Similarly, the meeting of 15 major emitters in Washington in September 2007 generated a strong consensus that the UNFCCC – of which the *Kyoto Protocol* is a part – should be the central forum for addressing climate change.[33] These statements are welcome, as policies adopted under the auspices of the UNFCCC are likely to have greater legitimacy than selective arrangements among coalitions of the willing. In addition, only a long-term global regime will send the signals that states and economic sectors need to make costly adjustments. The importance of strong signals – and strong action – was underlined yet again by the release of the International Energy Agency's *World Energy Outlook 2007* – projecting that current policies will cause oil and gas imports, coal use, and greenhouse gas emissions to grow dramatically through to 2030, which would threaten energy security and accelerate climate change.[34]

The apparent consensus on long-term goals and the need to address them within a legitimate global process[35] has not, however, been matched by anything close to a consensus on short- to medium-term steps to achieve them. Part of the reason is that, as scientific scepticism has decreased, the economic debate has heated up. In 2006 the UK government sponsored a report on the economics of climate change by former World Bank chief economist Sir Nicholas Stern. The resulting *Stern Review* applied cost-benefit analysis to the issue of climate change and found that delaying action will raise the costs from about one per cent per year of global GDP by 2050 to 5–20 per cent owing to climate-induced damage. It issued a strong call to action as 'the benefits of strong and early action far outweigh the economic costs of not acting.'[36]

The *Stern Review* was greeted with both acclaim and criticism. Much of the criticism focused on the assumptions underlying its conclusions. For example, one of the key criticisms was Stern's choice of a very low 'social discount rate' (a tool that economists use to compare the well-being of people at different times). Arguing that there is no ethical reason to place less value on future generations than current generations, Stern uses a very low discount rate, which considers the well-being of future generations equal to the well-being of those alive today. While this approach is shared by some prominent economists, it has sparked criticism on the basis that it is out of line with both how people actually view choices that have long-term implications and the discount rates typically used by economists. The critics argue that Stern's conclusions are driven by these assumptions and, as a result, weakened by them.[37]

In part, the controversy about the economics of climate change derives from the ethical choices that underlie the economic analysis. Some argue that discount rates are inappropriate or at least unhelpful in the case of climate change. According to Nobel laureate Thomas Schelling, for example, discounting should not be used for climate change as it is more like foreign aid – an ethical issue about the appropriate distribution of costs and benefits across time and different parts of the world.[38]

Given this controversy over the use of tools like the discount rate to make arguments that are essentially about values, our tolerance of risk, and the insurance we are willing to pay, it is not surprising that economics alone has not resolved the issue of what action is required. While the science of climate change points to the need for action, the choice of timing and scope of action by Canadian governments will not be found in a single discipline.

4. Why Canada Needs a Globally Integrated Climate Policy

It is against this general scientific and economic backdrop, together with the trajectory of Canadian policy that we sketched in section 2, that we must consider the political realities and constraints in finding feasible policy responses to the climate challenge. In this section, we consider the legal context in which Canada operates, as well as some key parameters for the globally integrated climate policy that we argue Canada must pursue.

Canada's failure to curb its emissions since it signed and ratified the UNFCCC in 1992 is a case in point. Rather than declining, greenhouse gas emissions in Canada increased by roughly 30 per cent by 2005.[39] While Canada faces particular challenges given its northern geography, relatively cold climate, and significant oil industry, it is hardly unique in facing structural challenges.[40] It does, however, stand apart among industrialized countries for its lack of policy development and implementation dating back to 1992, and, more recently, for backing away from its international legal commitment. Despite this record, the federal government noted in its October 2007 throne speech that, '[t]he world is moving on to address climate change and the environment, and Canada intends to help lead the effort at home and abroad.'[41] Taking this pledge of leadership at face value, we argue that only a more globally integrated climate policy can hope to regain opportunities lost by the policy trajectory of both the current and previous governments, and to chart a way forward.

The first step is to move away from the framing of a more globally integrated policy as a debate only about the *Kyoto Protocol*. A lesson can be learned here even from the United States. Despite the Bush administration's rejection of the *Kyoto Protocol*, it never abandoned an outward-looking policy, if for no other reason than to shift the discourse on global climate policy to one more amenable to its interests.[42] The apparent contradiction stems from a very successful campaign by those opposed to Kyoto to discredit it as unrealistic, unfair, harmful to the U.S. economy, inadequate for the task at hand, a threat to sovereignty, or all of the above. Yet, simultaneously, momentum built at both the state and national level for stronger policies. States like California and Massachusetts picked up where federal policies were lacking or counterproductive. They even engaged at times in their own foreign policy efforts in an attempt to counterbalance the U.S. federal government's recalcitrance.[43]

Meanwhile, the United States created the Asia-Pacific Partnership on Clean Development and Climate. For all its flaws, this initiative brought the world's fastest growing and probably largest future emitters of greenhouse gases, China and India, along with South Korea, into direct engagement with the United States, Japan, and Australia.[44] While the heavy focus on technology and voluntary action through private sector and public–private initiatives has drawn criticism from some quarters, as has the concern it will undermine UN processes in general and the legally binding *Kyoto Protocol* in particular, the initiative recognizes that engagement with major emitters in the North and South is absolutely essential for effective action on what is, after all, a global problem.

There is some irony, then, in the fact that a similarly fought debate in Canada over Kyoto has led to an inward-looking policy, which abandons Canada's long tradition of internationalism in its environmental policy and foreign policy more generally. While Kyoto has not been widely rejected in Canada, as it has in the United States, public opinion on the best approach to international climate policy has shifted considerably. In a 25 September poll on its website, the *Globe and Mail* asked Canadians: 'In terms of global warming, what approach do you favour? Strict adherence to the Kyoto Protocol? The "more flexible" approach suggested by Prime Minister Harper? A compromise between those two positions? No action is needed because global warming is not happening?' Roughly 86,000 respondents were almost equally divided between the first three options, with only 5 per cent opting for 'no action.'[45] Judging from these responses, Canadians are uncertain about climate policy, and only a minority now sees meeting the Kyoto targets as the preferred option. At the same time, polls also show that Canadians view the environment and climate change as their top priority.[46]

Perhaps in light of public opinion, the government has sent mixed signals on Kyoto. It has declared itself committed to global climate action, but not to the Kyoto targets. Further, it has opted to disregard its legal commitment under the treaty, while refusing to formally withdraw.[47] As we suggested in section 2, this rhetorical strategy arguably has been harmful to Canada's international standing at every level of its involvement in the international negotiations. It has also undermined the internal debate that Canada so urgently needs, as the effect has been to sidestep the real policy issue at stake: What can and will Canada do to seriously address the climate change challenge?

The *Kyoto Protocol* has always been only one piece of that puzzle – a piece meant to create momentum, reciprocal commitments to help promote collective action at the international level through its targets and compliance mechanism, and a set of incentives and opportunities to assist countries to reach their targets. It is important to note that the *Kyoto Protocol*, while it imposes a hard national cap on GHG emissions, does not dictate how countries are to achieve their targets. It also does not restrict states, individually or collectively, from moving beyond its mandate. Indeed, Kyoto should be understood as only one in a series of ongoing steps that are needed to achieve the ultimate goal of the UNFCCC to stabilize greenhouse gas emissions at a level that avoids dangerous human-induced climate change.

The *Kyoto Protocol* was negotiated under the UNFCCC, which sets a more general framework for global action. The UNFCCC provides a forum for regular meetings among its 191 parties and procedures for global decision making. It also provides the legal framework for an almost unlimited variety of new commitment tracks. The climate convention itself could be amended to include different sets of commitments for different groups of states. Alternatively, the *Kyoto Protocol* could be amended, or replaced by one or several new protocols. Through these devices, future commitments can be differentiated both between developing and industrialized states, and among developing and industrialized states themselves. In addition, the existing regime stipulates common methodologies for emissions inventories, rigorous reporting requirements, and elaborate performance-evaluation processes. It also contains the foundations for global emissions trading and a compulsory compliance assessment and enforcement procedure. All of these are key ingredients for an effective global approach.[48]

The way is open, then, for Canada to forge its own climate policy within this general global framework. As we noted above, Canada does face many challenges in reducing its greenhouse gas emissions, including its northern climate, its increasing population, a high rate of economic growth, significant regional differences, and, not least, an industrial structure encompassing many industries that currently emit large quantities of greenhouse gases. Its climate policy must take such challenges into account. But these challenges do not change the fact that Canada needs a *globally integrated* climate policy – a policy that reflects and engages with both the evolving global regime and national political, economic, and social realities. In concrete terms, a globally integrated policy has advantages along the four key dimensions that should serve as parameters for policy-making: effectiveness, efficiency, fairness, and political feasibility.

4.1. *Effectiveness*

In order to effectively stabilize and reduce greenhouse gas emissions worldwide, global cooperation is clearly needed. As Scott Barrett has written, addressing climate change is a global public good – a good from which no country can be excluded and the consumption of which by one country does not diminish enjoyment by others.[49] For this reason, it is apt to be underprovided, as countries attempt to 'free-ride' on the efforts of others. Moreover, climate change does not affect all countries equally nor are the costs of addressing climate change equally distributed across all countries and regions. If governments act from self-interest, climate change creates different incentives for action for national and sub-national governments.[50]

A globally integrated climate policy would connect Canada with the efforts of other countries, especially the efforts to build a long-term global regime. A globally integrated approach highlights the need both to contribute to the formation and further development of such a regime and to comply with its requirements. Not surprisingly, the willingness of states to participate in such collective efforts depends crucially on whether other key states are also involved.

Successive Canadian governments have understood this dynamic, which has guided their support for multilateral legal mechanisms and a rule-based order in a wide variety of international issue areas. A stable multilateral order has been good for Canadian prosperity, providing Canada leverage and opportunities that would otherwise be unavailable. It is curious, then, that when it comes to climate change, Canada should abandon the very kind of legal mechanism that has been so successful in facilitating its trade and security. As a middle power with an open economy, and geographically, economically, and politically in the shadow of the United States, Canada benefits immensely from rules that buffer direct U.S. influence while promoting predictability in world economic and political affairs.[51] This policy orientation also enables Canada to play a leadership role disproportionate to its economic and military power. Its environmental foreign policy has offered many such opportunities because leadership can rest on scientific expertise, and domestic financial and institutional support for such efforts, something Canadian governments have traditionally been able and willing to provide.[52]

The damage from Canada's retreat from serious engagement with international processes is already being felt in a number of areas, externally and internally. Externally, this approach has not only undercut Canada's leverage. It has also negatively affected global efforts to combat climate change. Canada's attitude towards its commitments has given greater credibility to rejectionists, undermining Kyoto's legitimacy and lessening the chances of negotiating a successor in time to allow continuity of a legal regime after 2012. Such a gap between Kyoto and any successor would mean uncertainty for businesses hoping to participate in world carbon markets. Arguably, it also undermines attempts to get the big developing-country emitters on board with binding commitments of their own, which is essential for long-term climate stability.

In their defence, Canada's national leaders claim they are now best placed to broker a way forward between the EU's call for stronger binding commitments and a U.S.-led focus on technology and voluntary action.[53] But the failure to make progress on such a way forward for global negotiations at the U.S.-led September meeting of major emitters demonstrated Canada's limited ability to genuinely bridge divides. Indeed, it could be said that this meeting simply reproduced entrenched positions and outcomes consistent with the much weaker Asia-Pacific Climate Partnership.

On one reading, the government's new fondness for the Asia-Pacific Partnership, which it recently joined,[54] might suggest a correction in an internationalist direction. Nonetheless, Canada must now reinvent discontinued programs that fostered international cooperation, learning, and technological innovation, such as the Canada-China Climate Cooperation Project. Canada lost not only the goodwill and leverage such programs might have offered, but also the opportunity to export clean technologies that they facilitated.

Internally, a globally integrated climate policy would also contribute to more effective responses to domestic challenges. The relationship between different countries in producing a global public good has obvious parallels with the relationship between provinces in meeting some form of national target. Canadian policy is potentially constrained by the constitutional division of powers between the federal and provincial governments. Federal powers can overcome a free-riding problem amongst provinces where such powers permit the federal government to take unilateral action. However, the federal government does not have complete power to address all issues related to climate change. Canadian climate policy, even if connected to the formation of a global regime, will not be effective – will not lead to addressing climate change either through adaptation or mitigation – without willingness at all levels to address this constraint. It will take cooperation both between provinces and between the provinces and the federal government.[55] A globally integrated climate policy incorporates and fosters such cooperation.

4.2. *Efficiency*

Only a globally integrated climate policy will be efficient. By efficient, we mean a policy that addresses climate change at the lowest cost both globally and domestically. Moreover, efficiency incorporates both the fostering of technological progress and the opportunity to take advantage of economic opportunities in the new policy climate. Timing is important to these notions of efficiency. A policy that in the short term is the least costly may not be sufficient to drive long-term cost minimization or technological progress.[56]

The *Stern Review* argues that setting a single price for carbon emissions permits the reduction of global emissions at the lowest cost.[57] This common price allows an equalization of costs across sectors and countries and can be set, in theory most easily, by either a carbon tax or an emissions trading system. Broadening the scope of the regime lowers the cost, at least in the context of trading.[58] What is true globally is also true nationally. Simpson, Jaccard, and Rivers, for example, call for emissions reduction policies that apply to the entire economy such as – potentially – an economy-wide carbon tax.[59]

Both taxes and emissions trading may drive efficiency in all three senses (least-cost reductions, promotion of technology, and fostering Canadian opportunities) if they are broadly based; although in some cases they may need to be supplemented with other measures such as subsidies for research and development.[60] In the context of emissions trading, John Drexhage has pointed out that

> economic model after economic model convincingly demonstrates that a global carbon trading mechanism will significantly reduce the costs of meeting our ultimate objective of delivering a safe climate system to

Introduction: A Globally Integrated Climate Policy *for* Canada 19

future generations. As it was put to me yesterday, 'in the case of climate change, just as emissions know no borders, neither do emission reductions.' Of course, we also need to focus on developing mechanisms and technologies here in Canada, but it is not an 'either-or' scenario and, in fact, if properly designed, Canada could take advantage of the carbon market as a way of launching and commercializing relevant carbon reduction technologies.[61]

Canada's current antipathy toward Kyoto has led it to squander opportunities to use market mechanisms to more easily achieve longer-term climate goals. Available international mechanisms include international emission trading and the clean development mechanism (CDM), through which Canadian companies could get credit for emissions reductions by investing in clean technology in developing countries. Several chapters in this volume examine the prospects for, limitations, and desirability of pursuing such mechanisms. Canadian business has already moved far ahead of government policy in seeking out international carbon markets and engaging with analytic efforts to explore ways to integrate national and international carbon markets.[62] Although Alberta is experimenting with a domestic trading system, only British Columbia has demonstrated a willingness to enter global markets even if the rest of Canada is unwilling to do so. On October 2007, it indicated it would sign onto the International Carbon Action Partnership, which aims at a global market in emission rights.[63]

In a recent about-face that shows some limited signs of re-engagement, the April 2007 Canadian climate plan, if implemented, would allow companies to invest in CDM projects to achieve up to 10 per cent of their emissions reductions – but the late Canadian start date (2010) and uncertainties over the fate of CDM after 2012 make it likely that demand for projects will be low.[64] The April plan also endorsed domestic emission trading and promised to 'explore opportunities' for linkages with future regional (U.S. and Mexican) trading systems, but not other international systems. Since the proposed emissions-trading regime would involve 'intensity-based' targets rather than firm emissions caps, however, there is real concern that this approach could undermine the kind of effective carbon pricing that will provide real incentives for emissions reductions. Canada's reluctance to encourage international market integration amounts to another missed opportunity.

While the Kyoto mechanisms are far from perfect – critics worry, for example, that they may disadvantage developing countries by crediting the least-costly changes to wealthy countries – no one disputes their economic benefits. As British prime minister Gordon Brown put it, 'In a globally competitive economy a multilateral approach is the only way forward.' Only such mechanisms 'make the economic opportunities of a climate friendly policy real and tangible.'[65]

4.3. Fairness

Talk of fairness can often be a deal-breaker behind which recalcitrant parties can hide to justify inaction.[66] This has been the case not only for major developing countries, but also for the United States and Canada, which have argued that they should not have onerous mitigation commitments when countries like India, Brazil, and China – potential competitors – face none. Only through a globally integrated policy can developed countries demonstrate that they take such concerns seriously enough to undercut the justification for inaction on the part of developing countries. Similarly, developing countries vulnerable to climate change – which include many of the large emitters – ultimately know that only global engagement will facilitate the technology transfer and cooperation needed to enable leapfrogging, foreign investment, and action on the part of the North to help prevent the worst consequences of climate change.

The drafters of the UNFCCC recognized this dilemma. Following difficult negotiations at the 1992 Rio conference on environment and development, the framework convention included the principle of 'common but differentiated responsibilities'[67] – recognizing that the largest share of historical and current global emissions of greenhouse gases stems from industrialized countries, and that per capita emissions in developing countries remain relatively low. While the precise implications of the principle remain contested,[68] it does acknowledge that eradication of poverty and promotion of economic growth are legitimate priorities for developing countries and asks industrialized countries to take the lead in combating climate change.

In light of emissions trends in some developing countries, it seems obvious that no climate change regime will be effective in the long run without their participation. It is also safe to say that no global arrangement will attract that participation unless it respects the principle of differentiation. As Lavanya Rajamani puts it, 'The Indian and Chinese negotiating stances, given the continuing stark differences in emissions levels between countries, fits squarely within the climate regime's burden sharing architecture, and is therefore legitimate.'[69] However, she adds, 'It is nonetheless not a sagacious position to hold. Poorer nations, and the poorest within them, will be the worst hit by climate change.' Canada can play a key role here as a traditional promoter of global norms of fairness as well as a possible broker between these positions, but only if it shows its willingness to participate fully in a global regime and in mechanisms like the CDM. Only a globally integrated policy allows Canada to be a model of diplomatic, technological, and investment-related programs to facilitate shifts in developing countries to low carbon economies.

While Canada's circumstances are obviously vastly different from those of European countries, it is the European Union that currently models all of these important policy moves. This fact should be of interest to Canada, if only from the standpoints of competitive and strategic advantage. While Canada looks to cast itself as a bridge builder and leader, the European Union is already practicing integrated climate policy, progressively decarbonizing its economies, and offering concrete proposals for a global regime that includes common but differentiated commitments by major players from North and South.[70] Again, to quote Rajamani, 'The challenge for the Bali negotiations will be to strengthen the existing confidence building architecture in the Convention and Protocol which will eventually persuade developing countries to undertake mitigation commitments.'[71]

What is more, similar principles and differentiation may provide lessons for the way that burden-sharing arrangements are implemented in Canada. From the start, a key principle of any national strategy has been that 'no region or sector [should be] asked to bear an unreasonable share of the burden.'[72] The costs of mitigation as well as the capacity to bear these costs vary from one region to another and one province to another. From 1990 to 2010, for example, the growth of emissions in Ontario and Quebec is expected to be about one-half that in Saskatchewan and Alberta.[73] The absence of a coherent national strategy has left provinces moving largely on their own in developing policies of variable rates of implementation and potential for integration. This situation militates against a fair and effective national policy.

4.4. *Political Feasibility*

According to recent polls, Canadians are concerned about climate change, but somewhat ambivalent about the *Kyoto Protocol* itself.[74] For this reason, it is not easy to translate public concern into concrete policies. Do Canadians really want costs imposed upon them? A globally integrated climate policy helps overcome reticence on the part of Canadians in three ways.

First, as noted above, reducing greenhouse gases is a global public good so that each country benefits from efforts by others to reduce emissions. There is understandable concern in such cases that some countries will free-ride on other countries' efforts, continuing to benefit economically from emissions while others take costly measures to avert climate change. The existence of a global regime can allay domestic concerns that Canada is taking action on the issue when other countries are not. Such a regime, which Canada could help build and maintain, would help counter the fear of some observers that Canada might be unfairly or naively bearing the costs of addressing climate change either directly (such as through government subsidies) or indirectly (through the loss of competitiveness of Canadian industries). In any case, assuming that Canada's policy goal is to avert dangerous climate change, domestic policy can only accomplish this goal in conjunction with collective efforts under an international regime.

Second, a globally integrated climate policy also increases the political acceptability of climate action by expanding the options for meeting international commitments. For example, if Canada can meet its obligations through steps that also help developing countries along the all-important path toward low-carbon development, the Canadian public might be more inclined to embrace options such as acquisition of international emissions reduction credits. Political acceptability would also be increased if Canadian policy were integrated with the global regime in a manner that leads to lower costs for meeting environmental goals or increasing markets for Canadian technology.

Third, a globally integrated climate policy that reflects domestic regional concerns helps overcome the sense that there is free-riding within the country, or that some provinces are bearing more than their fair share of the costs of addressing climate change. There are a range of potentially feasible methods to construct a policy that is both integrated with a global regime and reflects regional concerns. For example, Simpson, Jaccard, and Rivers propose that, in the event a carbon tax is adopted, any revenues be recycled back into the province from which it was generated. The hope is to overcome the reluctance of provinces such as Alberta to sign on to a system which appears to fiscally benefit other provinces.[75]

5. Canadian Policy Options and Obstacles

Quite obviously, in devising a globally integrated policy, policy choices must take into account not only the desirability of participation in a global regime, but also domestic policy options and attendant constraints. That is why we emphasize the need for a globally integrated climate policy *for* Canada. Thus, having canvassed the broad parameters that militate in favour of a globally integrated climate policy, we now consider what such an approach would mean in terms of policy choices for Canada, and what the constraints are on such choices.

Given that greenhouse gas emissions are a form of large-scale externality (where those who reap the benefits of an activity do not bear all of the costs), there are a range of potential policy measures that national or sub-national governments might adopt in order to reduce emissions, including the following:[76]

- Fostering *voluntary action* by individuals or industry such as through agreements or challenges, hoping that underlying norms or self-interest will provide the impetus for reductions in greenhouse gas emissions;
- providing *information or education* to either activate a latent concern by Canadians about the environment (if such a concern exists) or develop new norms or values around the environment;
- using *subsidies* to encourage environmentally friendly activity;
- introducing *taxes* which discourage greenhouse gas emissions, either on specific activities or more broadly on the emissions of carbon;
- establishing *cap-and-trade or obligation-and-certificate programs*, which would allow trading of requirements amongst participants; or
- *Prohibiting or regulating* activities which emit greenhouse gases (for example, set energy efficiency standards for appliances or emissions standards for vehicles).

Of course, Canada should also devise adaptation policies in conjunction with mitigation measures. Such policies would foster actions that reduce the harms from climate change either in Canada or in other countries, where the consequences of climate change are likely to be worse.[77]

Our point here is only to highlight the range of available policy options. They will be explored in more detail in the subsequent chapters of this book.[78] As those chapters serve to illustrate, there is a range of reasons for favouring different combinations of policies, at both the national and global levels. At the same time, there is a range of obstacles, or at least challenges, to the adoption of first best policies in Canada: the structure of the economy; the demands of federalism and regional differences; international competitiveness; and the underlying norms and values of citizens. We already noted in section 3 that these types of challenges are not unique to Canada; still, how they play out in the Canadian context will frame the appropriate choice of policies. Again, our goal here is only to flag some key factors, leaving a more in-depth discussion to other chapters in this book.

5.1. *Structure of the Economy*

The economic structure of Canada poses a challenge in addressing climate change. Oil and gas production, mining and transportation, along with fossil-fuel power generation, all contribute significantly to the Canadian economy but also to its greenhouse gas emissions.[79] Still, predictions of the costs of meeting the obligations of the *Kyoto Protocol* have varied widely. The Harper government, for example, released a report this year predicting high costs under the *Kyoto Protocol*.[80] The report estimated that attempting to meet the obligations under the *Kyoto Protocol* would cause the economy to shrink by 275,000 jobs and electricity and gas costs to jump 50 and 60 per cent respectively. Others point to the potential for technical innovation and an opportunity for Canadian industry to capture new markets, predicting that addressing climate change will not be overly costly.[81] Each of these estimates is based on modelling that contains controversial assumptions and have been part of the attempt to capture a political advantage.[82] In addition, most economic models focus on the costs of action as opposed to the costs of inaction, notwithstanding the *Stern Report* and the related controversy already noted. However, there is little controversy over one basic finding – the longer that action is delayed, the more expensive it becomes.

5.2. *Federalism*

The complexity of Canadian federalism in cross-cutting areas such as the environment further complicates the policy process.[83] The Canadian constitution does not assign the power over the environment to either the federal or provincial governments. Canadian courts have been will-

Introduction: A Globally Integrated Climate Policy *for* Canada   25

ing to find that the federal government has fairly broad powers to regulate environmental issues. However, there are limits on these federal powers. In particular, most of the powers cannot be used in a manner that infringes unduly on provincial areas of responsibility. The constitution grants the provinces jurisdiction over property and civil rights and non-renewable natural resources, with the result that provinces also have significant powers to regulate in the area of the environment.

The consequence of these potentially overlapping or conflicting powers is that the federal and provincial governments have in the past fought over who has control over environmental policy. Given the high stakes of climate policy, continued conflict seems likely. A key principle of any national strategy from the start had been that 'no region or sector [should be] asked to bear an unreasonable share of the burden.'[84] However, making this work in a federal system has proven very difficult.

5.3. *Competitiveness*

As noted above, concerns about competitiveness influence provincial action on climate change. They similarly limit what the federal government may be willing to do. Global competitive pressures work in both directions. Militating against aggressive action, Canadian manufacturing in energy-intensive sectors such as steel or auto manufacturing may compete against developing country manufacturers in, for example, South Korea or Brazil, while energy supply companies may compete against operations in OPEC countries, none of which face mandatory emissions targets. The fear of 'leakage' of industry across state borders may lead to less stringent policies.

There is, in addition, a growing threat from competitiveness on another front. France has threatened to impose a tax on imports from any country which has not signed on to a global climate change agreement. It could theoretically attempt to also impose such taxes on imports from countries which sign on to but fail to meet their obligations under such agreements. France and other members of the EU are concerned about the impacts on competitiveness to their industry from taking action when their major trading partners do not. There are significant limits on the ability of countries to impose such measures on imports under World Trade Organization rules, but they are possible in some forms.[85]

On the other hand, the evolving global regime may signal a broader shift to a less-carbon-intensive global economy. Such a shift may mean that the marketplace will favour more carbon-efficient, eco-efficient and energy-efficient companies – though such shifts are likely to occur far in the future. A globally integrated climate policy would take such a potential shift into account through the design of policies which maximize opportunities for Canadian companies to take advantage of potential new markets.

5.4. Values and Politics

Much has been made of the recent spike in interest by Canadians in environmental issues and, in particular, concern about climate change. It has not always been so. Kathryn Harrison has written about a series of 'waves' of public interest in the environment, each of which was followed by a decline in interest as other concerns such as about the economy took over.[86] Part of the reason why past climate change policies were so focused on largely ineffective combinations of subsidies and information may have been the low public salience of environmental issues.[87] The question is whether there is a new political situation as citizens become aware of the ramifications of continued greenhouse gas emissions. As noted in Kathryn Harrison's chapter in this volume,[88] the polls could be indicating a shift in public opinion and/or values, providing the basis for more stringent climate policy. Climate change has never been so salient. It is important to consider how to best take advantage of this opportunity.

6. Conclusion: There Is Still Time for Action – But It Must Be Now

Given the short time frame (10–15 years) that remains to shift emission trajectories, a more serious debate about trade-offs and values is required. While 'win-win' options may still be available, the need for effective action means that not all good things – short-term economic growth, low costs, and a stable environment – always go together. No one expects a government to commit economic or political suicide. However, successive governments have done little to prepare the public for the necessary choices for effective action, or to design policies that could be strengthened over time.

There is also an emerging consensus that any longer-term policy strategy needs to be based on some form of ramping up policies gradually, but with built-in reinforcing incentives to keep policies and actions headed in the right direction. For example, Mark Jaccard has proposed a schedule of gradually more intensive policies that range from a carbon tax (the toughest politically, but proven most effective in other jurisdictions) to regulations for fuel efficiency, small appliances, and buildings (arguably less controversial, but also less effective).[89] As indicated above, many have also proposed the creation of carbon markets with cap-and-trade systems for large industrial emitters, which could encourage the use of new carbon capture and storage technologies. The main challenge for Canada is to move from (by now seemingly endless) discussion of policy options to actual implementation. Thinking concretely about how to implement some policies *now* that will create their own momentum for future reductions is the key to policy success. Again, John Drexhage has put the challenge well: 'The first message that resonates strongly is that in implementing any sort of regulatory framework, start with a relatively simple system that begins gently. This is easier said than done, since an effective framework in Canada needs to balance international developments with the unique circumstances of Canada as a major energy-exporting developed country.'[90]

The good news is that, in the long run, investments, restraint, and channelling money to new technology are likely to have major pay-offs, as well as reduce adaptation costs, which are to some degree already inevitable.

While action is necessary at home, the idea of a totally 'made in Canada' policy is a myth that should be dispelled. A long-term effective strategy *must* include assistance and incentives to developing countries. A workable international framework, like the *Kyoto Protocol* or a successor agreement, would be helpful in providing such opportunities and incentives.

One popular line of thinking in Ottawa is that Canada is in Afghanistan to regain its international standing and prove its mettle. Taking back its traditional leadership on climate change would do even more for Canada, have a better chance of success, and is more likely to be remembered for generations to come.

Notes

1 *Kyoto Protocol to the United Nations Framework Convention on Climate Change*, 10 December 1997, U.N. Doc. FCCC/CP/1997/L.7/add. 1, reprinted in (1998) 37 I.L.M. 22 [hereinafter *Kyoto Protocol*].
2 On the growing trend towards alternative governance tracks, see M.J. Hoff-

mann, 'The Global Regime: Current Status of and *Quo Vadis* for Kyoto,' this volume and J. Drexhage, 'Climate Change and Global Governance: Which Way Ahead?' this volume.
3 E.A. Parson et al., 'Leading while Keeping in Step: Management of Global Atmospheric Issues in Canada,' in W. Clark et al., eds., *Social Learning in the Management of Global Environmental Risks* (Cambridge, MA: MIT Press, 2001) at 235; G. Toner and T. Conway, 'Environmental Policy,' in G.B. Doern et al., eds., *Border Crossings: The Internationalization of Canadian Public Policy* (Toronto: Oxford University Press, 1996) at 108.
4 The World Commission on Environment and Development, *Our Common Future* (Oxford: Oxford University Press, 1987).
5 See ICSU/UNEP/WMO, 'Report of the International Conference on the Assessment of the Role of Carbon Dioxide and Other Greenhouse Gases in Climate Variations and Assorted Impacts' (Villach, Austria, 9–15 October 1985). WMO document No. 661, 1986; and B. Bolin, B. Döös, J. Jäger, and R. Warrick, eds., *SCOPE 29: The Greenhouse Effect: Climate Change and Ecosystems* (Chichester: Wiley and Sons, 1986).
6 Other group members were Bert Bolin (later the first chair of IPCC), Gilbert White, Syukuro Manabe, Mohammad Kassas, Gordon Goodman, and Gueorgui Golitsyn. See S. Agrawala, 'Early Science-Policy Interactions in Climate Change: Lessons from the Advisory Group on Greenhouse Gases' (1999) 9(2) Global Environmental Change 157.
7 *United Nations Framework Convention on Climate Change*, U.N. Doc. A/AC.237/18 (Part II)/Add.1, reprinted in (1992) 31 I.L.M. 849 [hereinafter UNFCCC].
8 For a detailed analysis of these factors, see S. Bernstein, 'International Institutions and the Framing of Domestic Policies: The Kyoto Protocol and Canada's Response to Climate Change' (2002) 35(2) Policy Sciences 203 at 219–23.
9 Compare ibid., and K. Harrison, 'The Struggle of Ideas and Self-Interest: Canada's Ratification and Implementation of the Kyoto Protocol,' paper presented at the Annual Meeting of the International Studies Association, San Diego, 2006 (on file with authors).
10 D. Russell and G. Toner, 'Science and Policy when the Heat Is Rising: The Case of Global Climate Change Negotiations and Domestic Implementation,' paper presented to the CRUISE Conference on Science, Government and Global Markets: The State of Canada's Science-Based Regulatory Institutions, Ottawa, 1–2 October 1998, at 16.
11 Government of Canada, 'Statement by Environment Minister David Anderson on Climate Change,' Ottawa, 4 April 2001; online, https://

www.ec.gc.ca/press/2001/010404_n_e.htm; Bernstein, *supra* note 8 at 216.
12 For example, in 2002, roughly three-quarters of Canadians wanted the government to ratify the treaty. See M. Fox, 'Canada Mulls Kyoto Pull-Out,' *BBC News* (13 May, 2002); online, http://news.bbc.co.uk/2/hi/americas/1984427.stm.
13 See Harrison, *supra* note 9.
14 The 'Umbrella Group' is a loose negotiating alliance that includes Australia, Canada, Iceland, Japan, New Zealand, Norway, the Russian Federation, Ukraine, and the United States.
15 'Canada's Kyoto 'Energy Credits' Slammed by EU,' *BBC News* (15 April 2002); online, http://www.cbc.ca/canada/story/2002/04/14/kyoto_020414.html.
16 Harrison, *supra* note 9.
17 Ibid. at 17.
18 See 'Ambrose Blasted for Making Partisan Speech,' *CTV News* (15 November 2006); online, http://www.ctv.ca/servlet/ArticleNews/story/CTVNews/20061114/climate_conference_061115?s_name=&no_ads=.
19 For excellent reviews of Canadian climate policy, see Harrison, *supra* note 9; Bernstein, *supra* note 8; and, J. Simpson, M. Jaccard, and N. Rivers, *Hot Air: Meeting Canada's Climate Change Challenge* (Toronto: McClelland and Stewart, 2007).
20 For example, both independent and government reviews of the VCR concluded that despite some success stories, the program had a negligible impact on emissions and lacked sufficient incentives and penalties to be effective. Pembina Institute, *Corporate Action on Climate Change – 1997: An Independent Review* (Ottawa: Pembina Institute. 1998). Moreover, the 1999 updated Emission's Outlook had to revise figures upward by almost 60 million tonnes because the 1997 Outlook had overestimated the potential of the VCR: Analysis and Modelling Group, *Canada's Emissions Outlook: An Update*. (Ottawa: National Climate Change Process, 1999).
21 Harrison, *supra* note 9; Simpson et al., *supra* note 19.
22 Harrison, *supra* note 9; Simpson et al., *supra* note 19.
23 R. St Martin, 'The Made-in-Canada Climate Change Mystery,' *PoliticsWatch* (24 July 2006); online, http://www.politicswatch.com/climate-july24-2006.htm.
24 Government of Canada, *Turning the Corner: An Action Plan to Reduce Greenhouse Gases and Air Pollution*; online, http://www.ecoaction.gc.ca/turning-virage/index-eng.cfm?ecoaction_main.
25 Simpson et al., *supra* note 19 at 193.

26 Commissioner of the Environment and Sustainable Development, *2006 Report*; online, http://www.oag-bvg.gc.ca/domino/reports.nsf/html/c2006menu_e.html, at 11.
27 Stephen Harper, 'Speech from the Throne: Strong Leadership. A Better Canada' (16 October 2007); online, http://www.sft-ddt.gc.ca/eng/index.asp.
28 Contribution of Working Group I to the Fourth Assessment Report of the Intergovernmental Panel on Climate Change (AR4), *Climate Change 2007: The Physical Science Basis: Summary for Policymakers*, online, http://www.ipcc.ch/SPM2feb07.pdf, at 5.
29 T. Homer-Dixon, 'Positive Feedbacks, Dynamic Ice Sheets, and the Recarbonization of the Global Fuel Supply: The New Sense of Urgency about Global Warming,' this volume.
30 IPCC Working Group II, *Climate Change 2007: Impacts, Adaptation and Vulnerability*, online, http://www.ipcc-wg2.org/.
31 See, for example, Council of the European Union, Information note 7242/05 (11 March, 2005); online, http://register.consilium.europa.eu/pdf/en/05/st07/st07242.en05.pdf.
32 See *Joint Declaration by the G8–Presidency and Brazil, China, India, Mexico and South Africa*, 8 June 2007; online, http://www.g-8.de/Webs/G8/EN/G8Summit/ SummitDocuments/summit-documents.html.
33 See U.S. Department of State, 'Final Chairman's Summary: First Major Economies Meeting on Energy Security and Climate Change,' Washington, DC, 28–29 September 2007; online, http://www.state.gov/g/oes/climate/mem/93021.htm.
34 See International Energy Agency (IEA), *World Energy Outlook 2007 – China and India Insights* (OECD/IEA, 2007); online, http://www.worldenergyoutlook.org/.
35 On the legitimacy of the *Kyoto Protocol*, see R. Eckersley, 'Ambushed: The Kyoto Protocol, the Bush Administration's Climate Policy and the Erosion of Legitimacy' (2007) 44 International Politics 306.
36 N. Stern, *The Economics of Climate Change: The Stern Review* (Cambridge: Cambridge University Press, 2007) at xv.
37 See, for example, M. Weitzman, 'The Stern Review of the Economics of Climate Change' (2007) 45(3) Journal of Economic Literature 703; W. Nordhaus, 'The Stern Review on the Economics of Climate Change' (2007) 45(3) Journal of Economic Literature 686; and R. Tol, 'The Stern Review of the Economics of Climate Change: A Comment' (2006) 17 Energy and Environment 977.
38 T.C. Schelling, 'Intergenerational and International Discounting,' in T.C. Schelling, *Strategies of Commitment and Other Essays* (Cambridge, MA: Harvard University Press, 2006) at 53.

39 The figures provided by different sources vary somewhat. According to Environment Canada, Canada's emissions in 2005 placed it 33 per cent above its Kyoto target. See Environment Canada, 'Canada's 2005 Greenhouse Gas Inventory: A Summary of Trends,' online, http://www.ec.gc.ca/pdb/ghg/inventory_report/2005/2005summary_e.cfm. By contrast, the most recent UNFCCC numbers indicate an increase of 25.3 per cent (excluding emissions/removals from land-use, land-use change, and forestry – with these factors included, the increase is much higher). See UNFCCC, 'National Greenhouse Gas Inventory Data for the Period 1990–2005.' Note by the Secretariat, U.N. Doc. FCCC/SBI/2007/30 (24 October 2007) at 17–18; online, http://unfccc.int/resource/docs/2007/sbi/eng/30.pdf.
40 See, for example, Simpson, Jaccard, and Rivers, *supra* note 19.
41 Harper, *supra* note 27.
42 For a discussion see J. Brunnée, 'The United States and International Environmental Law: Living with an Elephant' (2004) 15 European Journal of International Law 617.
43 See D. Hunter, 'The Future of U.S. Climate Change Policy,' this volume.
44 See Government of Australia, Department of Foreign Affairs and Trade, 'Asia Pacific Partnership on Clean Development and Climate,' online, http://www.dfat.gov.au/environment/climate/ap6/.
45 On file with authors.
46 See 'Environment Leads Health as Issue: Poll,' Toronto *Star* (4 January 2007); online, http://www.thestar.com/News/article/167977.
47 See 'Canada Will Not Withdraw from Kyoto: Baird,' CBC (19 October 2007); online, http://www.cbc.ca/canada/story/2007/10/19/baird-kyoto.html?ref=rss.
48 See, for example, H. Ott, 'Climate Policy post-2012 – A Roadmap: The Global Governance of Climate Change,' Discussion paper for the 2007 Tällberg Forum, Tällberg Foundation, Stockholm, 2007; online, http://www.wupperinst.org/uploads/tx_wibeitrag/Ott_Taellberg_Post-2012.pdf. Specifically on the role of the Kyoto mechanisms, see M. Doelle, 'Global Carbon Trading and Climate Change Mitigation in Canada: Options for the Use of the Kyoto Mechanisms,' this volume.
49 S. Barrett, *Why Cooperate? The Incentive to Supply Global Public Goods* (Oxford: Oxford University Press, 2007).
50 Ibid. See also E. Posner and C. Sunstein, *Climate Change Justice* (University of Chicago Law School, Olin Law and Economics Working Paper No. 354, August 2007).
51 See, for example, J. Welsh, *At Home in the World: Canada's Global Vision for the 21st Century* (Toronto: Harper Collins, 2004) at 211.

52 Parson et al., *supra* note 3.
53 See A. Mayeda and N. Greenaway, 'Harper under Fire over Environment and African Aid,' *The National Post* (4 June 2007); online, http://www.canada.com/nationalpost/news/story.html?id=c6c0bd63-9703-470c-8818-d7eac9a22804.
54 See 'Kyoto Alternative - What Is This New Asia-Pacific Partnership All About?' *CBC News* (27 September 2007); online, http://www.cbc.ca/news/background/kyoto/asia-pacific-partnership.html.
55 See, for example, M. Winfield, 'Climate Change and Canadian Energy Policy,' this volume, discussing the lack of, and need for, a national energy policy in Canada.
56 See, for example, D.M. Driesen, 'Renewable Energy under the Kyoto Protocol: The Case for Mixing Instruments,' this volume, arguing that trading may reduce short run costs but may not be sufficient on its own to foster the technological change required to address climate change.
57 Stern, *supra* note 36 at 354.
58 Ibid. at 375.
59 Simpson et al., *supra* note 19 at 233, 256.
60 Driesen, *supra* note 56, Simpson et al., *supra* note 19 at 233, 256, and R. Posner, *Catastrophe: Risk and Response* (Oxford: Oxford University Press, 2004).
61 J. Drexhage, International Institute for Sustainable Development, 'Statement to the House of Commons Legislative Committee on Bill C-30,' 13 February 2007; online, http://www.iisd.org/pdf/2007/com_bill_c30.pdf, at 3.
62 For example, see the series of research and discussion papers prepared for a Pembina Institute sponsored conference on Carbon Pricing for a Sustainable Economy: Applying Market Forces to Climate Protection in Canada, 29–30 October 2007.
63 See International Carbon Action Partnership, *Political Declaration* (29 October 2007); online, http://www.icapcarbonaction.com/declaration.htm.
64 See Government of Canada, *Turning the Corner: A Plan to Reduce Greenhouse Gases and Air Pollution* (April 2007); online, http://www.ec.gc.ca/doc/media/m_124/agir-action_eng.htm.
65 See Speech by the Rt Hon Gordon Brown MP, Chancellor of the Exchequer, at the Energy and Environment Ministerial Roundtable, London, 15 March 2005; online, http://www.g7.utoronto.ca/environment/env_brown050315.htm.
66 See T. Roberts and B.C. Parks, 'Grandfathering, Carbon Intensity, Historical Responsibility, or Contract/Converge?' this volume.
67 UNFCCC, *supra* note 7 Article 3.1.

68 See, for example, L. Rajamani, 'From Stockholm to Johannesburg: The Anatomy of Dissonance in the International Environmental Regime' (2003) 12 RECIEL 23; S. Biniaz, 'Common but Differentiated Responsibility – Remarks,' in *Proceedings of the Annual Meeting* (American Society of International Law, 2002) at 359.
69 See L. Rajamani, 'China and India on Climate Change and Development: A Stance that is Legitimate but Not Sagacious?' this volume.
70 See J. Brunnée and K. Levin, 'Climate Policy beyond Kyoto: The Perspective of the European Union,' this volume.
71 Rajamani, *supra* note 69.
72 Environment Canada, 'Canada's Energy and Environment Ministers Agree to Work Together to Reduce Greenhouse Gas Emissions,' press release, 12 November 1997.
73 National Climate Change Process, Analysis and Modelling Group, *Canada's Emissions Outlook: An Update* (Ottawa, 1999) at 62.
74 See *supra* note 45 and accompanying text.
75 Simpson et al., *supra* note 19 at 212–13.
76 There are many sources describing the various options available to governments in addressing climate change including: Driesen, *supra* note 56; Simpson et al., *supra* note 19; Stern, *supra* note 36; and J.B. Wiener, 'Global Environmental Regulation: Instrument Choice in Legal Context' (1999) 108(4) Yale Law Journal 677.
77 See, for example, Stern, *supra* note 36 and Barrett, *supra* note 49.
78 See the contributions in this volume by D.M. Driesen, D.G. Duff and A.J. Green, A.J. Green, M.S. Winfield, and I.H. Rowlands.
79 See, for example, K. Harrison, 'Challenges and Opportunities in Canadian Climate Policy,' this volume and Simpson et al., *supra* note 19 discussing the impact of the structure of the economy on Canadian efforts to address climate change.
80 Environment Canada, 'The Cost of Bill C-288 to Canadian Families and Businesses,' (2007); online, http://www.ec.gc.ca/doc/media/m_123/report_eng.pdf.
81 Pembina Institute, 'Minister Baird: Canadians Want Leadership on Climate Change, Not a "Can't Do" Attitude' (20 April 2007); online, http://climate.pembina.org/media-release/1515.
82 Modelling costs and benefits is politically charged and filled with uncertainties, ranging from energy prices to technological innovations. For example, models that allow scope for cost-effective improvements in energy efficiency might translate into net benefits (i.e., no-regret measures). Whereas such models face criticism for underestimating hidden costs of

implementing new technologies, top-down models typically used in national forecasts tend to ignore such measures altogether, probably overestimating costs. See M. Grubb, C. Vrolijk, and D. Brack, *The Kyoto Protocol: A Guide and Assessment* (London: Royal Institute of International Affairs and Earthscan, 1999) at 163–65 and Appendix 2.

83 See A.J. Green, 'Bringing Institutions and Individuals into a Climate Change Policy *for* Canada,' this volume.
84 Environment Canada, *supra* note 71.
85 See, for example, J. Bhagwati and P. Mavroidis, 'Is Action against U.S. Exports for Failure to Sign Kyoto Protocol WTO-Legal?' (2007) 6(2) World Trade Review 299; and, A. Green and T. Epps, 'The WTO, Science, and the Environment: Moving towards Consistency' (2007) 10(2) Journal of International Economic Law 285.
86 K. Harrison, *Passing the Buck* (Vancouver: UBC Press, 1996).
87 Simpson et al., *supra* note 19; Harrison, this volume; and Harrison, *supra* note 9.
88 Harrison, this volume.
89 M. Jaccard, 'Canada's Kyoto Delusion: The Evidence is Finally Forcing Us to Admit We Have Done Nothing' (2007) 15(1) Literary Review of Canada 8.
90 Drexhage, *supra* note 61 at 1.

PART ONE

The Need for Action

2 Positive Feedbacks, Dynamic Ice Sheets, and the Recarbonization of the Global Fuel Supply: The New Sense of Urgency about Global Warming

THOMAS HOMER-DIXON*

I am delighted to be here to talk about my understanding of the current state of climate science. I should start by saying that I am not trained as a climate scientist, although I have been working in the area and reading the literature on climate science for twenty years, since my PhD studies at MIT. I have written articles with climate scientists, and I am deeply interested in the social and technological implications of climate change.

I think I have my finger on the pulse of the climate science community at the moment, and what I've noticed in the last several years is a shift in the perspective of leading scientists regarding the seriousness of the climate situation. A few years ago they regarded global warming as a matter of serious concern; now most appear to think that it's a matter of grave urgency – that we may be literally running out of time. The recent IPCC (Intergovernmental Panel on Climate Change) reports are increasingly viewed as out of date. Leading scientists perceive these reports as underestimating the degree and rapidity of climate change and the severity of its consequences.

We have to keep in mind that – around mid-2005 – the IPCC process brought a guillotine down on the scientific findings that were to be incorporated in the reports. These reports therefore do not reflect almost two years of extraordinarily important findings from multiple streams of scientific research. Indeed, immediately after the Working Group 1 report was released (in February 2007),[1] many climate scientists and geophysicists working on ice-sheet dynamics argued that it significantly underestimated potential sea-level rise this century. More recently, we've seen much higher carbon dioxide emissions than were anticipated by the IPCC, while the absorptive capacity of ocean and land-based carbon sinks appears to be decreasing more rapidly than anticipated.

Scientists working in this area are principally concerned about three issues: one concerns an underlying mechanism of climate change; another concerns key consequences of climate change; and the final issue concerns the nature of human energy systems. I'll talk about each of these issues today, and they're highlighted in the title of my presentation.

The first is destabilizing or 'positive' climate feedback. A positive feedback is a causal cycle – essentially a vicious circle – in which warming causes a series of changes that reinforce warming. There are two main kinds of positive feedback: the kind that operates more or less directly on temperature and the kind that operates on the carbon cycle. The feedbacks that operate on temperature are reasonably well incorporated into contemporary climate models. Those that operate on the carbon cycle are not, and it's becoming increasingly clear that they're the ones that could literally be deal-breakers for humanity. We may be quite close to creating circumstances in which the biosphere releases enormous quantities of carbon into the atmosphere. At that point, global warming could become its own cause, and it wouldn't really matter what we do in terms of mitigating our emissions of carbon dioxide – the global ecosystem would take over.

The next issue concerns ice-sheet dynamics: the nature of melting ice sheets, especially the Greenland ice sheet, and the rate at which they're melting. I'll talk more about this subject in a minute.

Finally there is the issue of the recarbonization of the global fuel supply. We have recently seen a reversal of a very important trend that had prevailed for about two hundred years – a progressive decarbonization of fuel supplies around the world. This trend meant that we released, over time, less and less carbon into the atmosphere for every unit of energy we produced. In the past five years, that trend has reversed, with potentially staggering implications for climate change.

Now, before going into these issues in more detail, I want to say a little bit about what I think is the decisive defeat of three main arguments that have been introduced over the years by climate sceptics. These are the arguments firstly about long-term temperature change, secondly about satellite data on tropospheric warming, and thirdly about radiation from the sun. I will make only brief remarks about each one of these sceptical arguments, because I think they have been pretty well demolished and shown to be invalid by careful research. The defeat of these arguments has great consequence for the larger debate about the policy implications of climate change.

The first argument concerns the long-term trend of Earth's average surface temperature. In 1999, Mann, Bradley, and Hughes released a paper that estimated average global temperature for the last millennium. This work was subsequently updated by Mann and Jones in 2003 to provide a temperature record from the years 200 to 2000 AD.[2] These researchers combined a number of different paleoclimatological records – like tree rings and coral growth rates – that are 'proxy' measures of atmospheric temperature during various historical epochs. They cobbled these proxy measures together to get a long-term record of the planet's temperature. Their graph famously showed a sharp uptick over the last half-century, which is why it was widely labelled the 'hockey stick' graph. It has been one of the most contentious pieces of evidence used to support the claim that we are experiencing an abnormally warm period.

You are probably familiar with this debate; it has been covered in the pages of the *Globe and Mail*. In response to criticism of the statistical methodology used to cobble these records together, the National Academy of Sciences in the United States created a panel to examine the Mann et al. methodology. The panel released its results last year, saying that, overall, while some questions remained about the methodology, the original study's conclusions were largely correct: the warming of the last 40 years very likely made Earth hotter than anytime in the last 1000 years, and it certainly made Earth hotter than anytime in the last 400 years. I think the National Academy of Sciences report dealt with the hockey stick issue; it's off the table now, except for some – and I use this word deliberately – crazies out there.[3]

The second argument concerns satellite data. There has been an enormous debate about an apparent discrepancy between data from satellites that show no warming in the troposphere and data from ground-level instruments that show warming. The argument was originally made by John Christy of the University of Alabama in Huntsville. But recent studies have looked very carefully at this apparent discrepancy between satellite and ground-level data and have shown that Christy and his colleagues made a number of methodological and statistical errors. Once these errors are corrected, the discrepancy disappears.[4] The satellite record actually shows tropospheric warming – in fact, it shows both tropospheric warming and, as we would expect from global warming theory, stratospheric cooling.

The third argument concerns radiation from the sun. The most common argument now put forward by climate sceptics is that the recent warming is a result of changes in the intensity of the sun's radiation. But a major review article last year in the journal *Nature* showed that it's virtually impossible to explain the warming we've seen in the last 40 years through changes in solar radiation.[5] This research is pretty well definitive, too.

So, these three arguments used by sceptics have been largely put to rest. We are now down to a hard core of climate change deniers who are essentially impervious to any evidence – and they write me all the time. Sometimes I engage in an amusing exercise just to see how detached from reality they can actually be. I send them scientific papers and reports on the latest climate research, and invariably the evidence in these reports makes absolutely no difference to their point of view.

This kind of psychological resistance points to something I think we need to confront directly: a process of denial of evidence that is quite powerful in some parts of our society and in some individuals. I think there are three stages of denial, which I talk about in my latest book.[6] The first is *existential* denial, where one denies the actual existence of the phenomenon. But existential denial is hard to sustain when the evidence becomes overwhelming, as is now the case with climate change. So, people tend to move away from existential denial and start engaging in what I call *consequential* denial, in which they deny that the consequences of the problem are going to be particularly serious. This is essentially the position taken by a lot of climate change sceptics now. They're saying, 'okay, there's climate change, but we can deal with it. It's basically a pollution problem that is not so serious. We can adapt as necessary.'

The evidence is also increasing, of course, that we won't be able to adapt adequately to the magnitude of the climate change that's likely even this century – or that the economic and social consequences of this change will be so great that, if we try to adapt, we'll still need to aggressively mitigate our output of carbon dioxide. So the final position, once it becomes impossible to support even consequential denial, is what I call *fatalistic* denial: one basically accepts that the problem is real and that it's going to hurt a lot, but then one simply says, 'there's nothing we can do about it.' In my future research I want to explore the larger social consequences of widespread fatalistic denial. I think they could be astonishingly bad.

Let me go on to quickly give you a sense of the three issues that I

Figure 2.1. Average global temperature increase

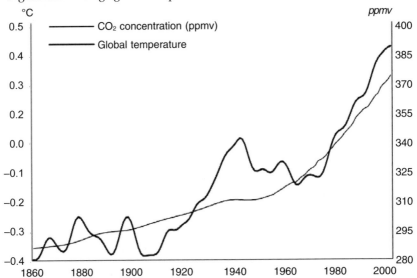

talked about before: positive feedback, ice-sheet dynamics, and recarbonization of the fuel system. Let's talk first about what the recent IPCC Working Group I report said about global warming to date – that the 'warming of the climate system is unequivocal, as is now evident from observations of increases in global average air and ocean temperatures, widespread melting of snow and ice, and rising global average sea level.' Figure 2.1 shows CO_2 concentration, which is the thinner line, and global temperature, which is the darker line. The period of particular interest to scientists is in the past 40 years when there appears to be, prima facie, quite a close correlation between CO_2 concentrations and global temperature.

From November 2006 through February 2007, large parts of Canada experienced warming in the neighbourhood of 2°C to 4°C. I'll give you a sense of the magnitude of that change: prior to this recent bout of warming over the last 40 years, Earth's average temperature had increased only about 5°C from the coldest period of the last ice age 15,000 years ago. Last winter's warming was most pronounced in the northern part of the planet – in the neighbourhood of 6°C – a fact that I'll return to shortly.

Using a range of scenarios with different configurations of technology, energy consumption, population growth, and trade relations between North and South, Working Group 1 estimated that warming in 2100 would fall between 1.1°C and 6.4°C, with the best estimate around 3°C. The scenario that I think represents the most likely modified 'business-as-usual' future is A1B, which predicts about 3°C by 2100. The Working Group also projected that 'climate sensitivity' – the amount of warming a doubling of pre-industrial levels of CO_2 (i.e., from 280 ppm to 560 ppm) will produce – will also be about 3°C, with a range from 2°C to 4.5°C. But they additionally said something very important that has not been widely reported – that values significantly higher than 4.5°C cannot be excluded. Computer models of the climate tend to break down at this upper range, but nonetheless substantial evidence now suggests we may see such extreme warming.

In the A1B scenario, by 2020 to 2029 Canada will warm between 1.5°C and 2.5°C; by 2090 to 2099, Canada will warm in the neighbourhood of 6°C to 7°C. Some people say that Canada is going to benefit from climate change. Well, let me challenge that assertion: we may have lower heating bills in the winter for a few years, but because we're a polar country, warming here will be twice as fast and the ultimate magnitude will be twice as great as the average warming for the planet. Optimistic comments about benefits to Canada neglect warming's staggering consequences for our flora and fauna, for our forests that can't adapt and will die en masse, for Canada's central grain-growing regions that could easily turn to desert, for the Great Lakes as their water levels fall, for transportation in the St Lawrence Seaway, and for northern permafrost that will melt. In actual fact, climate change may ultimately affect Canada as harshly as any country in the world.

Why are we warming more rapidly in the planet's northern reaches? The basic reason is the ice-albedo feedback. The sea ice floating on the surface of the Arctic Ocean is white, so it reflects a large proportion of the sun's radiation back into space. As this sea ice melts from global warming, it leaves behind open ocean water that absorbs about 80 per cent more of the sun's radiation. The ocean water becomes warmer. Then, after the summer passes in the north and fall comes, the water releases its heat back into the atmosphere, which impedes the refreezing of ice. So winter generates thinner ice, and this ice melts more easily the following summer. This is a positive feedback, a vicious circle.

Figure 2.2 was produced by the National Snow and Ice Data Center in Boulder, Colorado. It shows the minimum extent of Arctic sea ice

Figure 2.2. Minimum Arctic sea ice cover

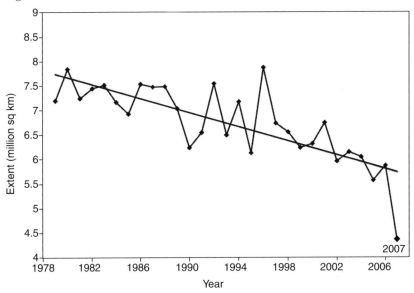

each year going back to 1978. The Arctic ice cap melts somewhat every summer – it always has and still does. Sea ice extent reaches its minimum sometime around the middle of September, and then as the days get shorter and cooler the ice starts to recover. A regression line has been plotted through the points up to 2006, showing a steady decline in ice extent, year by year. Notice that the melting in 2007 was much more severe and showed a sharp downward divergence from the trend. By 16 September 2007, we had lost about a third of the Arctic ice cap compared to the 1979–2000 average, and about 50 per cent compared to the 1950s average. The current expectation among scientists is that we will see a completely ice-free Arctic ocean in summer by the end of the next decade – perhaps even as early as 2013. I want to say a couple of words about the implications of this extraordinary change.

The last thing we should be worrying about or thinking about, I believe, is whether we're going to be able to run a lot of ships through the Arctic, or whether we'll be able to explore for oil and gas there as the ice vanishes. The area above the Arctic Circle makes up about nine per cent of the total surface area of Earth above the equator. The loss of ice will change the reflectivity of much of the polar region and therefore alter the energy balance of the northern half of the planet. We can't fully predict the consequences of this change, but they may be severe.

In the northern hemisphere, there are three important cycles of atmospheric circulation between the equatorial region and the pole. They are called Hadley cells. In each cell, warm air rises at the southern end of the cycle and flows northward at high altitude. Then, when the air cools, it descends to the surface and flows southward back to the starting point. There is a cell in the equatorial region, another at mid-latitude, and another over the Arctic. Some climate scientists think that loss of the Arctic sea ice could cause the Hadley cell in that region to break down, which would have consequences for the paths of jet streams further south. Jet streams influence storm tracks and precipitation patterns, which can in turn intimately affect our ability to grow food.

Let me now say a little bit more about some other feedbacks. This is one of the punch lines of my presentation today. I mentioned earlier that there are two general kinds of feedback: those that operate more-or-less directly on temperature, such as the ice-albedo feedback, and those that operate on Earth's carbon cycle, where warming produces a change in the amount of carbon in the atmosphere. We have a fairly good understanding of the former and not such a good understanding of the latter. One carbon feedback that worries scientists involves the melting of the permafrost in Siberia, Alaska, and Northern Canada. As the permafrost melts it releases large quantities of methane – a very powerful greenhouse gas that, in turn, causes more warming. Scientists are also concerned about the potential release of more carbon dioxide from forests: just yesterday researchers reported evidence that, as the climate has warmed, the Canadian boreal forest has gone from being a carbon sink to a slight carbon emitter.

And then there's the matter of pine bark beetles. As you likely know, we've lost wide swaths of pine forest in British Columbia and Alaska – huge areas of trees – to bark-beetle infestation. As the climate warms, bark-beetle populations reproduce through two generations during the summer, and beetle mortality is lower during the winter. Both these changes mean that beetle populations become much larger overall. If these larger populations cross the Rockies and get into the boreal forest that stretches from Alberta to Newfoundland, and if they kill that forest, the forest will be susceptible to fire that could release astounding quantities of carbon dioxide. I asked Stephen Schneider, a leading climate scientist at Stanford, about the implications of such a development. He just shrugged and said, 'well, we're talking about billions of tonnes of carbon.'

Other potentially destabilizing carbon-cycle feedbacks include the drying of the Amazon and the possibility that if it dries it will burn; the drying of peat bogs in Indonesia, which have already been susceptible to wide-spread burning; and the saturation of ocean carbon sinks. The Southern Ocean around Antarctica is no longer absorbing carbon dioxide to the extent it did in the past. Warming has produced much more vigorous winds closer to Antarctica. These winds have churned up the sea and brought to the surface deep carbon-rich water, which absorbs less carbon from the atmosphere. Also, higher levels of carbon dioxide in the atmosphere are acidifying the oceans, a change could reduce populations of molluscs and phytoplankton that absorb carbon into the calcium carbonate of their shells.

Our climate has both positive and negative feedbacks. The positive ones are self-reinforcing, and the negative ones equilibrate the climate and counteract the tendency towards self-reinforcing climate change. The big question for climate scientists then is: What is the balance is between the positive and negative feedbacks? A consensus has emerged over the last two years – a consensus again not reflected in the recent IPCC reports – that the positive feedbacks in the climate system are much stronger and more numerous than the negative feedbacks.

In a paper published last year in *Geophysical Research Letters*, Scheffer, Brovkin, and Cox carried out a comprehensive assessment of the feedback situation.[7] They wrote, '[we] produce an independent estimate of the potential implications of the positive feedback between global temperatures and greenhouse gasses.' In other words, these researchers focused specifically on carbon cycle feedbacks. They went on, 'we suggest that feedback of global temperature and atmosphere CO_2 will promote warming by an extra 15% to 78% on a century scale over and above the IPCC estimates.'

Let's turn to the issue of dynamic ice sheets. The Greenland ice sheet is the second largest mass of ice in the world, after that in Antarctica. If we melt Greenland entirely, we get seven metres of sea-level rise. If we melt the West Antarctic ice sheet, we get another five metres. If we melt the rest of Antarctica, we get an additional fifty or so metres. The Greenland ice sheet will probably be the first to melt, because it's the most vulnerable. During the last interglacial period 125,000 years ago, when temperatures were roughly what they're going to be at the end of this century, much of Greenland melted, and sea levels were four to six metres higher than they are right now.

We are probably already committed to temperatures in that range with the industrialization processes that are underway on the planet, especially in China and India. The estimate of the sea-level rise for this century that the IPCC produced was twenty to sixty centimetres – or somewhere around half a metre. Two independent studies of Greenland in the last two years, neither reflected in the IPCC reports, suggest that the ice sheet is now melting at the rate of 200 to 250 cubic kilometres a year, which is about 200 times the amount of water that Los Angeles consumes each year. According to the most recent study, that rate has doubled in the last ten years. That study used satellites to measure slight variations in the gravitational field around the planet; and based on these variations, the researchers estimated change in the mass of the Greenland ice sheet.[8] The two studies used very different methodologies, but their results correspond closely. So we can be confident that we're already seeing the Greenland ice-sheet disappear quite quickly.

Climate scientists now recognize that the models of ice sheet melting that the IPCC reports relied upon to estimate sea-level rise were radically inadequate. These models were 'static,' in that they assumed that atmospheric warming melts the ice, and the resulting water then runs off the surface of the ice sheet and down into the ocean. Scientists now know that these ice sheets have cracks in them. Water runs down the cracks, and as the ice melts the cracks can sometimes expand into gaps 10 to 15 metres across, with millions of tonnes of water flowing downwards. This flow creates pools underneath the ice sheets that lubricate the movement of glaciers and increase the speed of glacial movement into the ocean. Rates of movement are much higher than the IPCC reports expected. This phenomenon has truly scared scientists close to the subject.

Let me read some quotations of Robert Corell, chairman of the Arctic Climate Impact Assessment, the principle synthetic report on the state of the Arctic climate. Commenting on the Ilulissat glacier in northwest Greenland just a few weeks ago, he said 'we have seen a massive acceleration of the speed with which these glaciers are moving into the sea. The ice is moving at 2 meters an hour on a front 5 km long and 1,500 metres deep. That means that this one glacier puts enough fresh water into the sea in one year to provide drinking water for a city the size of London for a year.' He had flown over the glacier and seen 'gigantic holes in it through which swirling masses of melt water were falling. I first looked at this glacier in the 1960s and there were no holes. These so-called moulins, 10 to 15 meters across, have opened up all over the place. There are hundreds of them.'[9] The glacier is moving at 15 km a year into the sea, although it sometimes surges forward much faster. He measured one surge at 5 km in 90 minutes.

Big things are happening in Greenland – things that will affect sea level. The consensus emerging now among climate scientists is that we're going to see oceans rise by a metre this century and that we may even see two metres. A change of this magnitude would have enormous effects on coastal areas of Canada – on residential areas in Victoria and Vancouver (especially on the municipalities of Delta and Richmond in the Lower Mainland) and on the ports of Vancouver, St John's, and Halifax. With a two metre rise, concerns about rebuilding infrastructure and moving populations inland will – in a few decades – become real, even urgent.

With regards to global warming, changes are generally happening much faster than anticipated even a few years ago by the best scientific consensus as reflected in the IPCC reports. Faster change raises the issue of the relative balance – in terms of our policy response – between mitigation and adaptation. Some observers argue, myself among them, that we need to shift some of our policy resources to adaptation. We're going to see significant warming, with sometimes severe consequences, and we need to get ready for these consequences at the urban, municipal, and national levels. Of course, we can't neglect efforts to reduce carbon dioxide output. But in some respects the mitigation challenge we face is almost impossibly hard.

I'll give you an indication of what we're up against. Very soon humankind must cap and then ramp down global carbon emissions. We have very little room to warm: the estimated maximum safe warming from pre-industrial temperatures is around 2°C; beyond that point we get into a world where the positive feedbacks I've just discussed may develop great force. The warming to date has been about 0.8°C, and the warming in the pipeline – even if all emissions cease right now – is about 0.6°C. This leaves us with around 0.6°C room to warm.

Limited room to warm implies, in turn, that we have very little room to emit. The estimated carbon dioxide concentration that's likely to produce at least 2°C warming is about 450 ppm. (This is actually a conservative estimate; some people would put the threshold for carbon dioxide much lower. Notice, for instance, that I am talking about atmospheric carbon dioxide and not 'carbon dioxide equivalent.' In other words, these 450 ppm do not include chlorofluorocarbons, nitrous oxide, and a number of other powerful greenhouse gases. If they did, the actual limit for CO_2 itself would be much lower than 450 ppm.) The current concentration of CO_2 is about 380 ppm, so the room to emit, therefore, is about 70 ppm. The incremental annual increase is currently about 2 ppm and rising, so we have about 30 years left until we reach

Figure 2.3. Peak oil

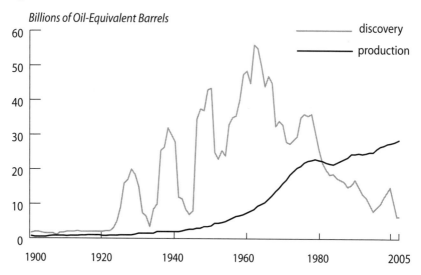

450 ppm. That doesn't mean we have 30 years before we have to start worrying about this problem: it means that in 30 years we'd better be heading south on carbon emissions really fast.

Indeed, we need to be heading towards an 80 to 90 per cent cut in carbon emissions by 2050. Scientists are talking about that kind of reduction, as are environmental activists, but in Canada it isn't even on the policy radar screen at the moment (notably, a number of U.S. Democratic and Republican presidential candidates have committed themselves to such reductions).

My last few comments concern the recarbonization of the global fuel supply. Figure 2.3 is a chart of world oil production and discovery. The lighter line is oil discovery and the darker line is oil production from 1900 and to 2000. We are very close to – if we aren't already at – a peak in the world's conventional oil production. Oil provides 40 per cent of the world's commercial energy and 98 per cent of its transportation energy. It's the stuff that the global economy literally runs on. And it's going to become more expensive, in terms of the energy cost of energy production. As we pass the mid-point of the amount of oil that's ultimately available on the planet, oil companies are finding that they have to go further and deeper into more hostile environments to find smaller pools of lower quality oil. This trend means that – at least when it

Figure 2.4. Energy return on investment (EROI) of various fuels

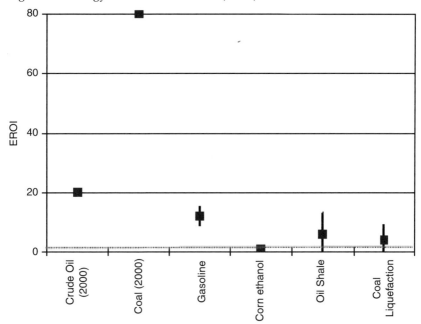

comes to conventional oil – we have to use a lot more energy to get energy. In the 1930s in Texas, drillers were rewarded with a return of about 100 barrels of oil for every barrel of oil of energy they invested to drill down into the ground and pump oil out. In the United States now, this energy return on investment (or *EROI*, as the concept is known among energy analysts) is around 17 to one. The Alberta tar sands give you an EROI of four to one. As we slide down the slope from 100 to one, past 17 to one, towards one to one, we're using a larger and larger fraction of the wealth and capital of our society simply to produce energy, and we've got less left over for everything else we want to do.

I believe the rising energy cost of energy is a very powerful binding constraint on economic development on the planet. We're entering a transition from a regime of abundant high-quality, and high-EROI energy to one of abundant, mixed quality, and often low-EROI energy.

Figure 2.4 compares the EROIs of various fuel systems. Crude oil has an EROI of around twenty to one, while corn ethanol and biofuel stand at about one to one – in other words, when it comes to ethanol we put in

Figure 2.5. Decarbonization trends

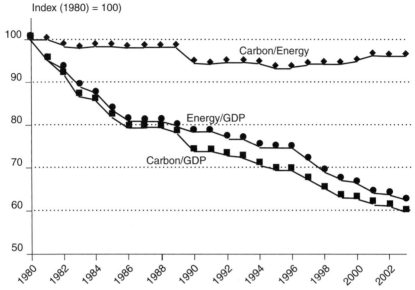

Intensity Ratios: Carbon/GDP, Carbon/Energy, and Energy/GDP
Source: Culter Cleveland, Boston University

about as much energy as we get out. Oil shale and tar sands have an EROI of around four to one or five to one. Creating diesel fuel from coal gives an EROI of about two to one or three to one. But coal by itself provides a very high EROI. In terms of an energy kick, it's great stuff.

The problem is that using more coal takes us in the wrong direction. In fact, it takes humankind in a direction that's radically different from the historical trend. In the last couple of centuries we've seen a steady decarbonization of our fuel supplies. We have moved from wood to coal to oil to natural gas as our main energy source, and with each of these transitions we have released less carbon into the atmosphere for each unit of energy produced. We're now seeing a reversal of this trend.

Figure 2.5 shows trends in three intensity ratios: from top to bottom, they are trendlines for carbon released per unit of energy, energy use per dollar of GDP, and carbon released per dollar of GDP, for the years from 1980 to 2002 in the United States. Although during this period the United States saw a decline of carbon output per unit of GDP, there was actually very little change in carbon released per unit energy. If the U.S.

situation is representative of the situation in wealthy countries, and on this issue it largely is, we actually haven't decarbonized our energy sources at all in the last 20 years. Almost all the gain in the decarbonization of GDP has been a product of increasing energy efficiency. That's an enormously important, and often overlooked, fact.

Since 2002, we've turned the corner. Decarbonization has stopped, and recarbonization has begun. Measurements of atmospheric carbon dioxide concentration taken at the Jubany station in Antarctica – a place where the atmosphere is very well mixed – show that from 1994 to 2001, the average annual addition to the atmosphere's carbon dioxide was about 1.64 ppm. Then from 2002 to 2006, the average jumped to about 2.1 ppm. The trend is heading upwards quickly towards 3 ppm annually. A significant component of this increase in the *rate* of increase (or what mathematicians call the second derivative) is a result of the higher carbon content of fuels. Basically, as oil has become more expensive, companies and economies have begun switching to more carbon-intensive fuel such as coal and oil derived from tar sands. Coal production, especially in China, is rising incredibly fast. China has doubled coal production from one tonne per person to two tonnes per person in the last six years, or from 1.3 to 2.7 billion tonnes of coal for the country as a whole. China is now a major driver of the increasing CO_2 concentration of the atmosphere.

A paper released last week in the *Proceedings of the National Academy of Sciences* provided a groundbreaking analysis of these trends. It looked at the acceleration in the magnitude of the annual addition of carbon dioxide to the atmosphere. The paper's authors break this acceleration into three parts. Sixty-five per cent is due to increasing global economic activity, in particular in India and China, and 17 per cent is due to increasing carbon intensity of the global economy, arising mainly from fuel switching to more carbon-intensive fuels like tar sands and coal. Together, these two factors explain a stark fact: global carbon emissions were increasing by about 1.3 per cent a year throughout the 1990s, but between 2000 and 2006 the rate rose to 3.3 per cent.

The third factor explaining the acceleration in the size of the annual increment of CO_2 in the atmosphere – explaining, in fact, the remaining 18 per cent of that acceleration – is an increase in the 'airborne fraction.' Normally, oceans and forests absorb about half of the carbon dioxide we emit, but that fraction seems to be declining, and the amount staying in the atmosphere – the airborne fraction – is now rising. The researchers note that the drop in carbon absorption, especially in the Southern

Ocean, is likely the result of global warming, which means we're starting to see positive feedbacks in the carbon cycle. The increase in airborne fraction is consistent with results of climate carbon cycle models, they continue, but 'the magnitude of the observed signal appears larger than that estimated by models.' All of these changes, they conclude, 'characterize a carbon cycle that has generated a stronger than expected and sooner than expected climate forcing.'[10]

I think we might actually be very close to self-reinforcing climate change – the situation where warming becomes its own cause. It is hard to say exactly when we will cross that threshold, but it could be closer than most experts anticipated even a few years ago, and certainly closer than implied by the IPCC reports.

Also, I expect that the first major socio-economic impact of climate change will be on our food supply. We'll see significant production shortfalls because of droughts and storms in major food producing areas. So keep a close eye on grain future prices, which are very high at the moment – the highest prices ever seen for corn and wheat. These high prices are significantly related to a drought-induced decline in grain production in Australia (they're also related to rising demand for grain, in particular corn, from ethanol producers).

My last remarks today concern a topic that – as recently as two years ago – I fervently hoped we would never have to discuss. That topic is geoengineering, the intentional human modification of the planet's climate. Geoengineering would involve, for example, putting sulphates into the atmosphere or putting mirrors into space to try to block a fraction of incoming solar radiation, or it would involve fertilizing the oceans to create plankton blooms to suck carbon out of the atmosphere.

Not only do I now think we have to discuss geoengineering, I believe we will almost certainly have to do it. Next week I'll be attending a meeting on the subject at the American Academy of Arts and Sciences in Cambridge, Massachusetts. Although the topic is at the margins of the public policy dialogue about climate change right now, I expect it will be at the centre of public discussion within four years. In 10 years, we will see demands from the public and many opinion leaders that we carry out geoengineering. And we'll probably start doing it within 20 years, likely when it becomes apparent that the Greenland ice cap is starting to collapse.

We will do it because we will be experiencing really large socio-economic impacts of climate change. We're going to look down the road and wonder about what kind of world we have created for our children and grandchildren. We will recognize that we're facing an emergency unlike anything humankind has ever faced before, and we will demand that our leaders do something, anything, to stop the slide.

I wish it weren't true, but the fact that some of the world's very best climate scientists are coming together to talk about the issue is a clear indication of the new sense of urgency about global warming.

Thanks very much for your time today.

Notes

* This chapter is an edited transcript of an address Professor Thomas Homer-Dixon gave to the conference A Globally Integrated Climate Policy *for* Canada, where the papers in this volume were originally presented. The address reports on the latest climate science, some of which has been published since the cut-off for inclusion in the latest Intergovernmental Panel on Climate Change Report (the fourth assessment report). These findings give added urgency to developing better climate policy and Professor Homer-Dixon places the policy challenge in the wider global context (the Editors).
1 Contribution of Working Group I to the Fourth Assessment Report of the IPCC, *Climate Change 2007 – The Physical Science Basis of Climate Change* (2007) online, http://ipcc-wg1.ucar.edu/wg1/wg1-report.html.
2 M. Mann, R. Bradley, and M. Hughes, 'Northern Hemisphere Temperatures during the Past Millennium: Inferences, Uncertainties, and Limitations' (1999) 26(6) Geophysical Research Letters 759; and M. Mann and P. Jones, 'Global Surface Temperatures over the Past Two Millennia' (2003) 30(15) Geophysical Research Letters.
3 Committee on Surface Temperature Reconstructions for the Last 2,000 Years, Board on Atmospheric Sciences and Climate, Division on Earth and Life Sciences, *Surface Temperature Reconstructions for the Last 2,000 Years* (Washington, DC: National Research Council of the National Academies, National Academies Press, 2006).
4 The original argument that satellite data show no warming of the lower atmosphere appeared in R.W. Spencer and J.R. Christy, 'Precise Monitoring of Global Temperature Trends from Satellites' (1990) 247(4950) Science 1558. For decisive evidence to the contrary, see B.D. Santer et al., 'Influence of Satellite Data Uncertainties on the Detection of Externally Forced Climate Change' (23 May 2003) 300(5623) Science 1280; and C. Mears and F. Wentz, 'The Effect of Diurnal Correction on Satellite-Derived Lower Tropo-

spheric Temperature' (2 September 2005) 309(5740) Science at 1548. On errors in interpreting weather balloon data, see S. Sherwood, J. Lazante, and C. Meyer, 'Radiosonde Daytime Biases and Late-20th-Century Warming' (2 September 2005) 309(5740) Science at 1556.
5 P. Foukal et al., 'Variations in Solar Luminosity and Their Effect on the Earth's Climate' (14 September 2006) 443 Nature 161.
6 T. Homer-Dixon, *The Upside of Down: Catastrophe, Creativity, and the Renewal of Civilization* (Toronto: Knopf, 2006) at 131–3.
7 M. Scheffer, V. Brovkin, and P. Cox, 'Positive Feedback between Global Warming and Atmospheric CO_2 Concentration Inferred from Past Climate Change' (2006) 33 *Geophysical Research Letters* L1072.
8 J.L. Chen, C.R. Wilson, and B.C. Tapley, 'Satellite Gravity Measurements Confirm Accelerated Melting of Greenland Ice Sheet' (2006) 313(5795) Science 1958.
9 As quoted in P. Brown, 'Melting Ice Cap Triggering Earthquakes,' *The Guardian*, 8 September 2007.
10 J. Canadell et al., 'Contributions to Accelerating Atmospheric CO_2 Growth from Economic Activity, Carbon Intensity, and Efficiency of Natural Sinks' (October 2007) *Proceedings of the National Academy of Sciences*, Early Edition.

PART TWO

Canada in the World

3 Climate Policy beyond Kyoto: The Perspective of the European Union

JUTTA BRUNNÉE AND KELLY LEVIN*

1. Introduction

A climate policy that is right *for* Canada must be built from the ground up, taking into account Canadian priorities, values, and constraints. It also requires taking into account the international policy context in which Canada operates, including the climate policies of the main players in the global arena and their likely priorities going forward.

The European Union (EU) is one such key player, not only because of its collective ecological and economic footprint, but also because it has been more determined than any other international actor to push for a comprehensive, ambitious, and legally binding global climate regime, extending beyond 2012 to new 2020 and 2050 benchmarks. The year 2012 has come to symbolize a major policy juncture because it marks the end of the first commitment period (2008–12) under the *Kyoto Protocol* to the *United Nations Framework Convention on Climate Change* (UNFCCC).[1] Although it is as yet unclear – and contested – what kind of global regime will follow the expiry of the Kyoto commitments, there is growing consensus that a long-term approach is required.

This paper begins with an outline of the salient features of the existing global regime, of the role that the EU has played in the development of the regime to date, and of the reasons why the regime is indeed at a turning point. It then considers the current state of European climate policy, including the European Emissions Trading System (ETS), and the EU's long-term priorities for international climate action. The paper closes with some reflections on the key features of the European approach to global regime-building and on their implications for Canadian policy. The scope of the paper is limited to mitigation of greenhouse gas emissions, although the EU is making headway on other climate-related activities, such as the advancement of adaptation policies.

2. Europe and the UN Climate Regime

The UNFCCC, adopted in 1992 and ratified by 191 states and the EU,[2] provides principles and objectives to guide global climate policy, as well as institutions and processes for further treaty development. The overarching objective of the convention is to stabilize greenhouse gas (GHG) 'concentrations in the atmosphere at a level that would prevent dangerous anthropocentric interference with the climate system.'[3] Among its foundational principles are the propositions that climate change is a 'common concern of humankind,'[4] and that states should protect the climate system 'on the basis of equity and in accordance with their common but differentiated responsibilities and respective capabilities.'[5] In light of the latter consideration, the convention also stipulates that 'developed country Parties should take the lead in combating climate change.'[6]

The *Kyoto Protocol*, adopted in 1997 and ratified by 175 states and the EU,[7] imposes binding GHG emission reduction commitments on developed country parties and countries with economies in transition (CEITs), but not on developing countries. Although this dimension of the protocol has become increasingly controversial, and was a key factor in the U.S. decision to abandon the treaty,[8] it is in keeping with the principles of the UNFCCC and with the idea that developed countries should take the lead in addressing climate change.

The protocol requires that emission reductions in comparison to 1990 emission levels are achieved during a 2008–12 'commitment period;' compliance with that target is assessed at the end of that period. Parties' specific commitments differ. For example, while Canada must reduce its emissions by 6 per cent below 1990 levels, the European Community (EC – the legal entity that is party to the protocol)[9] is committed to an 8 per cent cut.[10] In addition to the EC, its member states have their own commitments.[11] However, so long as the EC's collective commitment is met, member states will not be assessed for compliance with their individual commitments.[12] Indeed, an internal 'burden sharing agreement' has reallocated individual members' commitments within the so-called EU bubble.[13]

In addition to enshrining binding emission reduction commitments, the *Kyoto Protocol* provides for a non-compliance procedure that is designed, *inter alia*, to 'enforce' compliance with these commitments, as well as with related inventory and reporting commitments.[14] Finally, to give parties greater flexibility in meeting their commitments, the protocol establishes trading mechanisms through which parties (or legal entities under their jurisdictions) can exchange emission rights or emission reduction credits.[15]

Although it was adopted in 1997, the *Kyoto Protocol* entered into force only in 2005, for two main reasons. First, many of the details that needed to be in place for parties to assess whether or not to ratify (e.g. the rules governing the Kyoto mechanisms and the non-compliance procedure) were not fleshed out in the treaty itself. Second, the U.S. decision to abandon Kyoto threw the viability of the protocol into doubt, and made it more difficult to bring the treaty across the entry-into-force threshold of membership by developed-country or CEIT parties accounting for 55 per cent of at least 55 such parties' CO_2 emissions in 1990.[16]

The fact that the *Kyoto Protocol* did enter into force even though neither the United States (the single largest emitter of GHGs), nor those developing countries that will soon become major emitters (notably, China and India), have emissions reduction commitments under it is largely due to European determination not to see the treaty derailed. The EU made a conscious policy choice to press ahead with finalizing the protocol irrespective – even in spite – of the U.S. stance, and to lobby hard for sufficient buy-in to bring it into force.[17]

Many European policy-makers saw the U.S. decision not to join the protocol as an affront, and as further evidence of American 'unilateralism.'[18] Given the fraught trans-Atlantic relations after the U.S.-led invasion of Iraq, it is probably fair to say that European policy was motivated both by a desire to move the global climate regime forward, and a desire to prove that the 'single super-power' could not call all the shots. Either way, in the absence of U.S. initiative and in the face of U.S. reluctance, the EU emerged as the main force behind the UNFCCC and its *Kyoto Protocol*, and as the main laboratory for the integration of international and domestic climate policy.

Nonetheless, leading up to the third meeting of its parties in Bali in December 2007, the *Kyoto Protocol* is at a crossroads. Given that the protocol's commitment period expires in 2012, it is high time to settle whether or not there will be subsequent commitments and, if so, what these commitments will be and to whom they will apply. Not resolving these questions in favour of a Kyoto successor in the near future will have a number of consequences for global climate governance. At the most basic level, the legal result would be that Kyoto parties' emissions would be permitted to increase again at that point.[19] The absence of clear signals regarding subsequent commitments will also undermine the existing regime. A domino effect might ensue – more and more parties may be tempted to abandon their efforts to meet Kyoto commitments. Finally, even if the EU and its member states stay the Kyoto course – which is to be expected – industry might lose key incentives to step up climate action.

The barriers to speedy clarification of the post-Kyoto regime are formidable. First, since the treaty was negotiated more than ten years ago, emissions in many Kyoto parties have risen significantly, rendering their commitments much harder to meet than originally imagined. Some Kyoto parties, including Canada, have therefore begun to openly contemplate baselines and targets that deviate from the cornerstones established in the UNFCCC and the *Kyoto Protocol*, notably from the 1990 reference year and from the idea of an absolute emission reduction target.[20] Second, the United States remains opposed to a Kyoto-style approach to binding global action (as does Australia), especially in the absence of meaningful emission reduction commitments by the major developing economies. Third, developing countries remain entrenched in their resistance to taking on binding targets and their insistence on prioritizing action by industrialized states. Fourth, a number of other, non-binding, initiatives – such as the Asia-Pacific Partnership on Clean Development and Climate – have been launched to engage major emitters.[21] While these initiatives can be useful by helping to break the stalemate that stifles progress under the auspices of the UNFCCC, they also risk shifting the centre of gravity away from the UN framework and undercutting the basic ideas that underpin the convention: that climate change is of common concern and requires collective, albeit not uniform, action.

Three processes are currently underway to consider further actions under the UNFCCC umbrella.[22] Pursuant to Article 3.9 of the *Kyoto Protocol*, an Ad-hoc Working Group is tasked with negotiating new commitments for industrialized and CEIT parties. This track specifically excludes consideration of developing country commitments. Article 9 of the protocol would allow for a broader review of the adequacy of the protocol and its approach. However, the G7 and China have resisted the discussion of emissions-related commitments by developing countries. In turn, given this resistance, industrialized states have so far not been willing to discuss a concrete negotiating mandate under Art. 3.9 of the protocol. Lastly, under the auspices of the UNFCCC, a loose dialogue was launched to keep the United States and Australia in the global action tent. Both countries have expressed the desire to continue to global cooperation within the UN framework, but these gestures have not translated into concrete progress towards new commitments for more parties than those currently bound by the *Kyoto Protocol*.

Still, as noted above, the EU remains strongly committed to a Kyoto-style follow-up to existing commitments, and to expanding the reach of the climate regime to include all major industrialized and developing countries. Indeed, it appears that the EU is prepared to lead by example, and to take significant new action even without immediate buy-in by others.[23] To understand this policy stance, it is necessary to consider the current climate policy of the EU and its member states, as well as the rationale that underpins EU policy planning for the future.

3. European Climate Policy Today and the Agenda for the Future

3.1. *The EU and its Kyoto Commitments*

The European Union's contribution to global emissions is not insignificant: it emits roughly 14 per cent of global greenhouse gases,[24] 80 per cent of which is derived from the energy sector.[25] According to the Commission of the European Communities (EC Commission),[26] energy consumption is projected to grow in the coming years, boosting electricity demand by roughly 1.5 per cent per year; gas imports from 57 per cent of current consumption to 84 per cent by 2030; and oil imports from 82 per cent to 93 per cent by 2030. Overall, the EU's reliance upon energy imports will grow from 50 per cent at its current level to 65 per cent by 2030.[27]

The above being said, the EU is collectively on track to meet its Kyoto commitments.[28] In addition, European countries not covered by the 'EU bubble,' namely those that acceded to the EU more recently, are projected to meet their individual Kyoto targets.[29] The EU bubble's collective success masks uneven performance of EU member states. This pattern is hardly surprising, as the prospects of such uneven performance were the very reason for the EU's collective reduction commitment. Overall, the bubble concept has provided the desired flexibility for the EU member states. While some nations have greater abatement potential and can embrace steeper cuts of emissions, other countries can absorb additional reductions to make up for any shortfalls.

Specifically, considerable emissions reductions have been realized in the United Kingdom, which has already surpassed its target under the *Kyoto Protocol* 'burden sharing agreement,'[30] and is in the process of enacting an ambitious climate change legislation package with 2020 and 2050 targets.[31] Abatement has been achieved in large part because of adoption of fuel switching from coal to natural gas. Similarly, Germany has been a key player in advancing EU climate policy and will meet its target, in part due to economic restructuring,[32] shifting from coal to natural gas, and reductions in transportation and residential emissions.[33] By 2008–12, Germany's emissions are projected to be reduced by 21 per cent compared to 1990 levels, and by 40 per cent below 1990 levels by 2020.[34] The Netherlands has also achieved emissions reductions because of decreased reliance on heating due to warmer temperatures, as well as a move away from fossil fuel sources for electricity and heat.[35]

In contrast, countries such as Italy, Portugal, and Spain are struggling to meet their targets.[36] For example, in Spain and Portugal, hydroelectric power capacity has been compromised due to precipitation variability,[37] at times requiring a switch from renewable energy to more carbon-intensive energy sources. Countries that struggle to meet their targets under the burden sharing agreement may be able to address these problems by engaging in international emissions trading through the *Kyoto Protocol*'s flexibility mechanisms.

Throughout Europe, there has been a clear policy shift towards diversification of energy supply, carbon pricing, advancement of cleaner technologies and fuels, and promotion of behavioural changes through public information, education, and incentive programs. Indeed, Germany considers that a 'third industrial revolution' is underway: a transition from carbon-intensive energy sources derived from fossil fuels to a low carbon society based on renewable energies and energy efficiency.[38] Nonetheless, given the challenges inherent in meeting a regional target, the EU had the foresight to acknowledge that voluntary programs and scattered policies throughout its member states would likely not suffice. Thus, the EU created a comprehensive regulatory framework to help ensure member states collectively adhered to the EU commitment delineated in the *Kyoto Protocol*.

3.2. *The European Emissions Trading Scheme*

The architecture of this program is a cap-and-trade scheme: the EU Emissions Trading Scheme (ETS), which was launched in 2005 to support the individual targets assigned to each member state covered under the 'burden sharing agreement' mentioned above.[39] The ETS currently encompasses two phases: the first runs from 2005 to 2007, and the second from 2008 to 2012. The first phase was designed as a trial run prior to the *Kyoto Protocol* compliance period 2008–12, as the national targets are pinned to the protocol targets. During this first phase, the private sector and public officials were encouraged to 'learn-by-doing' in an effort to ease the member states into the compliance period. While the ETS is the centrepiece of climate policy in the EU, it is important to note that only 46 per cent of the region's emissions are covered under the scheme, and, thus, other policies have been adopted to complement the cap-and-trade program.[40]

Ironically, the design of the ETS was influenced greatly by the United States, despite the EU's initial distaste for the concept of emissions trading when the United States introduced it into the *Kyoto Protocol* negotiations.[41] Upon the U.S. withdrawal from the protocol, the EU championed the concept of emissions trading, modeling its emissions trading on the U.S. acid rain trading program for SO_2 emissions.[42] However, compared to the acid rain program, the ETS is characterized by greater decentralization, as state sovereignty concerns are highly relevant in the EU multi-jurisdictional trading program.[43] Thus, under the Emissions Trading Directive, the legal underpinning of the ETS, the member states are mandated to comply with some objectives (e.g. national targets) but are, at the same time, allowed great latitude in designing national policies for achieving these objectives (e.g. through national allowance allocation plans).[44]

A number of elements of the ETS, then, are uniform across the member states. First, while the ETS pilot phase is limited to CO_2 and covers only some of the major industrial sectors, subsequent phases of the scheme could expand sectoral and gas coverage.[45] Currently, the covered sectors are (a) iron and steel; (b) energy production, including electric power facilities above 20MW[46] and oil refineries; (c) pulp and paper; and (d) some mineral industries, including cement and glass production.[47] Altogether, 12,000 installations in these sectors are included in the ETS.[48] The transportation and residential sectors, which are highly energy intensive, are not included, although policymakers are currently considering the inclusion of aviation under future phases of the scheme.[49] In addition to required elements of source coverage, compliance fees are not left to the member state's discretion; a common fee of 40€/tonne of CO_2 exists for the first phase, and a 100€/tonne CO_2 for noncompliance during the second phase.[50] Also, member states must publicly disclose all operators that fail to comply with regulations.[51]

The majority of other design elements of the scheme are optional for member states. In fact, the ETS embraces such a high degree of flexibility that member states are required to draft National Allocation Plans in which they lay out their mitigation strategies for target attainment. Some elements of the scheme that are not rigid include options on banking/borrowing, opt-outs/opt-ins, set asides, auctioning, and the development of monitoring, reporting, and verification. For example, the ETS allows for the banking of excess allowances for future years, which can alleviate compliance costs.[52] Also, 'opt-outs' are allowed for individual installations during the first phase of the ETS if a firm has taken similar measures to reduce its carbon dioxide emissions elsewhere; member states are also allowed to include 'opt-ins' if they choose to augment source coverage.[53]

In addition, member states can opt to use a number of allowances to cover new entrants (i.e. firms that did not exist prior to the allocation of allowances), and draw from a regional pool of 'set asides' that account for 3 per cent of the total allowances allocated. Member states also have some latitude with regard to allowance auctioning. During the first phase of the ETS, up to 5 per cent of allowances can be auctioned, rising to 10 per cent during the second phase. This flexibility was designed in part to address concerns about perverse subsidies.[54] While a centralized electronic registry has been designed to catalogue emissions and allowances in the member states, each member state is required to design its own registry, or partner with others to design a joint registry, which connects with the centralized registry in a 'hub-and-spoke' design.[55] Although there are guidelines for monitoring and reporting, including the provision of emissions factors, member states have considerable flexibility in designing their own accounting, monitoring, and reporting techniques. And while third party verification is required, member states can specify verification protocols.[56] Also, while compliance fees are uniform, penalties for other acts of misconduct are determined by the member state.[57]

While the ETS has proven to be an important laboratory within which companies and member states can experiment with various abatement strategies, the EU has encountered some hurdles. First, price volatility has characterized the scheme as a result of both over-allocation of allowances and abatement timing. At one point the spot price of a carbon tonne dropped below €1,[58] as emissions were roughly 4 per cent lower than the amount of allowances distributed, a problem that may have arisen because of lack of installation data, which led to overestimation of emissions.[59] In addition to the problems of over-allocation, Ellerman and Buchner suggest that unanticipated abatement and banking contributed to the abundance of allowances.[60] In an effort to avoid another collapse of the carbon price, the EU announced in October 2007 that it would reduce the amount of allowances for the second phase of the ETS by 10 per cent to a total of 2.08 billion tonnes.[61] Some nations will have to contend with significant reductions: Luxembourg will have to cut allowances by 40 per cent; Hungary by 12 per cent; and Germany by 2.5 per cent. Other nations, such as the UK, Denmark, and France are not required to cut their amount of distributed allowances in the next ETS phase.[62]

These challenges notwithstanding, the ETS has already established itself as the largest contributor to the carbon market, comprising 62 per cent of the 2006 volume of credits and more than 80 per cent of the value.[63] The scheme is quickly integrating itself into the *Kyoto Protocol* regime, through trade with other Annex I and non-Annex I parties via the protocol's flexibility mechanisms. A Linking Directive was adopted, which allows for Clean Development Mechanism (CDM) and Joint Implementation (JI) credits to be converted into EU allowances.[64] Should such credits exceed 6 per cent of EU allowances during the second phase of the scheme, a review will be initiated. Forestry-related projects under the Kyoto mechanisms are currently not included in the ETS.[65]

In addition to linking through the Kyoto mechanisms, the EU Directive retains the option to link with non-Kyoto parties. Because the EU has greatly advanced the design of emissions trading, especially as mechanics relate to accounting for greenhouse gases rather than traditionally regulated pollutants, many other jurisdictions have turned to the EU not only for guidance but also for an invitation to link with their own emerging initiatives. For example, the state of California has discussed the prospect of linking its future abatement scheme with the ETS.[66] Similarly, the Regional Greenhouse Gas Initiative (RGGI) in the Northeast and Mid-Atlantic states will allow covered sources to purchase EU allowances for compliance.[67] In addition, in October of 2007, the EU decided to link with the schemes of Iceland, Liechtenstein, and Norway, which expanded the boundaries of the ETS beyond the EU.[68] Days later, on 29 October 2007, EU officials announced that it had joined several U.S. states, British Columbia, Norway, and New Zealand in a collaboration entitled the International Carbon Action Partnership.

The Partnership intends to create a forum to share best practices on design and implementation of emissions trading schemes, and to explore linkage potential and barriers.[69]

3.3. International Policy Priorities

National and sub-national activity on greenhouse gas abatement, as well as collaboration among different national programs, all help develop new approaches to international climate governance. Nonetheless, the EU contends that the global community must be guided by a global temperature target, consistent with the scientific evidence on impacts, in an effort to avert a 'dangerous' level of anthropogenic climate change. The EU has chosen a goal of limiting global temperature increase to no more than 2°C above pre-industrial levels.[70] The Earth has already warmed 0.8°C in the last century, having increased 0.2°C per decade within the last thirty years.[71] In addition, if concentrations of greenhouse gases were stabilized at today's levels, temperatures would still rise 0.6°C over time, as the Earth is absorbing more heat due to thermal inertia than it radiates back to space.[72] Thus, a temperature increase of roughly 1.4°C over pre-industrial levels is to be expected even if greenhouse gas concentrations were stabilized, which is highly unlikely given current trajectories. And, even if the global community were able to stabilize temperatures to no more than 2°C above pre-industrial levels, some projections indicate that we could witness a 10 per cent transformation of all ecosystems, millions displaced in coastal areas, and over a billion at risk of water stress, among other impacts.[73]

Policymakers must determine what level of emissions, and resultant concentrations of greenhouse gases, translates to a temperature target. There is a wide range of temperatures for a given concentration, which poses a challenge to policymakers in that they cannot be confident that a certain temperature target will be met by a chosen emissions target. Scientists have employed probability density functions of climate sensitivities to facilitate the inclusion of ranges of climate sensitivities into projections.[74] As a result, probabilities of exceeding temperature targets at a given concentration can be calculated. The risk of overshooting a 2°C target at 550 ppm CO_2equivalent (CO_2e)[75] is between 68 per cent and 99 per cent. At 450 ppm CO_2e, the mean risk of surpassing a 2°C target is 47 per cent, and the risk of overshooting 2°C is only 'unlikely' at 400 ppm CO_2e, which reduces the mean risk of exceeding the target to 27 per cent.[76] According to den Elzen and Meinshausen,[77] stabilization at 450 ppm CO_2e would require peaking of emissions around 2015 with global emissions reduced 15 per cent below 1990 levels by 2020 and 30 per cent below 1990 levels by 2050; stabilization at 400 ppm CO_2e requires reductions of 30 per cent below 1990 levels by 2020 and 50 per cent below 1990 levels by 2050. Accordingly, NASA climate expert Jim Hansen gives us only ten years to avert dangerous anthropogenic change.[78]

In light of this scientific evidence, the EC Commission now calls for action that ensures that global emissions of GHG peak in the next 10–15 years, and for a 50 per cent reduction of global emissions below 1990 levels by 2050.[79] To meet these emission reduction goals, all major emitters must agree to take measures to combat climate change. The EU proposes that the benchmark for a peak in global emissions can be met through emission reductions of 30 per cent below 1990 levels by developed countries by 2020. Developing countries would not be asked to make reduction commitments, but would be expected to begin reducing the *growth* of their emissions. However, since emissions by developing countries are projected to exceed the total emissions of industrialized countries by 2020, the EU proposes that developing countries would have to commit to emission reductions after that date, and that they should also take measures to avoid emissions from deforestation.[80] According to current EU proposals, major developing countries should commit to reductions of 50 per cent below 1990 levels by 2050, while developed nations would be required to reduce their emissions by 60–80 per cent. This staged and differentiated approach to emission reductions by developing countries, according to the EC Commission, reflects the historical contributions by industrialized countries to current GHG levels in the atmosphere and to deforestation, as well as their larger technological and financial capacity.[81]

In the absence of global policy developments post-2012, the now 27 EU member states have unilaterally agreed to reduce their emissions by at least 20 per cent below 1990 levels by 2020, and 30 per cent below 1990 levels if other industrialized nations join the effort.[82] Thus, a third phase of the ETS will cover 2013 to 2020.[83] While the third phase will benefit from the lessons learned during the previous phases, member states will no doubt encounter challenges, such as expanding the scheme to all 27 member states.[84] The 2020 target is of global significance, as it signals to the private sector that a carbon price will exist after the *Kyoto Protocol* expires in 2012, thus providing more security for investments into new technologies and research and development. In addition, the UK is set to adopt climate change legislation that would translate the 2020 target into law, and set a 2050 target of a reduction of carbon dioxide emissions by at least 60 per cent below 1990 levels.[85] The bill would create a climate change minister, as well as an independent climate change committee that would dedicate its activities to assessing long-term mitigation targets, and the scheme's coverage of gases and sectors, among other implications of climate policy.[86]

The EU has yet to draft regulations for the third phase of the ETS.[87] It currently projects that it will meet its new 2020 emissions target through a portfolio of activities,[88] including: increasing the share of energy derived from renewable sources to 20 per cent of total use by 2020;[89] boosting the reliance upon biofuels to 10 per cent of fuel use in the transportation sector by 2020; adopting efficiency standards; exploring carbon capture and storage; and, possibly banning incandescent light bulbs by 2010.[90] The EC Commission has also proposed to establish a single energy market, lifting trade barriers in an effort to encourage energy source diversification.[91] Recognizing that the ETS covers roughly 45 per cent of EU CO_2 emissions, the EC Commission suggests that the third phase of the ETS include mechanisms for curbing emissions from aviation, passenger cars, road freight transport and shipping, residential and commercial buildings, agriculture, forestry, and non-CO_2 greenhouse gases.[92]

4. Conclusion

Perhaps the most striking feature of European climate policy is the extent to which it is built around the UN climate regime. Several aspects of the EU agenda are worth highlighting in this respect. First, although the EU is not excluding the possibility of different commitment tracks (different speeds and approaches), its policy is clearly premised upon retaining the UNFCCC as core of the global climate regime.[93] The EU efforts to promote adoption of a multilateral post-2012 regime by 2009 also suggest that it is looking to maintain the basic architecture of the *Kyoto Protocol*.[94]

Second, the frame of reference for EU climate policy is the ultimate objective set out in the UNFCCC: averting dangerous climate change.[95] To this end, EU policy aims to keep global temperature increases to 2°C, which requires both shorter and longer term action to limit GHG concentrations in the atmosphere. It is this sequence of propositions that leads to the EU proposal that collective action be taken to ensure that global GHG emissions peak in the next 10–15 years, and are reduced by 50 per cent below 1990 levels by 2050.[96] While the 2°C goal and the time frames envisaged by the EU proposals are open to some debate,[97] it is noteworthy that EU policy is focused upon achieving what is considered to be necessary from an environmental protection standpoint. Of course, the approach is also deemed to be economically viable.[98] Still, in beginning with an environmental objective and building policy around that objective, the EU's approach to climate policy differs significantly from that of other global players.

Second, in specifying *how* global emission reduction goals are to be made, the EU policy is clearly guided by the core principles of the UNFCCC. In arguing for a global approach, EU policy is not merely pragmatic, but also builds upon the idea that climate change is a common concern of humankind. At the same time, EU policy statements explicitly acknowledge the greater historical contributions of industrialized countries to climate change, and their greater economic and technological capacity to address it. Accordingly, its policy proposals are crafted around the principle of common but differentiated responsibilities and the notion that developed countries must take the lead in combating climate change.[99]

These two UNFCCC propositions underpin the differentiated 2020 and 2050 emission goals advocated by the EU. In this explicit implementation of the UNFCCC's North-South equity principles, the EU proposals again differ quite significantly from other proposals. The EU evidently considers that significant action by industrialized countries, coupled with an approach that recognizes developing country concerns and grievances, is best suited to achieving the needed global climate action. Indeed, some EU member states have gone even further in signalling willingness to accommodate Southern policy priorities. For example, the German chancellor, Angela Merkel, recently indicated that states' GHG emissions could be measured in per capita terms, as demanded by many developing countries. Under Merkel's proposal the per capita emissions of industrialized countries would have to be reduced and those of developing countries could increase, but developing countries should not have higher per capita emissions than industrialized states.[100] Although the EU as a whole has not endorsed these proposals, they are not incompatible with the EU suggestions for differentiated 2020 and 2050 benchmarks.

These observations are not intended to suggest that the EU is engaged in an entirely idealistic policy-making exercise. Aside from the practical goal of designing a policy that can in fact avert dangerous climate change, there are other very pragmatic reasons for the European approach to global climate policy. Most notably, the EU has invested heavily in meeting its Kyoto commitments, investments that will not see a 'return' unless an assertive post-2012 approach is adopted at the global level. What is more, the EU has invested heavily in the UNFCC/Kyoto architecture, building its present and future targets around the 1990 baseline that underpins the UN regime, and launching an emissions trading system that is premised upon the existence of hard emission caps.

These policies have brought the EU considerable competitive advantages, flowing from accelerated technological innovation and accelerated conversion to a low-carbon economy. In addition, EU climate policies have entailed strategic advantages, especially in that any future global emissions trading system, as well as other national or regional systems, will likely be shaped with the ETS in mind. But, again, these advantages are best maintained by ensuring that a global climate regime builds on the existing foundations.

Finally, it should be noted that the EU's unilateral pledge of 20 per cent GHG emission cuts below 1990 levels by 2020, when examined more closely, is not as ambitious a commitment as it might appear at first glance. This pledge is rendered possible by what has been referred to as 'wall-fall profits,' resulting from the enlargement of the EU beyond its original membership of fifteen.[101] The 2020 emissions of the new Eastern European member states are projected to be 32 per cent below 1990 levels,[102] giving the EU's collective commitment a healthy cushion. For this reason, some commentators have called for a unilateral pledge of a 30 per cent cut by 2020.[103]

Nonetheless, there is an important normative dimension to European climate policy. The member states of the EU have long been comfortable with a supra-national approach to law and policy-making. It is no accident that the project of the 'constitutionalization' of international law is particularly popular in Europe.[104] The UN climate regime binds the EU and its member states, and its requirements have come to be enshrined in EC and national laws. As a result, the legal framework of the EU gives rise to continuous interactions between international, European, and domestic law. It is fair to say, therefore, that the EU and its member states have actually 'internalized' the goals, values and principles enshrined in the UNFCCC to a significant degree. They have become part of the legal and policy discourse within Europe, perhaps

Climate Policy beyond Kyoto: The Perspective of the European Union   71

even of the European identity as a member of a global climate action community. This pattern stands in some contrast to, for example, North American attitudes towards climate policy, which seem driven more by a search for (cost-)effective problem-solving and an attendant willingness to experiment with multiple modes of international governance.

Whatever the basic attitude towards the global climate regime, the EU provides a model for the kinds of significant policy and economic shifts that will be required of all countries – even countries with a carbon-based economy like Canada's. The EU is much further along the inevitable path towards a low-carbon economy than any other part of the world. It also provides a living laboratory for the integration of international and domestic climate policy. In the search for a globally integrated climate policy *for* Canada, this country would therefore be well advised to pay close attention to EU climate policy.

Notes

* The authors gratefully acknowledge the excellent research assistance provided by Nozomi Smith.
1 *United Nations Framework Convention on Climate Change*, U.N. Doc. A/AC.237/18 (Part II)/Add.1, reprinted in (1992) 31 I.L.M. 849 [hereinafter UNFCCC]; and *Kyoto Protocol to the United Nations Framework Convention on Climate Change*, 10 December 1997, U.N. Doc. FCCC/CP/1997/L.7/add. 1, reprinted in (1998) 37 I.L.M. 22 [hereinafter the *Kyoto Protocol*].
2 As of 22 August 2007, the UNFCCC has been ratified by 191 states and the EU. Online, http://unfccc.int/essential_background/convention/status_of_ratification/items/2631txt.php.
3 UNFCCC, *supra* note 1 Article 2.
4 Ibid., Preamble.
5 Ibid., Article 3.1.
6 Ibid.
7 As of 23 October 2007, the *Kyoto Protocol* had 176 parties. Online, http://unfccc.int/kyoto_protocol/background/status_of_ratification/items/2613.php.
8 See J. Brunnée, 'The United States and International Environmental Law: Living with an Elephant' (2004) 15 European Journal of International Law 617.
9 The EU, which currently has 25 member states, was established through the 1992 Treaty on European Union. The EU and the European Community

(EC) are legally distinct, but have the same member states and largely the same institutions. The EC rather than the EU is legally competent to enter into international agreements. On the salient distinctions, see L. Krämer, 'Regional Economic Integration Organizations: The European Union as an Example,' in D. Bodansky, J. Brunnée, and E. Hey, eds., *Oxford Handbook of International Environmental Law* (Oxford: Oxford University Press, 2007) 853 at 854–7.

10 See *Kyoto Protocol, supra* note 1, Article 3.1 and Annex B.
11 See ibid.
12 See ibid., Article 4.1.
13 For an overview, see J. Lefevre, 'The EU Greenhouse Gas Emissions Allowance Trading Scheme,' in F. Yamin, ed., *Climate Change and Carbon Markets: A Handbook on Emission Reduction Mechanisms* (London: Earthscan, 2005) 75 at 77.
14 See *Kyoto Protocol, supra* note 1, Article 18. For the text of the procedure, see UNFCCC, *Report of the Conference of the Parties serving as the meeting of the Parties to the Kyoto Protocol on its first session, held at Montreal from 28 November to 10 December 2005. Addendum. Part Two: Action taken by the Conference of the Parties serving as the meeting of the Parties to the Kyoto Protocol at its first session*, U.N. Doc. FCCC/KP/CMP/2005/8/Add.3 (30 March 2006) at 92–103.
15 See ibid., Articles 6, 12, and 17. For a detailed discussion of the Kyoto mechanisms, see M. Doelle, 'Global Carbon Trading and Climate Change Mitigation in Canada: Options for the Use of the Kyoto Mechanisms,' this volume.
16 See *Kyoto Protocol, supra* note 1, Article 25.
17 See, e.g., H. Ott, 'The Bonn Agreement to the Kyoto Protocol – Paving the Way for Ratification' (2001) 1 International Environmental Agreements: Politics, Law and Economics 469.
18 See, e.g., T. Karon, 'When it Comes to Kyoto, the U.S. Is the "Rogue Nation,"' *Time Magazine*, 24 July 2001; online, http://www.time.com/time/world/article/0,8599,168701,00.html.
19 See H. Ott, 'Climate Policy post 2012 – A Roadmap: The Global Governance of Climate Change,' Discussion paper for the 2007 Tällberg Forum (Stockholm: Tällberg Foundation, 2007); online, http://www.wupperinst.org/uploads/tx_wibeitrag/Ott_Taellberg_Post-2012.pdf.
20 See Government of Canada, Minister of Environment, *Regulatory Framework for Air Emissions* (2007); online, http://www.ec.gc.ca/doc/media/m_124/report_eng.pdf. The Canadian government has pegged its policy goals to a 2006 (rather than 1990) baseline and is advocating GHG intensity targets

(rather than absolute, Kyoto-style, targets). These moves amount to an attempt by Canada to 'grandfather' the significant emissions increases it has seen since 1990.
21 See Government of Australia, Department of Foreign Affairs and Trade, 'Asia Pacific Partnership on Clean Development and Climate,' online, http://www.dfat.gov.au/environment/climate/ap6/.
22 For an overview, see Ott, *supra* note 19 at 17.
23 See Commission of the European Communities, *Communication from the Commission to the Council, the European Parliament, the European Economic and Social Committee and the Committee of the Regions - Limiting Global Climate Change to 2 Degrees Celsius – The Way ahead for 2020 and Beyond*, COM (2007) 2 final (1 October 2007) at 2, 9; online, http://ec.europa.eu/environment/climat/future_action.htm [hereinafter EC Commission, *Limiting Global Climate Change*].
24 European Parliament, Press Service, *A New Industrial Revolution: Parliamentarians Debate EU Responses to Climate change*, Ref. 20071001PR11004; online, http://www.europarl.europa.eu/news/expert/infopress_page/064-11025-277-10-40-911-20071001IPR11004-04-10-2007-2007-false/default_en.htm.
25 See Commission of the European Communities, *Communication from the Commission to the European Council and the European Parliament – An Energy Policy for Europe*. COM (2007) 1 final (1 October 2007) at 3; online, http://ec.europa.eu/energy/energy_policy/doc/01_energy_policy_for_europe_en.pdf [hereinafter EC Commission, *Energy Policy for Europe*].
26 The Commission is charged with making proposals for the adoption of EU regulations and directives and with ensuring member state compliance with these measures. See Krämer, *supra* note 9 at 856–7.
27 See EC Commission, *Energy Policy for Europe*, *supra* note 25 at 3.
28 M. Xuequan, 'Barroso Says EU Will Meet Kyoto Targets,' *China View* (29 October 2007); online, http://news.xinhuanet.com/english/2007-10/29/content_6972705.htm.
29 European Environment Agency, *Greenhouse Gas Emission Trends and Projections in Europe 2006*, EEA Report No 9/2006 (2006) at 23; online, http://reports.eea.europa.eu/eea_report_2006_9/en/eea_report_9_2006.pdf [hereinafter European Environment Agency 2006].
30 David Suzuki Foundation, *Backgrounder: Who's Meeting Their Kyoto Targets?* (2006) at 1; online, http://www.davidsuzuki.org/files/climate/Kyoto_Progress.pdf [hereinafter Suzuki Foundation *Backgrounder*].
31 The draft *Climate Change Bill* and related documents are available at http://www.defra.gov.uk/environment/climatechange/uk/legislation/index.htm.

32 See Suzuki Foundation *Backgrounder, supra,* note 30, at 2. And see Pew Center on Global Climate Change, *The European Union Emissions Trading Scheme (EU-ETS) Insights and Opportunities* (2005) at 2; online, http://www.pewclimate.org/docUploads/EU-ETS%20White%20Paper.pdf [hereinafter *EU ETS Insights and Opportunities*].

33 European Environment Agency, 'EU Greenhouse Gas Emissions Decrease in 2005 – Press Release' (15 June 2007); online, http://www.eea.europa.eu/pressroom/newsreleases/eu-greenhouse-gas-emissions-decrease-in-2005 [hereinafter European Environment Agency 2007].

34 Federal Ministry for the Environment Nature Conservation and Nuclear Safety, *Taking Action Against Global Warming: An Overview of German Climate Policy* (31 October 2007) at 5; online, http://www.bmu.bund.de/english/international_environmental_policy/downloads/doc/40014.php [hereinafter *Taking Action*].

35 See European Environment Agency 2007, *supra* note 33.

36 'No More Hot Air [Editorial]' (2007) 446(7132) *Nature* 109 at 109.

37 See European Environment Agency 2006, *supra* note 29 at 42.

38 See *Taking Action, supra* note 34 at 5.

39 See European Community, *Directive 2003/87/EC of the European Parliament and of the Council of 13 October 2003 Establishing a Scheme for Greenhouse Gas Emission Allowance Trading within the Community and Amending Council Directive 96/61/EC (Text with EEA Relevance),* OJ L 275, 25.10.2003, at 32–46; online, http://ec.europa.eu/environment/climat/emission/implementation_en.htm.

40 J. Kruger, W.E. Oates, et al., *Decentralization in the EU Emissions Trading Scheme and Lessons for Global Policy,* RFF DP 07–02 (Washington: Resources for the Future, February 2007) at 5; online, http://www.rff.org/Documents/RFF-DP-07–02.pdf.

41 D. Ellerman and B. Buchner, 'The European Union Emissions Trading Scheme: Origins, Allocation, and Early Results' (2007) 1 Review of Environmental Economics and Policy 66 at 67; online, http://reep.oxfordjournals.org/cgi/content/abstract/1/1/66.

42 J. Kruger and W.A. Pizer, 'Greenhouse Gas Trading in Europe: The New Grand Policy Experiment' (2004) 46(8) Environment 8; online, http://www.rff.org/rff/News/Features/loader.cfm?url=/commonspot/security/getfile.cfm&PageID=16480.

43 See Kruger et al, *supra* note 40 at 24. And see Ellerman and Buchner, *supra* note 41 at 68.

44 For the directive, see *supra* note 39. For a discussion, see Kruger and Pizer, *supra* note 42 at 11.

45 See *EU ETS Insights and Opportunities, supra* note 32 at 7.
46 J. Vogler and C. Bretherton, 'The European Union as a Protagonist to the United States on Climate Change' (2006) 7(1) International Studies Perspectives, at 6.
47 See Kruger et al., *supra* note 40 at 5.
48 Ibid.at 5.
49 G. Schwarze, 'Including Aviation into the European Union's Emissions Trading Scheme' (January 2007) 16(1) European Environmental Law Review 10. This move has proven to be contentious internationally and has met with the resistance, most notably, of the United States. See United States Mission to the European Union, 'U.S. Seeks Comprehensive International Approach to Aircraft Emissions Reductions' (22 September 2007); online, http://useu.usmission.gov/Dossiers/Aviation/Sep2207_Steinberg_ICAO.asp.
50 See *EU ETS Insights and Opportunities, supra* note 32 at 8.
51 See Kruger and Pizer, *supra* note 42 at 16.
52 See *EU ETS Insights and Opportunities, supra* note 32 at 7.
53 Ibid. at 8.
54 See Kruger et al., *supra* note 40 at 8. However, in practice only 0.13 per cent of emissions have been assigned to auctions. Only Denmark, Hungary, Lithuania, and Ireland have auctioned a percentage of their allowances, and with the exception of Denmark, with an auction of 5 per cent of total allowances, the other three countries auction 2.5 per cent or less. See Ellerman and Buchner, *supra* note 41 at 73.
55 See *EU ETS Insights and Opportunities, supra* note 32 at 8.
56 Ibid. at 8. Also, see Kruger et al., *supra* note 40 at 9.
57 Ibid. at 16.
58 D. Gow, 'Smoke Alarm: EU Shows Carbon Trading Is Not Cutting Emissions,' *The Guardian* (Brussels, 3 April 2007); online, http://business.guardian.co.uk/story/0,,2048733,00.html.
59 D. Ellerman and B. Buchner, *Over-Allocation or Abatement: A Preliminary Analysis of the EU ETS Based on the 2005 Emissions Data* 06–016, Center for Energy and Environmental Policy Research (2006) at 1; online, http://web.mit.edu/ceepr/www/2006-016.pdf.
60 Ibid. at 32.
61 'Commission Tightens Screw on Carbon Market,' *EurActiv.com* (29 October 2007); online, http://www.euractiv.com/en/climate-change/commission-tightens-screw-carbon-market/article-167966.
62 J. Mason, 'EU and U.S. at Odds over Airline Emissions Trade,' *Reuters* (21 September 2007); online, http://www.enn.com/pollution/article/23271.

63 PointCarbon, 'EU ETS Now Significantly Reducing Emissions,' Press Release (13 March 2007) at 1; online, http://www.pointcarbon.com/getfile.php/fileelement_103925/13_March_2007_EU_ETS_now_significantly_reducing_emissions.pdf.
64 See European Community, *Directive 2004/101/EC of the European Parliament and of the Council of 27 October 2004 Amending Directive 2003/87/EC Establishing a Scheme for Greenhouse Gas Emission Allowance Trading within the Community, in Respect of the Kyoto Protocol's Project Mechanisms*, Official Journal L 338, 13.11.2004, at 18–23; online, http://ec.europa.eu/environment/climat/emission/implementation_en.htm.
65 See International Institute for Sustainable Development, *Summary of the Seminar on Linking the Kyoto Project-Based Mechanisms with the European Union Emissions Trading Scheme* (19 September 2005) 115(1) Earth Negotiations Bulletin, at 2; online, http://www.iisd.ca/download/pdf/sd/ymbvol115num1e.pdf.
66 'California, U.K. to Strike Emissions Deal,' Toronto *Star* (1 August 2006); online, http://www.ieta.org/ieta/www/pages/index.php?IdSitePage=1166.
67 Pew Center on Global Climate Change, *Q & A: Regional Greenhouse Gas Initiative*; online, http://www.pewclimate.org/what_s_being_done/in_the_states/rggi/rggi.cfm.
68 'EU to Link Emissions Scheme with 3 Countries,' *Reuters* (26 October 2007); online, http://www.reuters.com/article/environmentNews/idUSL268295520071026.
69 International Carbon Action Partnership, *Political Declaration* (29 October 2007); online, http://www.icapcarbonaction.com/declaration.htm.
70 See, e.g., Council of the European Union, Information note 7242/05 (11 March 2005); online, http://register.consilium.europa.eu/pdf/en/05/st07/st07242.en05.pdf.
71 J. Hansen, M. Sato, et al., 'Global Temperature Change' (2006) 103(39) Proceedings of the National Academy of Sciences 14288.
72 J. Hansen, L. Nazarenko, et al., 'Earth's Energy Imbalance: Confirmation and Implications' (3 June 2005) 308(5727) Science 1431.
73 R. Warren, 'Impacts of Global Climate Change at Different Annual Mean Global Temperature Increases,' in H.J. Schellnhuber et al., eds., *Avoiding Dangerous Climate Change* (Cambridge: Cambridge University Press, 2006) 93.
74 M. Webster and A. Sokolov, 'Quantifying the Uncertainty in Climate Predictions,' Report #37 (MIT Global Change Joint Program, July 1998); online, http://web.mit.edu/globalchange/www/rpt37.html.

75 CO_2e is a measure of the contribution of key greenhouse gases.
76 B. Hare and M. Meinshausen, 'How Much Warming Are We Committed To and How Much Can Be Avoided?' (2006) 75 Climatic Change 111 at 131.
77 M.G.J. den Elzen and M. Meinshausen. *Meeting the EU 2°C Climate Target: Global and Regional Emission Implications*. Report 728001031/2005 (Netherlands Environmental Assessment Agency, 2005) at 2; online, http://www.mnp.nl/en/publications/2005/Meeting_the_EU_2_degrees_C_climate_target_global_and_regional_emission_implications.html.
78 J. Hansen, 'The Threat to the Planet' (13 July 2006) 53(12) New York Review of Books; online, http://www.nybooks.com/articles/19131.
79 See EC Commission, *Limiting Global Climate Change*, *supra*, note 23 at 9–10.
80 Ibid. at 10.
81 Ibid. at 9. See also Energy Information Agency, *Greenhouse Gases, Climate Change and Energy* (United States Department of Energy, 2004); online, http://www.eia.doe.gov/oiaf/1605/ggccebro/chapter1.html.
82 'The European Union Thinks It Can Be a Model for the World on Climate Change: Can It?' *The Economist* (15 March 2007) [hereinafter 'EU as Model for the World']. Also, it should be noted that, according to den Elzen and Meinshausen (see, *supra* note 77 at 2), a delay of a global effort to stabilize, and subsequently decrease emissions, by as little as a decade could require a doubling of the rates of abatement, with concomitant costs.
83 J. Mason, 'UPDATE 2-Third Phase of EU Carbon Trading to Go through 2020,' *Reuters UK* (30 October 2007).
84 I. Melander, 'EU Commission Delays Emissions Trading Proposals,' *Reuters* (19 October 2007); online, http://www.reuters.com/article/environmentNews/idUSL1956762720071019?feedType=RSS&feedName=environmentNews.
85 See draft *Climate Change Bill*, *supra* note 31.
86 'Benn Pledges Tougher Climate Bill,' *BBC News* (29 October 2007); online, http://news.bbc.co.uk/2/hi/uk_news/politics/7066735.stm.
87 It is expected to do so in January of 2008. See Melander, *supra* note 84.
88 See 'EU as Model for the World,' *supra* note 82.
89 This goal represents a significant improvement over the less than 7 per cent today; enhancing energy efficiency by 20 per cent by 2020 – an initiative that would translate to 780 tonnes of CO_2 saved annually. See EC Commission, *Energy Policy for Europe*, *supra* note 25 at 11.
90 'EU Agrees Renewable Energy Target' *BBC News* (25 October 2007); online, http://news.bbc.co.uk/2/hi/europe/6433503.stm.
91 Q. Schiermeier, 'Europe Moves to Secure Its Future Energy Supply' (18 January 2007) 445 Nature 234.

92 See EC Commission, *Limiting Global Climate Change, supra* note 23 at 6, 7.
93 See, e.g., Government of Germany, 'Speech by Federal Chancellor Angela Merkel at the "Mitigation" Panel of the UN Secretary-General's High-Level Event on Climate Change' (24 September 2007); online, http://www.bundesregierung.de/Content/EN/Reden/2007/09/2007-09-24-rede-bk-high-level-event.html.
94 On the goal of adoption by 2009, see 'Statement by Miguel Silvestre, First Secretary, Permanent Mission of Portugal to the United Nations, on Behalf of the European Union, United Nations 62nd General Assembly, 2nd Committee: Sustainable Development and Human Settlements' (29 October 2007); online, http://www.europa-eu-un.org/articles/en/article_7459_en.htm.
95 See *supra* note 2 and accompanying text.
96 See 'Submission from Portugal on Behalf of the European Community and Its Member States,' Dialogue Working Paper 16, UNFCCC, Dialogue on long-term cooperative action to address climate change by enhancing the implementation of the Convention, Fourth Workshop, Vienna, 27–31 August 2007, at paras. 4–6; online, http://unfccc.int/files/meetings/dialogue/application/pdf/wp16-eu.pdf [hereinafter *Dialogue*].
97 See *supra* notes 73–78 and accompanying text.
98 See UK Treasury, *Stern Review on the Economics of Climate Change* (30 October 2006); online, http://www.hm-treasury.gov.uk/independent_reviews/stern_review_economics_climate_change/stern_review_report.cfm.
99 See *supra* notes 80–82, and accompanying text. See also *Dialogue, supra* note 96 at paras. 10–11.
100 See Government of Germany, 'It Makes Economic Sense to Take Ecological Action,' Press Release (31 August 2007); online, http://www.bundesregierung.de/nn_6562/Content/EN/Artikel/2007/08/2007-08-31-f_C3_BCnfter-tag-bundeskanzlerin-in-japan_en.html. And see 'Moving beyond Kyoto: Merkel Takes the Climate Fight to Asia,' *Der Spiegel* (31 August 2007); online, http://www.spiegel.de/international/world/0,1518,503155,00.html.
101 See Ott, *supra* note 19 at 26.
102 See Suzuki Foundation *Backgrounder, supra* note 30 at 2.
103 See Ott, *supra*, note 19, at 36.
104 See, e.g., E. de Wet, 'The International Constitutional Order' (2006) 55(1) International and Comparative Law Quarterly 51.

4 The Future of U.S. Climate Change Policy

DAVID B. HUNTER

Being asked to predict the future of U.S. climate change policy is virtually impossible, particularly given the uncertainty that comes with a Presidential election year. Indeed, we can no sooner predict future U.S. climate policy than can we predict who will be the next president. On the other hand, so much uncertainty makes it impossible to be wrong.

Despite the potentially paralyzing uncertainty, some general statements about the future of U.S. climate change policy can be made. First, U.S. federal climate change policy is going to change; the status quo is under siege. Profound changes have taken place outside Washington, including a fundamental shift in public opinion regarding the severity of the threat from climate change. Growing public concern has already been translated into real action in state and local climate policy that now stands in stark contrast to the lack of movement at the federal level. The ongoing debate in Congress over climate legislation is one manifestation of the new reality, as is the increasing attention climate change is receiving from the presidential candidates. Even the Supreme Court has recently added its voice to the climate change debate.

Second, future U.S. federal climate change policy will almost certainly include some version of a cap-and-trade system as well as other policies and measures. Virtually all of the nearly dozen proposed climate change bills in the U.S. Congress include a cap-and-trade system. Other policy measures such as renewable portfolio standards, fuel efficiency standards, and investments in research and development have broad support and will also likely be part of any federal climate policy package.

Third, whatever legislation is passed by the U.S. Congress will shape the international negotiation position of the United States. Thus, future

U.S. negotiators will likely be supportive of whatever cap-and-trade limits are passed by the Congress. In the short term, it is likely that such legislation (and thus the U.S. negotiating position) will not be as strong as the targets and timetables championed by the European Union.

Fourth, the next president of the United States will likely not be as hostile to addressing climate change. Although several of the Republican candidates have platforms virtually indistinguishable from the Bush administration's climate policy, none of the current candidates appear to be as closely tied to the oil and gas industry as is President Bush nor as seemingly allergic to multilateral processes. Each of the leading Democratic candidates adopted relatively strong approaches to climate change. Internationally, this means that the United States will likely take a stronger position, perhaps even a leadership position, in future negotiations. Domestically, it means that any federal climate legislation will likely be implemented with at least some greater degree of seriousness.

Finally, regardless of what happens in Washington, future U.S. climate change policy may continue to be shaped by what is happening at the state and local levels, and by private industry. Climate change is already having significant impacts across the United States, which is causing political realignments and strong actions at subnational levels.

This chapter discusses these and other aspects of current and future U.S. climate policy. Part 1 describes current and future U.S. climate policy as it relates to international relations and negotiations. The future U.S. negotiating position will depend significantly on the outcome of the next presidential election and thus this part briefly reviews the current climate platforms of the leading U.S. presidential candidates. Part 2 discusses current and future federal domestic policy. This part also discusses the impact of the recent U.S. Supreme Court case, *Massachusetts v. EPA*, as well as summarizes draft legislation that has been introduced in the U.S. Congress. Part 3 analyzes recent activities at the state and local levels that are collectively having a major impact on U.S. climate policy. Part 4 concludes by identifying some of the implications of future U.S. climate policy for Canada and Canadian climate policy.

1. Current and Future U.S. International Climate Policy

Externally, the U.S. repudiation of Kyoto, its effort to undermine post-Kyoto talks, and its failure to commit to targets and timetables are viewed rightly as the most salient aspects of U.S. international climate policy. Although the United States remains a party to the *UN Framework Convention on Climate Change* (UNFCCC), that commitment has limited policy implications apart from reporting. While continuing to participate in the UNFCCC, the U.S. international strategy largely ignores that and all other global forums. Rather, the U.S. strategy has been to launch a series of research-oriented initiatives both multilaterally and through bilateral climate agreements, primarily aimed at promoting technology transfer for clean development and joint research initiatives.

International Cooperation in Research

The United States has promoted bilateral and multilateral cooperative agreements to encourage research into clean energy technologies and to climate-related science. The most important of these agreements is the Asia-Pacific Partnership on Clean Development and Climate (APP), a multilateral initiative launched in 2006.[1] In addition to the United States, the APP initially included Australia, China, India, Japan, and Korea. Canada just joined the APP on October 15, 2007. The United States has also developed bilateral partnerships with approximately twenty countries, including Brazil, India, China, South Africa, Italy, and Mexico.

None of these agreements require any specific emission reductions or the adoption of specific climate mitigation policies. The APP's Vision Statement exemplifies the approach:

> The Partners will collaborate to promote and create an enabling environment for the development, diffusion, deployment and transfer of existing and emerging cost-effective, cleaner technologies and practices, through concrete and substantial cooperation so as to achieve practical results. The Partners will also cooperate on the development, diffusion, deployment and transfer of longer-term transformational energy technologies that will promote economic growth while enabling significant reductions in greenhouse gas intensities. In addition, the Partners will share experiences in developing and implementing our national sustainable development and energy strategies, and explore opportunities to reduce the greenhouse gas intensities of our economies.[2]

The partners have since identified eight key market sectors to try to expand clean energy technologies through public-private task forces. These market sectors include aluminium, buildings and appliances, cement, clean fossil fuel use, coal mining, power generation and transmission, renewable energy, and steel. The bilateral cooperative agreements are essentially the same, reflecting general technology co-operation on energy efficiency, clean energy, and distributed energy technologies, as well as shared scientific research on topics such as forest carbon assessment, climate and environmental observations, economic and environmental modelling, and integrated environmental strategies.[3]

The United States also developed a number of international science and technology initiatives, most notably hosting the Earth Observation Summit in 2003, which helped to launch the Global Earth Observation System of Systems (GEOSS) a new international, integrated, sustained, and comprehensive Earth observation system that will greatly advance our understanding of climate change.[4] The Bush administration also launched the Carbon Sequestration Leadership Forum, a multilateral effort to advance technologies that capture and store carbon emissions.[5] Along the same lines, the United States participated in the International Partnership for a Hydrogen Economy, which is aimed at reducing the technological, financial, and institutional barriers to hydrogen and fuel cell technologies and to improve their energy security, environmental security, and economic security.[6]

All of these initiatives are similar in that they emphasize research and development, and technology transfer and dissemination as long-term solutions to global climate change. Yet these initiatives include no significant, measurable obligations against which to measure progress, and most outside observers have discounted them as essentially window-dressing to obscure the administration's lack of any meaningful international climate policy.

Post-Kyoto Negotiations

Because it is not a party to the *Kyoto Protocol*, the United States has not formally joined in the negotiations for post-2012 mandatory reductions. Initially, the Bush administration's strategy was both to ignore and seek alternatives to the post-Kyoto negotiations. For example, although the APP was ostensibly presented as consistent with efforts under the UNFCCC and complementary to the *Kyoto Protocol*, most observers believed that the United States and Australia were promoting the non-binding and flexible approach in the APP as an alternative attractive to developing countries who might otherwise be expected to adopt binding targets in future reporting periods under the Kyoto regime. The United States and Australia, as the only two major industrialized countries outside of the Kyoto regime, hoped they could gain the support of developing countries to oppose the Protocol. If the United States, Australia and, for example, China, India and Brazil all repudiated the Protocol's approach, it would then have been obvious that Kyoto was dead – or at least would be after the first reporting period.

This strategy came to a head at the first meeting of the Kyoto parties in Montreal in 2005, where the parties were expected to launch the beginnings of the negotiations for the post-Kyoto reporting periods. Ultimately, the United States could not entice developing countries to join them in repudiating the post-Kyoto negotiations. The United States reluctantly agreed to participate in a dialogue on strategic approaches for long-term cooperative action as long as it was clearly understood that the dialogue was not a first step toward future negotiations on new binding emission reductions.[7]

In the two years since the 2005 Meeting of the Parties, public opinion in the United States has shifted dramatically, leaving the Republican Party vulnerable to the perception that the administration was virtually alone in its strong anti-climate position. At the same time, the push by European leaders to make climate change a high profile issue at the G8 meetings beginning in Gleneagles left the United States even more isolated internationally. In response, the Bush administration has had to adjust its approach to international negotiations, although the changes have been mostly cosmetic.

First, the administration reached a compromise with the Europeans at the 2007 G8 Summit in Heiligendamm, Germany, in which the United States agreed that combating climate change was 'one of the major challenges for mankind' but failed to endorse explicitly the commitment promoted by the rest of the G8 to cut global emissions in half by 2050.[8] Moreover, the United States launched its push for a two-track approach which, while agreeing to 'actively and constructively' participate in the UNFCCC process for a post-2012 agreement, also prevailed in arguing that future talks must include all 'major emitters.' According to the G8 Communiqué, 'it is vital that the major emitting countries agree on a detailed contribution for a new global framework by the end of 2008 which would contribute to a global agreement under the *UNFCCC* by 2009.'[9]

The Bush administration viewed this as at least a public relations victory because by hosting the 'major emitters conference' the United

States was arguably bringing China, India, and Brazil to the table. This was in keeping with the U.S. position that large developing-country emitters have to be part of any multilateral climate negotiations. On the eve of the major emitters conference in September, 2007, China insisted on the conference being changed to 'major economies' so that they would not be lumped together as a major emitter with the United States.[10]

All of the invited major emitters participated in the first meeting in September 2007, but many of them sent mid-level bureaucrats and few policy advances were made. In keeping with its general emphasis on research and voluntary incentives, the Bush administration announced a proposal for an 'international clean technology fund' to help finance clean energy projects in developing countries.[11] Other governments and many observers view the major emitters' process as a U.S. attempt to assume a cloak of climate leadership in the administration's last year and to blunt momentum around the post-Kyoto talks.[12] Little reason exists to doubt this, and the reality is that little more can be expected from the United States in the post-Kyoto process before the Bush administration leaves office in January 2009.

This is not to say that the Bush administration is completely unwilling to take some bolder steps in the international arena as long as they do not force them to take stronger positions in the Kyoto/UNFCCC negotiating processes. Thus, the United States took a leadership role in the recent twentieth anniversary meeting of the *Montreal Protocol*. The United States, which is historically supportive of the *Montreal Protocol*, proactively pushed for the accelerated phase-out of hydrochlorofluorocarbons (HCFCs). Although negotiated under the *Montreal Protocol* regime, the primary purpose for accelerating the phase-out of HCFCs was to mitigate future climate change. Perhaps because the accelerated HCFC phase-out would primarily affect developing countries, did not implicate the oil and gas industry, and did not require any change in position vis-à-vis a global emission cap on carbon, the Bush administration saw a win-win situation. It could answer critics back home who want to see leadership on climate change, while not harming friends of the administration in the oil and gas industry. Whatever the reasons, the United States was an important positive voice in the parties' decision to accelerate HCFC phase-out schedules.[13] The resulting adjustment to the *Montreal Protocol* could result in as much as a billion tons of CO_2 reductions in the atmosphere.[14]

The U.S. position on HCFCs did not signal a major shift in U.S. climate change policy generally. One week after completing the *Montreal Protocol* negotiations, the United States hosted its major emitters conference and once again appeared outside the mainstream of international negotiations, at least with respect to emissions targets. As noted above, this position is unlikely to change at least until a new president takes office in 2009, or until Congress enacts national legislation introducing an emissions cap. The positions of the major presidential candidates are discussed below, followed in the next section by a discussion of U.S. national climate policy, including prospects for major legislation in the U.S. Congress.

Future Foreign Policy on Climate: The Next President

U.S. foreign policy is virtually the exclusive domain of the president, and thus the external U.S. position on climate change will be significantly affected by who prevails in next year's presidential elections. In general, the major presidential candidates' positions on climate change reflect partisan divides on the issue.[15]

All of the major Democratic candidates, including Hillary Clinton, John Edwards, and Barack Obama, have committed publicly, either by co-sponsoring legislation or in their presidential platforms, to a cap-and-trade system and to exhibiting 'leadership' in international negotiations. Both Clinton and Obama co-sponsored the *Global Warming Protection Act* of 2007, which would commit the United States to reducing emissions 80 per cent by 2050, and Senator Edwards has made the same commitment as part of his climate policy platform.[16] Although Senator Clinton has provided virtually no details regarding her climate platform,[17] both Edwards and Obama have complex, detailed positions, with some unique proposals. Edwards, for example, has called for a moratorium on the construction of any coal-fired power plants that do not have carbon capture and sequestration technology,[18] and has proposed a $13 billion fund, paid for by greenhouse gas (GHG) emitters, to be used for investing in renewable energy.[19] Obama would implement his cap-and-trade system by auctioning 100 per cent of the pollution credits, and use $150 billion of the resulting revenue to invest in clean energy technologies.[20] He also suggested that the federal government should assist auto manufacturers with their health care costs in exchange for one-half of those savings being invested in fuel efficiency research.

With one exception, John McCain, the major Republican candidates have essentially made no commitments to address climate change. McCain has been a leader on climate policy in the U.S. Senate and has co-sponsored the *Climate Stewardship Act* that would call for a cap-and-trade system reducing emissions by 30 per cent by 2050.[21] The other leading candidates, Rudy Giuliani and Mitt Romney, while acknowledging that climate change poses a threat, have made no specific commitments on climate change, preferring instead to discuss the issue in terms of energy independence.[22] All three leading Republican candidates also agree that the United States should have rejected the *Kyoto Protocol* and that the United States should not act 'unilaterally' without China, India, and the developing countries. Fred Thompson, another leading Republican candidate, claims uncertainty over what is causing climate change, but believes 'it makes sense to take reasonable steps to reduce CO_2 emissions without harming the economy.'[23]

This brief survey of the presidential candidates suggests, not surprisingly, that who wins the presidency matters with respect to U.S. leadership on climate policy. Clearly the Democratic candidates have seen climate change as an important issue for their electoral base and one that can help them in their primary election. Just as clearly, Republican candidates do not see the issue as one that is important to their base, and have for the most part avoided taking any clear positions on the issue. Particularly with respect to the Republicans who have yet to take a position, we can expect that whoever wins the Republican primary may need to move back toward the electoral centre. That centre in the United States now seems to be squarely behind some greater federal action on climate policy.

2. U.S. National Domestic Climate Policy

The current administration's domestic approach largely mirrors its international policy, which emphasizes a reduction in the growth of emissions by improving energy intensity, supporting voluntary initiatives by the private sector, and expanding funding into scientific research and research into alternative energy.

President Bush set a national goal of reducing U.S. greenhouse gas intensity (GHG emissions per dollar of GDP) by 18 per cent over the next 10 years. According to the administration, this increase in the economy's efficiency would represent a nearly 30 per cent improvement over business-as-usual and result in approximately 500 million metric tons in cumulative carbon-equivalent emissions reductions from business-as-usual estimates through 2012. The administration expected to achieve those reductions, however, through a series of voluntary and research initiatives.[24]

Environmental critics of the approach have pointed out that the administration inflated their estimates. In fact, it would provide only modest, if any, changes over business-as-usual, because GHG intensity in the United States has been declining over the past few decades. More importantly, this approach minimizes economic impact by allowing emissions to rise or fall with economic output – but it provides no assurance that a given level of environmental protection will be achieved. In fact, according to the Pew Center on Climate Change, even if the president's announced goal was met, total GHG emissions in the United States would *increase* 12 per cent by 2012, because of growth in the GDP.[25] The government's reliance exclusively on voluntary initiatives has also been heavily criticized. Even the Government Accounting Office has questioned the effectiveness of voluntary measures championed by the administration, noting that nearly half of the companies enrolled in the programs had not even set emissions reduction goals.[26]

Current Federal U.S. Climate Legislation and Policy

On the federal level, no legal requirement exists to reduce carbon dioxide or other GHGs for purposes of climate change. Current U.S. federal policy does promote research relevant to monitoring climate change and to promoting some clean technologies. The United States has also implemented its obligations to create a baseline of emission sources under the UNFCCC. Beyond those approaches, the U.S. federal approach has been to reach its goal of an 18 per cent reduction in GHG intensity by supporting a series of voluntary initiatives.[27] Finally, the U.S. Supreme Court has clarified that the U.S. Environmental Protection Agency (EPA) has the authority to regulate CO_2 under the *Clean Air Act* but until now the EPA has chosen not to do so. Each of these approaches is described below.

PROMOTING RESEARCH AND DEVELOPMENT

The first U.S. legislation promoting research relating to climate change was enacted as early as 1978. The legislation created a National Climate Program, located in the Department of Commerce, to coordinate research regarding 'natural and man-induced climate processes and their implications.'[28] Later, the Congress passed the *Global Change*

88 David B. Hunter

Research Act of 1990, which included global climate change within its scope. That Act requires the interagency development of a ten-year National Global Change Research Plan to 'advance scientific understanding of global change and provide usable information on which to base policy decisions relating to global change.'[29] With those two acts as background, the Bush administration announced the creation in 2002 of the U.S. Climate Change Science Program, which in turn led in 2003 to the release of a ten-year strategy for climate research. That strategy has five goals:

1. Improve knowledge of the Earth's past and present climate and environment, including its natural variability, and improve understanding of the causes of observed variability and change;
2. improve quantification of the forces bringing about changes in the Earth's climate and related systems;
3. reduce uncertainty in projections of how the Earth's climate and related systems may change in the future;
4. understand the sensitivity and adaptability of different natural and managed ecosystems and human systems to climate and related global changes;
5. explore the uses and identify the limits of evolving knowledge to manage risks and opportunities related to climate variability and change.[30]

This climate science strategy remains the primary federal initiative in climate science research, although as noted above the United States also participates actively in international science cooperation.

In addition to basic research the Bush administration has also supported research aimed at developing and promoting technological solutions to climate change. In 2002, the president established a cabinet-level Committee on Climate Change Science and Technology Integration (CCCSTI) that in turn oversees the Climate Change Technology Program (CCTP). The CCTP's goal 'is to assist the United States, including its R&D collaborators, at home and abroad, in developing advanced technologies eventually needed to achieve the long-term UNFCCC goal [of stabilizing GHG concentrations].'[31]

The *Energy Policy Act* of 2005[32] also emphasized research and development with a special focus on technologies that could reduce GHG intensity, in keeping with the President's overall policy goal of reducing GHG intensity. The *Energy Policy Act* calls for the creation of a Committee on Climate Change Technology, a national inventory of GHG-intensity reducing technologies under development and an overall national plan to deploy such technologies. The Act also made the export of GHG intensity reducing technologies an integrated part of U.S. foreign policy towards developing countries, directing the secretary of state to survey the 25 largest emitters among developing countries to determine market opportunities for the transfer, deployment and commercialization of U.S. GHG-intensity reducing technologies. Promoting these technologies has since been a visible part of U.S. bilateral trade and development agendas with several developing countries.

U.S. INVENTORY OF EMISSION SOURCES

As a party to the UNFCCC, the United States is obligated to prepare and maintain an inventory of 'anthropogenic emissions by sources and removals by sinks of all [GHGs] not controlled by the Montreal Protocol.'[33] The *Energy Policy Act* of 1992 was passed to meet this obligation and directs the Energy Information Agency to 'develop [and update annually] ... an inventory of the national aggregate emissions of each [GHG] for each calendar year of the baseline period of 1987 through 1990.'[34] Under the Act, the United States also established a voluntary GHG Registry where private industry and others could report the results of voluntary efforts to reduce, avoid, or sequester GHG emissions. For the year 2004, 226 companies reported that they had undertaken 2,154 projects to reduce or sequester approximately 480 million metric tons carbon dioxide equivalent (million $MTCO_2e$) greenhouse gases in 2004.[35]

VOLUNTARY APPROACHES

In announcing the U.S. policy goal of reducing GHG intensity by 18 per cent, President Bush also made it clear that the U.S. would achieve that goal through a combination of investments in research and development along with voluntary measures. The primary program for reaching the GHG intensity goal is called the Climate VISION (Voluntary Innovative Sector Initiatives: Opportunities Now) program, which is a public-private partnership launched in 2003 explicitly to contribute to the president's GHG intensity goal. Thus far, trade associations representing fourteen energy-intensive industrial sectors and the Business Council on Climate have joined Climate VISION and vowed to reduce their GHG intensity. According to the Climate VISION mission statement, the program helps its members:

- Identify and implement solutions for reducing GHG emissions that are cost-effective today;
- develop and utilize the tools to calculate, inventory, and report GHG emissions reduction, avoidance, and sequestration;
- develop strategies to speed the development and commercial adoption of advanced technologies;
- develop strategies across the commercial and residential sectors to help energy consumers reduce GHG emissions; and
- recognize voluntary mitigation actions.[36]

Despite the high profile of Climate VISION and the central role that it plays in implementing U.S. climate change policy, a recent Government Accounting Office (GAO) report found that it was not effective in meeting its goals. In particular, the GAO reported that the government did not track implementation of the promised commitments by industry and thus had no way of either encouraging implementation or measuring the program's actual effectiveness.[37] The GAO report found that, as of November 2005, only eleven of the fifteen participating trade groups had set targets under the program and only five had reported on their emissions.

The report was no more encouraging regarding the Climate Leaders program, the Administration's second highest profile voluntary initiative. Under the Climate Leaders program, private companies are encouraged to set ambitious GHG emission reduction targets, develop a plan for achieving the targets, and report annually on their progress.[38] According to the GAO report, however, as of November 2005 only 38 of the 74 firms had even set reduction goals.[39] The number of firms participating in Climate Leaders has grown since to nearly 150, but still less than half have set reduction targets.

In addition to Climate VISION and Climate Leaders, the U.S. government has launched a number of more specific voluntary initiatives aimed at specific aspects of climate change.[40] The Energy Star program, launched in 1992, is a multi-faceted effort to improve energy efficiency and thus reduce GHG emissions. The program provides the most widely recognized energy efficiency label for appliances, light bulbs and other products and has had a significant impact on improving energy efficiency in the United States. EPA claims that with Energy Star's help U.S. consumers reduced their emissions by the equivalent of 25 million cars and saved $14 billion in energy costs in 2006.[41] EPA has also had some success with public-private partnerships targeting specific sectors or GHGs, including a partnership aimed at reducing methane, particularly from landfills, and industrial gases, such as perfluorocarbons and HCFCs, with high global warming potential.[42]

THE *Clean Air Act* AND *Massachusetts v. EPA*

Substantial controversy has surrounded the issue of whether and to what extent the U.S. *Clean Air Act* empowers the Administration to regulate GHGs for purposes of addressing climate change.[43] The Bush Administration's position has been that without new amendments the government has no power to regulate carbon dioxide and other GHGs under the Act. Frustrated with the slow pace of federal climate policy, nineteen environmental organizations petitioned the EPA in 1999 to regulate GHG emissions from new motor vehicles under Section 202 of the *Clean Air Act*. EPA denied the petition in 2003, arguing that the Act 'does not authorize EPA to issue mandatory regulations to address global climate change ... and (2) that even if the agency had the authority to set GHG emission standards, it would be unwise to do so.'[44] EPA argued that GHGs were not 'air pollutants' as defined by the *Clean Air Act* and, alternatively, that climate change was so important that we would expect Congress to address it explicitly if they intended for the Act to cover the issue. The environmental groups, joined by several states and localities, sued the EPA claiming that they had misinterpreted the *Clean Air Act*. The case eventually reached the Supreme Court.

In *Massachusetts v. EPA*, the Supreme Court addressed climate change for the first time. A critical jurisdictional issue was standing of the parties to raise a case involving injury from climate change. To have standing, a party must allege that it has suffered some injury caused or potentially caused by the defendant's actions. The Court found (5 votes to 4) that at least the State of Massachusetts in its quasi-sovereign status had alleged a potential threat to its coastline and other interests, and thus had standing to bring a climate change-related action.

Turning then to the merits of the case the Court first found that the *Clean Air Act* had a sufficiently broad definition of 'air pollutant' to cover carbon dioxide and the other GHGs at issue.[45] Although Congress in passing the *Clean Air Act* did not explicitly address climate change, the Court found that it had deliberately provided EPA with sufficient flexibility to address new air pollution threats that might arise over time.[46] Having found that EPA had the regulatory authority to address GHGs, the Court then turned to EPA's argument that, in its discretion, now was not the time to regulate GHGs. There the Court found that EPA had not properly evaluated the issue. Section 202(a)(1) of the *Clean Air Act* states that '[t]he [EPA] Administrator shall by regulation prescribe ... standards applicable to the emission of any air pollutant from any class or classes of new motor vehicles or new motor vehicle engines, which in his judgment cause, or contribute to, air pollution which may reasonably be anticipated to endanger public health or welfare.'[47]

The Court then focused on the requirement that EPA must come to a judgment on whether GHGs endanger public health or welfare. 'Under the clear terms of the Clean Air Act, EPA can avoid taking further action only if it determines that GHGs do not contribute to climate change or if it provides some reasonable explanation as to why it cannot or will not exercise its discretion to determine whether they do.'[48] Because EPA 'offered no reasoned explanation for its refusal to decide whether GHGs cause or contribute to climate change,' EPA must revisit the issue and make an endangerment judgment. Thus, the Court did not order EPA to regulate GHGs, but it does require the Administration to make a reasoned judgment as to whether GHGs endanger public health or welfare. If it does make an endangerment finding, then the Court implied that EPA would still have broad latitude to decide what regulatory approach it would take.[49]

Massachusetts v. EPA could have far-reaching implications. The case was remanded to EPA for further action, but EPA has yet to respond. Environmental groups have petitioned EPA to try to get them to respond and have also filed a new petition to go beyond motor vehicles and have EPA regulate GHG emissions from maritime ships.[50] Moreover, although *Massachusetts v. EPA* applied only to mobile sources, the ruling that GHGs are 'air pollutants' may open the door to regulating carbon dioxide from stationary sources. Several cases seeking to force EPA to regulate GHGs from stationary sources had been held in abeyance pending the Supreme Court decision and those cases will now move ahead.[51] The initial result will likely not be immediate changes in the regulatory approach by the Bush administration, but the cases do clear the way for future administrations to regulate GHGs, *without* any new Congressional legislation.

Future U.S. Climate Legislation

Significant action in the U.S. Congress on climate change is probably not far off. Serious attention to climate change began in the 2005 Congress, when Senators John McCain (R-AZ) and Joe Lieberman (D-CT) introduced the *Climate Stewardship and Innovation Act* (S. 1151). The bill would have introduced a cap-and-trade mechanism as well as subsidies, construction loans, and a reverse-bid program to encourage such technologies as nuclear power, solar power, biofuels, and integrated gasification combined-cycle (IGCC) coal plants with geological carbon sequestration. Other bills introduced in both the House and Senate would have set even stronger emission limits, but few gained significant political support. The Senate did pass a Sense of the Senate Resolution that Congress should enact a national program of mandatory, market-based limits and incentives to slow, stop, and reverse the growth of GHG emissions. The Resolution came shortly after a cross-section of industry leaders called on the White House to support clearer federal standards on GHG emissions.[52]

The mid-term elections in 2006, which placed leadership of both houses of Congress in Democratic hands, were widely viewed as significantly improving the future prospects for sweeping climate change legislation. According to the Pew Center on Climate Change, as of mid-July 2007 no fewer than 125 bills, resolutions, and amendments specifically addressing global climate change and greenhouse gas (GHG) emissions had been introduced into the Congress this year – compared with 106 proposals submitted in the entire two-year (2005–6) term.[53]

Among those 125 proposed bills are 11 different relatively comprehensive bills that include economy-wide, cap-and-trade schemes for GHG emission reductions. Summarizing and evaluating even those eleven bills in detail is beyond the scope of this paper.[54] The proposed bills differ in the details about how initial pollution credits will be allocated, whether and how to include a safety valve for the price of carbon, and what the ultimate targets should be.[55]

The most stringent proposed bill is the *Safe Climate Act*, H.R. 1590, introduced by Congressman Henry Waxman. The *Safe Climate Act* would freeze U.S. GHG emissions in 2010 at the 2009 levels, and have an interim target of achieving 1990 levels by 2020. Thereafter, the Act would cut emissions by roughly 5 per cent per year so that U.S. GHG emissions would be 80 per cent lower than 1990 levels by the year 2050. Other proposed bills are not far behind. The proposal probably having the most support at this point is the bipartisan Lieberman-Warner *Climate Security Act*, S.2191, which was unveiled fully on October 18, 2007. That bill, which is the direct descendant of the McCain-Lieberman bill introduced in 2005, would achieve 1990 levels by 2020 and implement a full cap-and-trade system. The bill has garnered support from many Senators as well as mainstream environmental groups, such as Environmental Defense, National Wildlife Federation and Natural Resources Defense Council.

Although it is impossible to tell from the current proposals what the precise details of future climate legislation will be in the United States, it is probably fair to say that such legislation will likely include commitments to achieve 1990 levels of GHG emissions by 2020 and will either be silent about future periods or will set a goal of greater reductions by 2050 (although most likely those reductions will be less than the 80 per cent reductions set forth in the strongest proposals). Most observers also believe that, despite growing support for the Lieberman-Warner proposal, comprehensive climate legislation is unlikely to pass through the full Congress until after the next elections in 2008. Although too early to be certain, most analysts now believe that the Democrats will have larger majorities in both the House and the Senate, making passage of stronger climate legislation more likely.

3. Subnational Efforts

While little action took place at the federal level in the decade following the negotiation of the *Kyoto Protocol*, the same cannot be said for efforts at the state and local levels. Many regions, states, and localities around the country have moved to fill what they see as a void at the federal level and have taken significant actions to address climate change.

Some of these subnational initiatives have come in the form of emission targets and timetables. In August, 2006 California became the first state in the nation to set binding limits on GHGs when it passed the *Global Warming Solutions Act*.[56] That Act commits California to implementing a regulatory program that will reduce its emissions to 1990 levels by the year 2020, an estimated 25 per cent reduction over today's emission levels. The Act is expected to result in a state-wide emissions trading program as part of the strategies for achieving the emissions levels (assuming that it is not pre-empted by a federal program). Other states that have committed to emissions reductions targets include New Mexico, New Jersey, Maine, Massachusetts, Connecticut, New York, Washington, and Oregon.

States have also adopted a wide variety of policies and measures addressing climate change.[57] Nearly half of all of the states, for example, have adopted some type of renewable portfolio standard (RPS), which requires that a certain portion of a utility's energy comes from renewable sources by a given date. Thus, for example, Minnesota's RPS requires that 25 per cent of its utilities energy be from renewable sources by the year 2025. Also noteworthy is that California has taken advantage of its special status under the *Clean Air Act* to set its own automobile emissions standards for carbon dioxide. Vermont and nine other states have agreed to follow those standards as well.[58] Examples of other state policy initiatives have included the establishment of funds to support energy efficiency and renewable energy, support for 'green' premium pricing at public utilities, requirements that new and in some cases existing power plants offset some of their GHG emissions, and efficiency standards for appliances not covered by federal standards.

Significant movement on climate has also been seen at the local level in the United States. As of October 2007 nearly 700 mayors representing cities in every state had endorsed the U.S. Mayors Climate Protection Agreement,[59] which in addition to calling for bipartisan federal response to climate change also commits their municipalities to meeting the reduction goals of the *Kyoto Protocol* and beyond. Many U.S. cities have gone further, announcing substantial commitments to GHG reductions, including the following:

- Portland, Oregon: 10 per cent below 1990 levels by 2010
- Chula Vista, California: 20 per cent below 1990 levels by 2010[60]
- Oakland, California: 15 per cent below 1990 levels by 2010
- Berkeley, California: 15 per cent below 1990 levels by 2010[61]
- San Jose, California: 10 per cent below 1990 levels by 2000
- Seattle, Washington: 7 per cent to 40 per cent below 1990 levels by 2010[62]

The International Council for Local Environmental Initiatives estimates that municipalities from North America have collectively reduced GHG emissions by an equivalent of 60 million tons of carbon dioxide.[63]

In addition to state and local efforts, several regional initiatives have been launched. Most advanced is the Regional Greenhouse Gas Initiative (RGGI – a strategy for controlling emissions that has now been adopted by ten New England and mid-Atlantic states. The centrepiece of the RGGI is a cap-and-trade system that is conceptually similar to the *Kyoto Protocol*. Under the agreement, each of the participating states has committed to a mandatory 10 per cent reduction in CO_2 emissions from electric power generators by 2019. When the Initiative takes effect in

2009, each state will allocate or auction emission allowances to each power generator in their state. The emission reduction allowances can be traded between the participating states.[64]

These subnational initiatives are significant for several reasons. First, many U.S. states are significantly larger than many foreign countries, particularly when it comes to GHG emissions. California, for example, would rank 12th in emissions if it were a country, New York would rank 23rd, and Texas would rank 6th. Second, state legislation and other initiatives are providing lessons for the future development of federal climate legislation and policy. States have been experimenting with many policies and measures as discussed below, which both provide a testing ground for federal laws and build political will for federal policymakers. In part, because of the activity at the state level, many observers believe that future federal climate policy will also incorporate a renewable portfolio standard. Finally, bold action at the state and local level has served to isolate the Bush administration internationally and domestically. The leadership position taken by California's Republican governor, Arnold Schwarzenegger, for example, including the governor's outreach to international leaders such as Britain's Prime Minister Tony Blair and even Canadian provincial leaders, has provided a stark and unflattering contrast to the president's approach.[65]

4. Conclusion: Implications of U.S. Policy for Canada

From the outside, it is easy to view U.S. climate policy as being only the president's foreign policy – in this case, the repudiation of the *Kyoto Protocol* and an emphasis on research and non-binding, voluntary initiatives. That view, however, misses much of the complexity within the United States and the substantial changes occurring outside of the executive branch. These trends have isolated the administration domestically every bit as much as it is isolated internationally. Given that the presidential election is now only one year away, the influence of the Bush administration on future climate policy is even further diminished.

This also means that the specifics of future U.S. climate policy are uncertain. Nonetheless, we can say with some confidence that, in the future, the United States will likely join in more constructive participation within the UNFCCC/Kyoto framework. For Canada, this may mean that much of the criticism aimed at the United States may be redirected elsewhere, and specifically to countries that have not complied with the international obligations they have accepted. To the extent that the derelict position of the United States has muted international criticism regarding Canada's climate policy, this may no longer be the case. We might therefore expect that the international community will demand more from Canada.

Although the United States is likely to participate more fully in international climate negotiations, the U.S. positions on targets and timetables may continue to be less stringent than the European proposals, at least for near-term goals. Given that the United States has not been subject to any international or domestic obligation to curb its GHG emissions, the U.S. GHG budget has expanded steadily. According to the U.S. 2005 National Inventory, total U.S. greenhouse gas emissions were 7,260.4 Tg CO_2 eq. – an overall increase of 16.3 per cent in total U.S. emissions from 1990 levels. Just in the year 2004–5, emissions increased nearly 1 per cent (0.8 per cent). Although GHG intensity has gone down, the facts suggest fundamental changes have not occurred in the economy, and even meeting U.S. Kyoto commitments in the short term would be difficult.[66] Over the longer term (that is, by 2050 or later), U.S. commitments could come into alignment with those in Europe, but in the short-term the United States may continue to provide a dampening effect on international targets and timetables.

Regardless of whether the United States participates more constructively in the international climate talks, it is now likely that the United States will create a carbon market with linkages to regional or global markets. This may occur through international negotiations and federal sanction of cooperation in post-Kyoto markets, through the continued growth of subnational and state markets, or through the growing participation by U.S. firms in private carbon markets.[67] Just the scale of the U.S. economy ensures that this will fundamentally change the carbon market within which Canada operates. The particulars are hard to predict, however, because the details of the regulatory structure underlying the U.S. market are not yet known. Still, it seems relatively certain that the United States will become a more significant player in international carbon markets in the future. How the U.S. market is formed will undoubtedly impact how Canadian firms can participate, but in general we can expect further integration of a North American carbon market, either formally or informally.

Integration of a carbon-regulated U.S. economy into a global or regional carbon market will have uncertain impact on the price, but overall and over time it would seem to drive the price of carbon downward as the U.S. regulation drives technology innovation. This could help Canadian firms to the extent that less expensive technologies are made available in the Canadian market, but it may also lead to a competitive advantage for some industry sectors. Canada's window to benefit from the 'first movers' advantage is closing and is likely only a few years at most. Creating incentives now for Canadian firms to innovate would allow the chance for a vibrant clean technology industry to emerge in Canada that can participate effectively in the U.S. market once carbon is regulated there. In general, we can expect a larger North American market for GHG-reducing technologies in the future and how Canada prepares for that eventuality now may dictate whether it can shape the market and profit from it.

Finally, the public opinion shifts in the United States are not likely to reverse any time soon. Worsening climate impacts will continue to keep climate change policy a major political issue, particularly for those sectors of the economy that live close to the land. For farmers, ranchers, indigenous peoples, and fishermen, for example, climate change is simply a label for the profound changes they have been seeing for the past decade. Calls for action will likely get louder, and climate-related activists in the United States will likely share strategies and energy with their Canadian counterparts. Not only may this mean that U.S. strategies of litigation may find their way north of the border, but it also may be that continental campaigns will target Canadian activities, such as the development of Alberta's oil sands.[68] To be sure, these activists will ensure that climate change remains high on the North American policy agenda and a more significant component of bilateral U.S.-Canada relations in the future.

Notes

1 See online, http://www.asiapacificpartnership.org.
2 About the Asia-Pacific Partnership on Clean Development and Climate, available at http://www.asiapacificpartnership.org/About.htm#Purposes.
3 See, for example, remarks of Dr Harlan L. Watson, senior climate negotiator and special representative, at the U.S.-India Cooperation on Climate Change briefing, New Delhi, India, 11 November 2003.
4 See Interagency Working Group on Earth Observations, NSTC Committee on Environment and Natural Resources, *Strategic Plan for the US Integrated Earth Observation System* (2005) online, http://usgeo.gov/docs/EOCStrategic_Plan.pdf.

5 See *Charter for the Carbon Sequestration Leadership Forum (CSLF): A Carbon Capture and Storage Technology Initiative* (June 2003) online, http://www.cslforum.org/documents/CSLFcharter.pdf. Canada is also a member of the CSLF.
6 See online, http://www.iphe.net/. Canada is also a member of this partnership.
7 UNFCCC, 'Dialogue on Long-Term Cooperative Action to Address Climate Change by Enhancing Implementation of the Convention,' Decision-/CP11, online, http://unfccc.int/files/meetings/cop_11/application/pdf/cop11_00_ dialogue_on_long-term_coop_action.pdf.
8 *Chair's Summary,* G8 Summit 2007 Heiligendamm (8 June 2007).
9 Ibid.
10 R. Harribin, 'Manipulating the Climate Message,' *BBC News,* 7 October 2007, online, http://news.bbc.co.uk/2/hi/science/nature/7027887.stm.
11 J.M. Broder, 'Bush Outlines Proposal on Climate Change,' *N.Y. Times,* 28 September 2007, online, http://www.nytimes.com/2007/09/28/world/28cnd-climate.html.
12 Ibid.; see also Harribin, *supra* note 10.
13 'Historic Agreement Safeguards Both Climate and Ozone Layer,' *Env't. News Service,* 22 September 2007, online, http://www.ens-newswire.com/ens/sep2007/2007-09-22-01.asp.
14 'Nations to Ramp up Phase-out of Ozone-Killers,' *Associated Press,* 23 September 2007, online, http://www.msnbc.msn.com/id/20925073/.
15 For a summary of the Presidential candidates' positions on climate change, see K. Bennett and F. Hossain, 'N.Y. Times Election Guide 2008: The Presidential Candidates on Climate Change,' online, http://politics.nytimes.com/election-guide/2008/issues/climate/index.html.
16 John Edwards for President, 'Achieving Energy Independence & Stopping Global Warming through a New Energy Economy,' online, http://johnedwards.com/issues/energy/new-energy-economy/. Note that Senator Clinton has co-sponsored cap-and-trade legislation, but has not made specific targets a part of her platform. See Hillary for President, 'Promoting Energy Independence and Fighting Global Warming,' online, www.hillaryclinton.com/issues/energy/.
17 See ibid., Hillary for President.
18 John Edwards for President, 'Press Release: Edwards Calls for Cleaner Use of Coal as Part of Fight against Global Warming,' 26 March 2007, online, http://johnedwards.com/issues/energy/20070326-cleaner-coal/.
19 Bennett and Hossain, *supra* note 15.
20 See B. Obama, 'Meeting Energy Needs,' online: http://www.barackobama.com/issues/energy/.

21 See Bennett and Hossain, *supra* note 15; see also http://johnmccain.com/Informing/Issues/.
22 See Bennett and Hossain, *supra* note 15; On Giuliani generally, see online, http://www.joinrudy2008.com/issues/; on Romney generally, see online, http://www.mittromney.com/Issue-Watch/index.
23 See 'Issues: Energy Security,' online, http://www.fred08.com/Principles/PrinciplesSummary.aspx. Apparently Thompson's uncertainty over whether climate change is caused by humans is because he believes solar activity may be the cause of climate change. See Bennett and Hossain, *supra* note 15.
24 See The White House, 'Addressing Global Climate Change,' online, http://www.whitehouse.gov/ceq/global-change.html.
25 Pew Center on Climate Change, 'Analysis of President Bush's Climate Change Plan,' 12 February 2003.
26 See U.S. Government Accountability Office, *Climate Change: EPA and DOE Should Do More to Encourage Progress Under Two Voluntary Programs*, GAO-06-97 (April 2006) online, http://www.gao.gov/new.items/d0697.pdf.
27 The United States also has substantial amounts of energy-related legislation, or legislation in other sectors that has a clear climate change impact. This article does not discuss energy legislation generally or other laws that do not directly and explicitly aim at addressing climate change. For a lengthier discussion of US policies relating to climate change, see J. Dernbach, 'US Climate Policy,' in M. Gerrard, ed., *Global Climate Change and US Law* (Chicago: ABA Book Publishing, 2007).
28 15 U.S.C. s. 2902.
29 15 U.S.C. s. 2934(b)(1); see also Dernbach, *supra* note 27 at 75–6.
30 The U.S. Climate Change Science Program and the Subcommittee on Global Change Research, *The US Climate Change Program: Vision for the Program and Highlights of the Scientific Strategic Plan* (July 2003) at 3.
31 *Charter of the US Climate Change Technology Program Office*, online, http://www.climatetechnology.gov/about/charter.htm.
32 42 U.S.C. 15801.
33 UNFCCC, Art. 4(1)(a).
34 42 U.S.C. 13385(a); see also U.S. Environmental Protection Agency, *Inventory of US Greenhouse Gas Emissions and Sinks: 1990–2005*, USEPA #430-R-07-002 (April 2007), online, http://www.epa.gov/climatechange/emissions/usinventoryreport.html [hereinafter US GHG Inventory].
35 2004 is the last year for which data is available. Energy Information Administration, U.S. Department of Energy, 'Executive Summary,' *Voluntary Reporting of Greenhouse Gases, 2004*, DOE/EIA-0608(2004) (March 2006) at

ix. Online, http://www.eia.doe.gov/oiaf/1605/archive/vr04data/pdf/execsum.pdf.
36 Climate VISION, 'Program Mission,' online, http://www.climatevision.gov/mission.html.
37 GAO Report, *supra* note 26.
38 For a list of the participating entities in Climate Leaders, see online, http://www.epa.gov/climateleaders/partners/index.html.
39 GAO Report, *supra* note 26 at 3.
40 For a more complete review of voluntary programs, see T. Kerr, 'Voluntary Climate Change Efforts,' in M. Gerrard, ed., *Global Climate Change and US Law* (Chicago: ABA Book Publishing, 2007) 591.
41 U.S. Environmental Protection Agency, *Energy Star and Other Climate Protection Partnerships, 2006 Annual Report* (September 2007) at 1. Online, http://www.energystar.gov/ia/news/downloads/annual_report_2006.pdf.
42 Ibid.
43 Other federal environmental laws may also impact climate change indirectly. For example, the Center for Biological Diversity and other groups have petitioned to list the polar bear and other species under the *Endangered Species Act* because of impacts to their habitat caused by climate change, and Friends of the Earth has sued under the *National Environmental Policy Act* to force the Overseas Private Investment Corporation to assess climate impacts of its operations. See Center for Biological Diversity, *Petition to List the Polar Bear (Ursus maritimus) as Threatened under the Endangered Species Act before the Secretary of the Interior* (16 February 2005), online, http://www.biologicaldiversity.org/swcbd/SPECIES/polarbear/petition.pdf; Center for Biological Diversity, *Petition to List 12 Penguin Species under the Endangered Species Act before the Secretary of the Interior* (28 November 2006), online, http://www.biologicaldiversity.org/swcbd/SPECIES/penguins/Penguin Petition.pdf [hereinafter Penguin Petition]; *Friends of the Earth v. Mosbacher*, No. C02–4106 JSW, 2007 WL 962955 (N.D. Cal. Mar. 30, 2007) (order denying plaintiffs' motion for summary judgement and granting in part and denying in part defendants' motion for summary judgement).
44 *Massachusetts v. EPA*, 127 S.Ct. 1438, 1450 (2007).
45 Ibid. at 1460.
46 Ibid. at 1462.
47 42 U.S.C. 7521(a)(1).
48 *Massachusetts v. EPA* at 1462.
49 Ibid. at 1463.
50 Common Dreams, Press Release, 'Environmental Advocates Urge the EPA

to Reduce Global Warming Pollution from Ships,' 3 October 2007, online, http://www.commondreams.org/news2007/1003–03.htm.
51 *Coke Oven Environmental Taskforce et al. v. EPA*, Nos. 06–1131 and consolidated cases (D.C. Cir. filed 7 April 2006); *New York v. EPA*, No. 02–1387 (D.C. Cir. filed 31 December 2002).
52 See Bloomberg News, 'Companies Call on Bush for Emissions Guidelines,' *International Herald Tribune* (7 June 2005).
53 *Legislation in the 110th Congress Related to Global Climate Change*, online, http://www.pewclimate.org/what_s_being_done/in_the_congress/110th congress.cfm.
54 For detailed comparison of proposed climate legislation, see Resources for the Future, *Summary of Market-Based Climate Change Bills Introduced into the 110th Congress*, online, http://www.rff.org/rff/News/Releases/2007Releases/July2007Climate ChangeBillsinCongress.cfm.
55 Ibid.
56 AB 32 (2006). California's Governor Arnold Schwarzenegger has made even further commitments, issuing an executive order setting GHG reduction targets at 80 per cent below 1990 levels by 2050.
57 See generally Pew Center on Global Climate Change, 'Learning from State Action on Climate Change,' March 2007 Update, online, http://www.pewclimate.org/docUploads/States%20Brief%20Template%20 _March%202007_jgph.pdf (regularly updating a summary of state climate policies).
58 See *Green Mountain Chrysler-Plymouth-Dodge-Jeep et al. v. Crombie et al.*, No. 05-cv-302 (D. Vt. 12 Sept. 2007). For any of these emissions standards to become effective the U.S. EPA has to grant California a waiver, which historically the Agency has always done.
59 See the map and list of mayors online, http://usmayors.org/climateprotection/ClimateChange.asp
60 'Chula Vista CO_2 Reduction Plan' (14 November 2000) at 75; online, http://www.chulavistaca.gov/City_Services/Development_Services/Community_Development/PDFs/ChulaVista-CO2ReductionPlan.pdf.
61 'City of Berkeley Resource Conservation and Global Warming Abatement Plan' (January 1998) at 1; online, http://www.ci.berkeley.ca.us/sustainable/government/CO2%20Reducti on%20Plan.pdf.
62 West Coast Governors' Global Warming Initiative Staff Recommendations to the Governors, Appendix I: 'Setting Global Warming Pollution Reduction Targets' at 3, online, http://www.ef.org/westcoastclimate/I_Targets.pdf; see also, Chula Vista, *supra* note 61, and Berkeley, *supra* note 62.
63 International Council for Local Environmental Initiatives, *International*

Progress Report: Cities for Climate Protection (2006) at 4; online, http://www.iclei.org/documents/USA/documents/CCP/ICLEI-CCP_International_Report-2006.pdf.
64 For more information on the Regional Greenhouse Gas Initiative, see online, http://www.rggi.org/about.htm.
65 Adrian Croft and Timothy Gardner, 'Britain, California to Work Together on Global Warming,' *Reuters*, 1 August 2006, online, http://www.planetark.com/dailynewsstory.cfm/newsid/37457/story.htm.
66 U.S. GNP rose 55 per cent from 1990 to 2005. U.S. GHG Inventory, Executive Summary, *supra* note 34 at 3.
67 Interesting in this respect is the non-binding agreement California's Governor made with then-prime minister Tony Blair; see Croft and Gardner, *supra*, note 66.
68 In this regard, see for example the Natural Resource Defense Council's Comments to the Alberta Department of Energy's Oil Sands Consultations, 3 October 2006, online: http://docs.nrdc.org/international/int_06100301A.pdf.

5 China and India on Climate Change and Development: A Stance That Is Legitimate but Not Sagacious?

LAVANYA RAJAMANI

Climate change may well be the most significant environmental problem of our time – significant not just for its portended consequences, which are severe, but for its ability to test the extent of humankind's collective conscience to take moral responsibility for a problem of its own making. If the ongoing international negotiations are any indication, the international community is yet to come of age. Since their inception, the climate negotiations have witnessed intense bickering between and within the industrial and developing worlds over who should take responsibility, in what measure and under what conditions, to avert climate change.

China and India, with other large developing countries, have consistently argued that it is inequitable to ask developing countries, who have played little part thus far in creating the problem, to take on greenhouse gas reduction commitments.[1] There may well be legitimacy to this position vis-à-vis the industrialized world, when countries like the United States, which with 4% of the world's population is responsible for 23% of the world's emissions,[2] has rejected reduction commitments under the *Kyoto Protocol*.[3] India, 126th on the Human Development Index,[4] with 16% of the world's population, is responsible for 5.1% of the world's emissions.[5] China, 81st on the Human Development Index, with 21% of the world's population, is responsible for 16% of the world's emissions.[6] China's per capita emissions are 3.2 metric tons annually and India's are 1.2 – low compared to most industrialized countries.[7] It is in recognition of this powerful rhetoric of equity, enshrined in the climate change treaties as the principle of common but differentiated responsibility,[8] that developing countries do not as yet have reduction commitments.

Yet China and India are together responsible for a fifth of the world's emissions and their energy use is on the rise. In the last decade China's economy has n growing at an annual rate of 10.2%, its greenhouse gas emissions at 4%, and its energy consumption at 5.6% per year.[9] India's economy, which grew at 9.2% in 2006,[10] is fast catching up with China's. If India's current growth rate continues, energy demand will more than double by 2020.[11] In addition, if India is to meet its targets on poverty, unemployment, and literacy in its 11th five year plan,[12] with some of these targets being more ambitious than the Millennium Development Goals,[13] and to also provide energy to the estimated 44% of the population without access to electricity,[14] it will require much greater energy use. Thus, China and India will soon be significant contributors to the problem.

While the rhetoric of equity may serve these countries well in international forums, lack of serious domestic action will hamper the ability of the international community to tackle climate change. Besides, climate change will have significant impacts – economic, social and environmental – in both these countries. The Intergovernmental Panel on Climate Change (IPCC)[15] and the *Stern Review*[16] underscore this expectation. Even a small change in temperature could result in significantly lower agricultural yield, desertification, loss of arable land, and an escalating refugee crisis.[17] Climate change will critically impair China and India's economic growth and their ability to meet development goals.

In the face of increasing international pressure and mounting domestic concern, China and India's responses to the climate challenge have remained consistent and conservative – at least in the international arena. In international forums, China and India, both parties to the UN Framework Convention on Climate Change (UNFCCC)[18] and its *Kyoto Protocol*,[19] refuse to take on mitigation commitments. Domestically, these countries have set up systems to facilitate the operation of the Clean Development Mechanism (CDM). They are exploring action in areas where synergies exist between development and climate goals, for instance in energy efficiency, energy conservation, exploitation of renewable energies, and sustainable transport. China has, in addition, a dedicated climate policy – the National Climate Change Programme.[20] India does not yet have dedicated climate or adaptation laws or policies.[21]

Mitigation Commitments – 'Not Now – Not Ever'[22]

China and India are parties to the UNFCCC and its *Kyoto Protocol*. They participate in the G8+5 Gleneagles dialogue,[23] are members of the Asia Pacific Partnership on Clean Development and Climate,[24] and have bilateral relationships with the United Kingdom, the United States, and the European Union on climate research and technology. India has also entered into an agreement with the United States on civil nuclear energy, an agreement touted to have significant climatic benefits.[25]

India's position in these international forums as well as its response to Nicholas Stern, who visited India in 2006,[26] and the release of the IPCC Reports has been consistent. The National Environment Policy, 2006,[27] lists the elements that comprise India's response:

- to adhere to the principle of common but differentiated responsibilities and respective capabilities;
- to prioritise the right to development;
- to base discussion on equal per capita entitlements of all countries to global environmental resources;
- to engage in multilateral approaches;
- to participate in voluntary partnerships consistent with the UNFCCC.

Drawing from these elements, India argues that, given its limited role in contributing to the problem thus far, its overriding development needs that will lead to an increase in energy use, and the historical responsibility of industrialized countries and their current lack of leadership, it cannot be expected to take on mitigation commitments. India made this position clear at the inception of the international dialogue on climate change,[28] and has repeated it often since.[29] Indian negotiators claim that the burden-sharing architecture of the UNFCCC, in particular the language of the principle of common but differentiated responsibilities and the linking clause in Article 4(7), was crafted by Indian diplomats.[30] In keeping with this rhetoric of equity, some also present an interesting variant of the inter-generational equity argument – that is, unless the current generation generates and sustains high levels of economic growth, future generations will inherit an earth that is highly vulnerable to climate change. Inter-generational equity would therefore demand that the current generation prioritise development as a matter of urgency.[31]

China's response to climate change is driven by a set of similar principles:

- to address climate change within the framework of sustainable development;

- to follow the principle of common but differentiated responsibilities;
- to place equal emphasis on mitigation and adaptation;
- to integrate climate policy with other interrelated policies;
- to rely on the advancement and innovation of science and technology;
- to participate in international cooperation actively and extensively.[32]

Building on these elements, China argues that industrialized countries should face up to their historical responsibility and their current high per capita emissions, strictly abide by their emissions reduction targets in the *Kyoto Protocol*, and continue to take the lead in reducing emissions after 2012.[33] China believes that climate change is a development issue,[34] the UNFCCC and *Kyoto Protocol* should be the main channels for addressing climate change, and the principle of common but differentiated responsibilities and respective capabilities should guide efforts to address climate change.[35]

Following from their similar positions, both China and India reject developing country mitigation commitments in any way, shape, or form.

Given the congruence in climate policy drivers and imperatives, China and India often join forces to oppose discussion on developing country mitigation commitments.[36] Together they opposed a discussion of climate change at the UN Security Council in April 2007, arguing that the Security Council did not have any competence to deal with climate change, and opposed any suggestion that climate change, 'an uncertain long term prospect,' may have security implications.[37] They also stalled talks at the meeting to release the IPCC Working Group III Report on the mitigation of climate change,[38] part of the Fourth Assessment Report to be released in November 2007.[39] China and India insisted on a formal quantification of the historical responsibility of industrialized countries.[40]

The Chinese and Indian negotiating positions are critical to shaping the G77/China[41] consensus on various issues in the climate negotiations, in particular the issue of mitigation commitments. In its latest version the issue of developing country commitments (or rather the prospect of discussing it) is being discussed in the context of the Russian proposal on voluntary commitments.[42] Russia is seeking to offer procedures and mechanisms for those wishing to voluntarily reduce their emissions, and those not in Annex I but wishing to join it. In doing so it is seeking recognition for 'the sovereign right' of states to voluntarily undertake commitments. India is deeply resistant to this idea for fear that the notion of voluntary commitments is merely a trigger to negotiated commitments. There is some justification for this concern. Since the price of carbon depends on the stringency of the commitments undertaken, if unilaterally determined generous baselines are selected by developing countries, the supply of carbon will increase significantly and lead to a crash in the price of carbon. Unmonitored self-selection of mitigation commitments and baselines will provide considerable room to manipulate the carbon market. Therefore it is unlikely, and perhaps inadvisable, for the regime to permit truly 'voluntary' commitments.

In the discussions at the UNFCCC workshop on the Russian proposal in May 2007, India advocated voluntary 'actions' rather than commitments. Some Indian diplomats argue that at issue is not the desirability of moderating emissions growth but the question of incremental costs and who bears them.[43] The conversation about commitments, it is believed, is directed at persuading developing countries to bear part of the incremental costs.[44] China, at this workshop, questioned the utility of the notion of voluntary commitments, stressed existing commitments, and stated its opposition to any follow-up process to the workshop.[45]

At a more pragmatic level, India argues that its emissions will not have a significant impact on the global climate change trajectory since there are significant polluters outside the regime.[46] There are also some perverse incentives built into the system for developing countries to stay energy inefficient and environmentally unsound – for only then will they be able to demonstrate the 'additionality' necessary for CDM projects.

The Alternative: Voluntary Actions to Decarbonize the Economy Coupled with Technology Transfer, Financial Assistance, and Capacity Building

India advocates equal rights to the atmosphere. It has done so since the negotiations at the Intergovernmental negotiating committee in the lead up to the UNFCCC.[47] In India's view, while developing countries, with less than equal current use of the atmosphere, cannot be expected to take on absolute emission limitation commitments, they can, nevertheless, take voluntary practical actions to decarbonize their economies. Indian negotiators are willing, however, to commit that India will not on a per capita basis allow its emissions to exceed those of OECD countries.⁴⁸ Decarbonization, according to India, refers to lower carbon intensity of the economy over time. It does not refer to a reduction in the absolute level of greenhouse gas emissions over time, or to a reduction in the rate of GDP growth.⁴⁹ Decarbonization includes:

- enhanced energy efficiency;
- shift in primary energy use from fossil fuels to renewables (including hydropower) and nuclear energy; and
- changes in production and consumption patterns.⁵⁰

In India's Initial National Communication,⁵¹ presentations at the Dialogue,⁵² and other submissions to the UNFCCC, it lists actions that it has taken to contribute to decarbonization. These include initiatives to promote renewable energy,⁵³ energy efficiency,⁵⁴ and energy conservation,⁵⁵ ensure cleaner transport (including through imposition of European norms and conversion of all public vehicles to compressed natural gas in New Delhi⁵⁶), develop fuel-efficient appliances, and implement afforestation and land restoration.⁵⁷ India also identifies opportunities for emissions reductions in various sectors, the 'CDM potential' as it were, and quantifies the investment outlay needed to transition to a low carbon path in the 2007–2012 plan period (US$25.1 billion).⁵⁸

In addition, India proposes the establishment of a Clean Technology Acquisition Fund. At Gleneagles 2005,⁵⁹ at the UN General Assembly in 2006,⁶⁰ and at the fifteenth session of the UN Commission on Sustainable Development 2007,⁶¹ India argued that the technologies that can help it decarbonize are out of reach because of intellectual property rights and prohibitive costs. It asserts that the international property rights issue was successfully addressed in the HIV/AIDs context in developing countries, and should be tackled in the same way in the field of sustainable development.⁶²

China's effort to decarbonize also contains similar elements. In China's Initial National Communication,⁶³ presentations at the Dialogue,⁶⁴ and other communications to the UNFCCC, it lists its laws on energy efficiency, conservation, and renewable energy, actions promoting forestry, and efforts to develop low carbon energy solutions by developing *inter alia* hydropower, nuclear power, thermal power, and bio-energy.⁶⁵ In its 11th five-year plan China targets to increase forest cover by 20% and decrease pollution by 10% and energy intensity by 20%.⁶⁶ Like India, China also emphasizes the need for effective technology transfer and capacity building.⁶⁷

China and India's membership in the Asia Pacific Partnership on Clean Development and Climate, which focuses on 'voluntary practical measures,' and on 'national strategies, experience-sharing, and technology development and deployment,' is in keeping with these countries' preferred strategy.[68] So too are India and China's bilateral partnerships with the European Union,[69] the United States,[70] and the United Kingdom[71] on climate research and technology.

A recent development, in keeping with India's policy priorities, is the controversial India–U.S. nuclear agreement.[72] Under this agreement, India will gain access to U.S. civil nuclear technology in return for opening up its facilities for inspection.[73] There is continuing domestic political opposition to this agreement for reasons ranging from national security and sovereignty to nuclear non-proliferation concerns, but one forgotten element of this agreement is its carbon implications. The U.S. Senate Committee on Energy and Natural Resources considered testimony that suggested that the annual carbon savings from this agreement could be nearly as large as the entire commitment of the EU to meet the *Kyoto Protocol*.[74] Although it does not have a similar agreement with the United States, China also considers nuclear power as a central element of its strategy to address climate change.[75]

Evaluating China and India's Stance

As the chair of the Nobel Prize–winning IPCC, R.K. Pachauri, noted 'while the [Indian and Chinese] position may well be legitimate, whether it is sagacious is questionable.'[76]

A Legitimate Stance

The stance taken by India and China is arguably legitimate for it is firmly positioned within the burden-sharing architecture of the UNFCCC and its *Kyoto Protocol*. This burden-sharing architecture contains three central elements which would impart these countries' positions with legitimacy: redistribution of the ecological space, common but differentiated responsibility, and the linking clause.

REDISTRIBUTION OF THE ECOLOGICAL SPACE
One of the central goals of the climate regime is the redistribution of the ecological space. Evidence for this exists in the language of the UNFCCC and *Kyoto Protocol*. The UNFCCC preamble contains a recog-

nition that 'the share of global emissions originating in developing countries will grow to meet their social and development needs.'[77] Elsewhere in the preamble the UNFCCC adds that in order for developing countries to progress towards sustainable social and economic development 'their energy consumption will need to grow.'[78] These preambular provisions do not provide developing countries with a carte blanche to increase their emissions. The phrase 'share of global emissions' is critical. It implies that the UNFCCC countenances growth of emissions in developing countries *relative* to the emissions of industrial countries, not in itself.[79] Further, the recognition of the need for increased energy consumption in developing countries is buttressed by references to 'greater energy efficiency' and 'application of new technologies.'[80] Despite the boundaries within which growth in developing countries' emissions is countenanced, there is a clear signal that one of the objects of the UNFCCC is the redistribution of the ecological space.

This goal is in keeping with UNGA Resolution 44/228 referred to in the UNFCCC preamble, which mandates that 'the protection and enhancement of the environment must take fully into account the current imbalances in global patterns of production and consumption.'[81] The recognition that the share of developing countries' emissions will grow is to be read in conjunction with the objective in UNFCCC Article 2 to 'stabilize greenhouse gas emissions,' and the emphasis in the common but differentiated responsibility principle and elsewhere[82] that the industrial world is responsible for the largest share of historical and current greenhouse gases and must assume a leadership role in rising to the climate challenge. It follows that industrial countries are required under the climate regime to reduce the ecological space they occupy in favour of developing countries.

COMMON BUT DIFFERENTIATED RESPONSIBILITY AND INDUSTRIAL COUNTRIES' LEADERSHIP

In the 1980s, in the process leading up to Rio[83] and at Rio, in particular in the climate negotiations,[84] there was a growing[85] albeit not universal acknowledgment[86] of industrial country contributions to the global environmental crisis. This acknowledgment was articulated as the principle of common but differentiated responsibilities.[87] The principle establishes unequivocally the common responsibility of states for the protection of the global environment. Next, it builds on the acknowledgment by industrial countries that they bear the primary responsibility for creating the global environmental problem by taking into account the contributions of states to environmental degradation in determining their levels of responsibility under the regime. In doing so it recognizes broad distinctions between states, whether on the basis of economic development or consumption levels. It also, by its clear terms, assigns a leadership role to developing countries.

The principle of common but differentiated responsibilities, contained in Rio Principle 7[88] and UNFCCC Article 3, notwithstanding much legal wrangling,[89] is the overarching principle guiding the future development of the climate regime. It is found in two operational paragraphs of the UNFCCC, a binding treaty with near universal participation, and reiterated in the preamble of the *Kyoto Protocol*. It is also frequently referred to in UNFCCC conference of parties' decisions and ministerial declarations.[90] Even though this principle does not assume the character of a legal obligation in itself, it possesses a 'species of normativity' implying a certain legal gravitas.[91] It is still the *context* within which international environmental law functions such that this principle, among others, forms the bedrock of the burden sharing arrangements crafted in different environmental treaties. And, it is a fundamental part of the conceptual apparatus of the climate regime such that it forms the basis for the interpretation of existing obligations and the elaboration of future international legal obligations within the regime in question.[92]

THE LINKING CLAUSE

UNFCCC Article 4(7), oft quoted by developing countries,[93] termed here as the 'linking clause' reads:

> The extent to which developing country Parties will effectively implement their commitments under the Convention will depend on the *effective implementation by developed country Parties of their commitments under the Convention related to financial resources and transfer of technology* and will take *fully into account* that *economic and social development and poverty eradication* are the *first and overriding priorities* of the developing country Parties.[94]

This provision, like similar provisions in the *Montreal Protocol*,[95] the *Convention on Biological Diversity*,[96] and the *Stockholm Convention*,[97] is a significant innovation in that by linking developing countries' participation and implementation to industrial countries' commitments, it underpins and reinforces the compact between developing and industrial countries with respect to international environmental protection.[98]

There are various elements to this provision,[99] but for current purposes, it is worth noting that the recognition that economic and social development and poverty eradication are the first and overriding priorities of the developing countries could be read as a limited exception to the link between industrial and developing countries' implementation. In other words, even if industrial countries fulfil their assistance commitments, developing countries could claim, provided it did not constitute a breach of customary obligations, that overriding priorities had come in the way of the implementation of their commitments.

But Not Sagacious

The Indian and Chinese negotiating stances, given the continuing differences in per capita emissions levels between countries, fits squarely within the climate regime's burden sharing architecture, and is therefore legitimate. It is nonetheless not a sagacious position to hold.

Poorer nations, and the poorest within them, will be the worst hit by climate change.[100] This is indeed the fundamental inequity at the heart of the climate change problematic – that those who have contributed the least to causing climate change will bear the real brunt of it. 'Like the sinking of the Titanic, catastrophes are not democratic,' and 'a much higher percentage of passengers from the cheaper decks' will be lost.[101] A vast majority of the occupants of the cheaper decks are Chinese and Indians. The Human Development Indicators estimate that 16.6% of China's 1.3 billion, and 34.7% of India's 1 billion live on less than US$1 a day.[102] A vast majority of China and India's poor are in rural areas and are dependent directly on climate-sensitive natural resources.[103] China estimates that 23.65 million of its people in rural areas are poverty-stricken.[104] The poor have the least adaptive capacity.[105] And, climate change is predicted to have severe impacts in China and India.[106] Climate change will increase the severity of draughts, land degradation, and desertification, the intensity of floods and tropical cyclones, the incidence of malaria and heat-related mortality, and decrease crop yield and food security.[107] In addition, rising sea-levels will displace coastal populations and lead to an escalating refugee crisis.[108] Melting Himalayan glaciers will initially increase flood risk and eventually threaten water shortages for the one-sixth of humanity primarily in the

Indian sub-continent.[109] The impacts in China are likely to be equally severe: 27.4% of China suffers from desertification, and it has a continental coastline of over 18,000 kilometres, as well as more than 6,500 islands larger than 500 square metres. China is also intensely vulnerable to climate change.[110] The Chinese National Assessment Report on Climate Change released in December 2006 estimates that the negative impact of a warming climate on agricultural productivity could reach 37% by the second half of the 21st century, and seriously impair food security in China.[111]

The *Stern Review* highlights the toll that climate change could take on the Indian economy. Even a small change in temperature could have a significant impact on the Indian monsoon, resulting in up to 25% lower agricultural yield.[112] A 2–3.5-degree centigrade temperature increase could cause as much as 0.67% GNP loss, and a 100-centimetre increase in sea level could cause 0.37% GNP loss.[113] A quarter of the Indian economy is dependent on agriculture, and any impact on this sector will fundamentally impair India's ability to meet its development goals. Similar impacts are expected in China. Weather-related natural disasters already cause a direct economic loss of crops of 2% of GDP.[114]

It is of critical importance that climate change concerns are mainstreamed into development planning and concrete actions are taken to transition to a low carbon or beyond carbon development pathway. It is also important that commitments are undertaken at a global level, for it is only cumulative global not regional or national emissions reductions that will impact the trajectory of climate change. Both India and China are taking action – to differing extents and with differing emphases – at the domestic level. Recent reports indicate that China, India, and Brazil are already making significant unilateral greenhouse gas reductions. Seventy per cent of these reductions are occurring outside the CDM in Brazil and China, and 30% in India.[115] It is politically unlikely, however, that they will be willing to accept commitments in any shape or form at the international level.[116] The challenge for the Bali negotiations will be to strengthen the existing confidence-building architecture in the convention and protocol, which will eventually create the conditions necessary for developing countries to undertake mitigation commitments without economic hardship.

The Climate Negotiations at Bali

In the last two years countries have, as part of discussions in the Dialogue,[117] identified essential elements of a post-2012 climate regime. These include, *inter alia*:

- a long term goal consistent with science with short term reduction objectives for industrialized countries, and strategies for cleaner development paths for developing countries;
- national climate change strategies covering mitigation and adaptation;
- market mechanisms that can reduce the cost of mitigation and provide incentives for cleaner development;
- urgent attention to adaptation needs; and
- efforts to stimulate diffusion, transfer, and deployment of lower-emissions technology.[118]

It has become clear over time that poverty eradication and economic growth continue to function as the overriding priorities of developing countries. In their positions going into the Bali negotiations most developing countries are likely, in addition to reiterating their opposition to mitigation commitments, to highlight the need to

- address, on an urgent basis, adaptation needs;
- address energy security and energy efficiency concerns;
- address lack of access to climate-friendly technologies and investment;
- identify and exploit co-benefits[119] of development and climate actions; and
- provide continuity and certainty to the carbon market, and incentives for equitable geographical distribution of the benefits.

Therefore if developing countries are to be integrated more 'meaningfully' into the climate regime going forward, the post-2012 agreement must

- not include binding emissions reduction targets for developing countries;
- provide incentives for developing countries to limit their emissions growth;
- facilitate the adoption of a cleaner development path which exploits co-benefits and does not limit economic growth; and
- address competitiveness concerns.

The agreement must, however, further strengthen and expand mitigation commitments for industrialized countries. The Ad Hoc Working Group on Further Commitments for Annex I Parties under the *Kyoto Protocol* (the AWG)[120] is mandated to consider future commitments for industrialized countries under the *Kyoto Protocol*, for the period after 2012. Parties are currently considering a range of 25–40% below 1990 emissions levels, in keeping with latest scientific input, for further emissions reductions by Annex I parties.[121]

There are promising signals that the necessary political will for ambitious mitigation targets may exist among some industrialized nations. The G8 summit in Heiligendamm in June 2007 promised to 'seriously consider' at least a halving of global emissions by 2050.[122] The European Union has announced ambitious unilateral targets to cut emissions by at least 20% below 1990 levels by 2020.[123] Japan, through its 'Cool Earth 50' strategy, has proposed setting a long-term global target of halving emissions from current level by 2050.[124] And even the United States, by bringing together the 'major economies' to discuss energy security and climate change,[125] is endorsing the value of multilateral approaches.

Conclusion

Popular consciousness of climate concerns has never been higher. An evident warming of the globe, coupled with high-profile endorsement and nurturing media interest has ensured that climate concerns are seldom far from the surface of public discourse. And, rightly so. The scientific consensus is overwhelming. The IPCC warns that the warming of the climate system is unequivocal and accelerating.[126] The global average temperature has increased by 0.74 centigrade in the last century, the largest and fastest warming trend in the history of the Earth.[127] And, the economics favour action over inaction.[128] There is, as a result, gathering political momentum. The UN, the G8, the EU, the APEC, and others have prioritised climate change in their meetings this year, and set the scene for ambitious climate actions.

The intergovernmental climate negotiations, however, have yet to be instilled with the appropriate sense of urgency. At the root of this – across and within the industrialized and developing world – lies an enduring fear of economic hardship. Economic growth and climate protection are not mutually exclusive. But this fear, albeit misplaced, is creating a disturbing political drag on the intergovernmental negotiations. Unless the fear of economic hardship is squarely addressed in the post-2012 agreement, the future of the international climate regime, and indeed of the world, will hang in balance.

Notes

1 Preamble, 'Chinese Non-Paper & Statement by the Leader of the Indian Delegation,' 19 June 1991 (Washington Session of the Inter-governmental Negotiating Committee), cited in Chandrashekar Dasgupta, 'The Climate Change Negotiation,' in I.M. Mintzer and J.A. Leonard, eds., *Negotiating The Climate Convention* (Cambridge: Cambridge University Press/Stockholm Environment Institute, 1994) 129, 133–4. At the IPCC Working Group III session in May 2007, India, China, and Brazil sought language to formally recognize and quantify the historical responsibility of industrialized countries. See 'India, China Hold Up Climate Talks,' *The Times of India*, 3 May 2007.
2 Indian negotiators claim that they were instrumental in crafting this principle. Interview with Chandrashekar Dasgupta, member of the Indian delegation to the UNFCCC and Kyoto Protocol Conferences and Meetings of Parties, 16 April 2007.
3 In March 2001, U.S. president George Bush rejected the *Kyoto Protocol, 1997*, in part because it 'exempts 80 percent of the world, including major population centers such as China and India, from compliance.' See 'Text of a Letter from the President to Senators Hagel, Helms, Craig, and Roberts,' The White House, Office of the Press Secretary, 13 March 2001, online, http://www.whitehouse.gov/news/releases/2001/03/20010314.html.
4 *Human Development Report 2006*, online, http://hdr.undp.org/hdr2006/statistics/.
5 Ibid.
6 Ibid.
7 The United States has a per capita emissions rate of 19.8, Australia of 18 and Canada of 17.5. Ibid.
8 For an extended analysis and references on this principle see Lavanya Rajamani, *Differential Treatment in International Environmental Law* (Oxford: Oxford University Press, 2006), chapter 5.
9 Presentation by Gao Guangsheng, Director General, Office of National Coordination Committee on Climate Change, National Development and Reform Commission, P.R. China, at the Major Economies Meeting on Energy Security and Climate Change, 26 September 2007, Washington.
10 India's growth rate for 2006–2007 is 9.2% and has averaged just above 8%

118   Lavanya Rajamani

in the last four years. See Economic Surveys, Ministry of Finance, Government of India, online, http://finmin.nic.in/index.html.
11 See India Country Presentation to the UNFCCC, 'Dialogue on Long Term Cooperative Action to Address Climate Change by Enhancing Implementation of the Convention,' May 2006, online, http://unfccc.int/meetings/dialogue/items/3669.php.
12 See Planning Commission, Government of India, 'Towards Faster and More Inclusive Growth: An Approach to the 11the Five Year Plan' (2006) at 98, online, http://www.planningcommission.nic.in/ plans/planrel/app11_16jan.pdf .
13 See Ministry of Environment and Forests, Government of India, 'India's Initial National Communication to the United Nations Framework Convention on Climate Change,' 2004, Table 6.1 at 192–3, comparing the *Millennium Development Goals* to the related Indian targets under the 10the five year plan, online, http://unfccc.int/national_reports/non-annex_i_natcom/items/2979.php.
14 See National Electricity Policy, 2005, s. 2, online, http://powermin.nic.in/whats_new/national_electricity_policy.htm
15 IPCC Working Group II Report, 'Impacts, Adaptation and Vulnerability,' 2007; Summary for Policy Makers, online, http://www.ipcc-wg2.org/.
16 *Stern Review on the Economics of Climate Change* (2006) online, http://www.hm-treasury.gov.uk./independent_reviews/stern_review_economics_clima te_change/sternreview_index.cfm.
17 Ibid.
18 *United Nations Framework Convention on Climate Change*, 29 May 1992, U.N. Doc. A/AC.237/18 (Part II)/Add. 1, reprinted in (1992) 31 I.L.M. 849 [hereinafter UNFCCC].
19 *Kyoto Protocol to the United Nations Framework Convention on Climate Change*, 10 December 1997, U.N. Doc. FCCC/CP/1997/L.7/add.1, reprinted in (1998) 37 I.L.M. 22 [hereinafter the *Kyoto Protocol*].
20 China's National Climate Change Programme, National Development and Reform Commission, People's Republic of China (June 2007) online, http://en.ndrc.gov.cn/newsrelease/P020070604561191006823.pdf.
21 For an extended discussion of India's approach to energy, climate and development goals, see Lavanya Rajamani, 'The Indian Way: Exploring Synergies between Energy, Development and Climate Goals,' in Catherine Redgwell et al., eds., *Beyond the Carbon Economy* (Oxford: Oxford University Press, forthcoming Spring 2008).
22 In conversation with Prodipto Ghosh, erstwhile Secretary, Ministry of Environment and Forests, Government of India, and member of the Indian dele-

gation to the UNFCCC and Kyoto Protocol Conferences and Meetings of Parties, 17 April 2007.
23 For further details see online, http://www.g8.gov.uk.
24 In 2005 the United States joined forces with Australia, China, India, Japan, and South Korea to initiate the Asia-Pacific Partnership on Clean Development. The partnership focuses on 'voluntary practical measures' to 'create new investment opportunities, build local capacity, and remove barriers to the introduction of clean, more efficient technologies.' It disavows climate solutions that put the world on an 'energy diet' and expresses solidarity with developing countries for, in President Bush's words, 'like us, developing countries are unlikely to join in approaches that foreclose their own economic growth and development.' The White House, Office of the Press Secretary, Fact Sheet, 'President Bush and the Asia-Pacific Partnership on Clean Development,' 27 July 2005, online, http://www.state.gov/ g/oes/ rls/fs/50314.htm. See also *Vision Statement of Australia, China, India, Japan, the Republic of Korea and the United States of America for a New Asia-Pacific Partnership on Clean Development and Climate*, 28 July 2005, online, http:// www.asiapacificpartnership.org/vision.pdf. The Asia-Pacific partners met in a ministerial meeting in Sydney in January 2006 to release a charter, a communiqué and a work plan, available on the website of the Australian Government's Department of Foreign Affairs and Trade, http://www .dfat.gov.au/environment/climate/ap6/index.html. The partners met in a second ministerial meeting in New Delhi in October 2007. Canada joined the partnership on 15 October 2007 in New Delhi. See for further details, http://www.asiapacificpartnership.org/.
25 See *infra*, text accompanying notes 72–74.
26 Sonu Jain, 'Climate Report Says India Vulnerable but Is Delhi Listening?' *Indian Express*, 3 February 2007.
27 *National Environment Policy, 2006*, available on the website of the Ministry of Environment and Forests, http://envfor.nic.in/nep/nep2006e.pdf.
28 See *supra* note 1.
29 For a flavor of these see Saifuddin Soz, 'India Rejects Incorporation of New Environmental Commitments for Developing Countries,' Address to the UNFCCC, 8 December 2007, online, http://www.indianembassy.org/ policy/Environment/soz.htm; Katie Mantell, 'Indian PM Says No to Emissions Targets,' *SciDev.Net*, 31 October 2002, online, http://www.scidev.net/ content/news/eng/ indian-pm-says-no-to-emissions-targets.cfm; and more recently, 'Ministry Hints Tough Stand on Climate; PM Meet Today,' *The Indian Express*, 16 May 2007, and 'India to Take Firm Stand on Global Warming,' *The Hindustan Times*, 22 May 2007.

30 Interview with Chandrashekar Dasgupta, 16 April 2007.
31 Ibid.
32 See *supra* note 20 at 24–6.
33 See *supra* note 20 at 58–62. See also 'Hu Jintao Expounds China's Climate Stance in APEC Meeting,'*Xinhua*, 8 September 2007.
34 Ibid.
35 Ibid.
36 'China and India Agree to Work More Closely,' *China Daily*, 8 June 2007.
37 Indrani Bagchi, 'Climate Change: India, China and Pak Join Forces,' *The Times of India*, 20 April 2007.
38 Online, http://www.mnp.nl/ipcc/pages_media/ar4.html
39 'India, China Hold Up Climate Talks,' *The Times of India*, 3 May 2007.
40 Ibid.
41 The G77 is a group of 132 developing countries. Further details are available online, http://www.g77.org/.
42 'Proposal of the Russian Federation to Develop Appropriate Procedures for the Approval of Voluntary Commitments,' FCCC/KP/CMP/2006/MISC.4, online, http://unfccc.int/resource/docs/2006/cmp2/eng/misc04.pdf. Updated information note, Russian submission and presentation at the May 2007 workshop, online, http://unfccc.int/meetings/workshops/other_meetings/items/3971.php.
43 Interview with Chandrashekar Dasgupta, 16 April 2007.
44 Ibid.
45 'SB 26 Highlights, Friday 11 May 2007' (2007) 12 Earth Negotiations Bulletin 327. It is worth noting one of the early negotiating drafts of the *Kyoto Protocol* contained a proposed article 10 on voluntary commitments. The strongest opposition came from China and India who questioned the legality of creating a new category of parties under the Convention and the 'voluntary' nature of the commitments under discussion. 'Report of the Third Conference of Parties to the UN Framework Convention on Climate Change, 1–11 December, 1997' (1997) 12 Earth Negotiations Bulletin 76 at 13.
46 Interview with Ajay Mathur, Director General, Bureau of Energy Efficiency, India, 13 April 2007.
47 Interview with Chandrashekar Dasgupta, 16 April 2007.
48 Remarks by Pranab Mukherjee, Minister of External Affairs of India, Meeting of Major Economies on Energy Security and Climate Change, Washington, DC, 27 September 2007.
49 *Dealing with the Threat of Climate Change*, India Country Paper, the Gleneagles Summit, 2005.
50 Ibid.

51 See *supra* note 13.
52 See *supra* note 11.
53 For further details see Ministry of New and Renewable Energy, Government of India, online, http://mnes.nic.in/.
54 For further details see Bureau of Energy Efficiency, Ministry of Power, Government of India, online, http://www.bee-india.nic.in/.
55 *Supra* note 51 at 196–202.
56 For a detailed study of this initiative see Lavanya Rajamani, 'Public Interest Environmental Litigation in India' (2007) 19(3) *Journal of Environmental Law* 293.
57 Some of these are clearly identified ex-post facto, as for instance the New Delhi CNG transport initiative. It was imposed by the Supreme Court at tremendous cost, and with significant resistance from the Delhi Government. See ibid.
58 See *supra* note 52.
59 See *supra* note 49.
60 Statement by Rahul Gandhi, Member of Parliament, Second Committee of the 61st session of the UN General Assembly, 25 October 2006.
61 Statement by Prodipto Ghosh, High-level segment, Fifteenth Session of the Commission on Sustainable Development, 10 May 2007.
62 Ibid. As Indian foreign secretary Shiv Shankar Menon observed, India views climate change 'in the context of the promises made by the international community for technology transfer and additional financing since Rio, which have remained unfulfilled.' See Indrani Bagchi, 'Climate Policy Must Address Third World Needs,' *The Times of India*, 11 April 2007.
63 See China's 'Initial National Communication on Climate Change,' 2004, chapters 4 and 7, online, http://unfccc.int/resource/docs/natc/chnnc1e.pdf
64 See *supra* note 11.
65 See *supra* note 9.
66 Ibid.
67 Ibid.
68 See *supra* note 24.
69 'India-EU Strategic Partnership Joint Action Plan,' along with further details available online, http://ec.europa.eu/environment/climat/montreal_05.htm. Also, 'EU and China Partnership on Climate Change,' available online, http://europa.eu/rapid/pressReleasesAction.do?reference=MEMO/ 05/298&format=HTML&aged=1&language= EN&guiLanguage=en.
70 See 'Overview of the US-India Climate Change Partnership,' U.S. Depart-

ment of State, online, http://www.state.gov/g/oes/rls/rm/2003/26110.htm. Also 'US-China Working Group on Climate Change,' U.S. Department of State, online, http://www.state.gov/g/oes/climate/c22933.htm.
71 Government of the UK, Department for Environment, Food and Rural Affairs, 'Working with Developing Countries – India,' online, http://www.defra.gov.uk/environment/climatechange/internat/ devcountry/india.htm. See also 'Working with Developing Countries – China,' online, http://www.defra.gov.uk/environment/climatechange/internat/devcountry/china.htm.
72 For full coverage of this deal see 'Indo-US Nuclear Deal,' *Indian Express*, online collection of all stories published on this issue. Online, http://www.indianexpress.com/fullcoverage/42.html.
73 Ibid.
74 David G. Victor, 'The India Nuclear Deal: Implications for Global Climate Change,' testimony before the U.S. Senate Committee on Energy and Natural Resources, 18 July 2006. The Indian PM, Manmohan Singh, estimates that India will increase nuclear energy by 40GW by 2015 – which will result in 300 million tons of CO_2 reductions. Cited in ibid.
75 Feng Gao, Ministry of Foreign Affairs, P.R. China, 'China's View on Future Climate Change Negotiation and Measures to Address Climate Change,' presentation to seminar of governmental experts, UNFCCC, 16–17 May 2005, Bonn, Germany; online, http://unfccc.int/files/meetings/seminar/application/pdf/ sem_pre_china.pdf. See also *supra* note 9.
76 Interview with Rajendra Pachauri, chair, IPCC, 20 April 2007.
77 Paragraph 3, Preamble, UNFCCC, 1992.
78 Paragraph 22, Preamble, UNFCCC, 1992.
79 The UNFCCC does not specifically accept or reject absolute growth in emissions.
80 See *supra* note 78.
81 U.N. Doc. G.A. Res. 44/228 (1989).
82 Preamble, UNFCCC, 1992.
83 The discussions at the second meeting of the parties to the Montreal Protocol, London, 1990, formed part of the backdrop to Rio. At the London Conference, whilst arguing for financial assistance and technology transfer, developing countries led *inter alia* by then Indian Minister, Maneka Gandhi, took the position, '[w]e did not destroy the ozone layer, you have done that already. Don't ask us to pay the price.' See Maneka Gandhi, 'A Lesson for Humanity: The London Meeting,' excerpted in Stephen O. Anderson and K. Madhava Sarma, *Protecting the Ozone Layer: The United Nations History* (London: Earthscan, 2002) at 126.

84 At the Second World Climate Conference participating countries recognizing that the 'principle of equity and common but differentiated responsibility of countries should be the basis of any global response to climate change' and postulated that 'developed countries must take the lead.' Ministerial Declaration of the Second World Climate Conference, Geneva, 6–7 November 1990, at paragraph 5. Earlier, the Noordwijk Declaration on Climate Change, 1989 recognized that, '[i]ndustrialized countries, in view of their contribution to the increase of greenhouse gas concentrations, and in view of their capabilities have specific responsibilities.' 'Noordwijk Declaration on Atmospheric Pollution and Climatic Change, Noordwijk, The Netherlands, 7 November 1989' (1990) 5 Am. U.J. Int'l L. & Pol'y at 592. A similar sentiment was expressed in Hague Declaration on the Environment, 11 March 1989, reprinted in (1989) 28 I.L.M. 1308.

85 G.A. Res. 2849 highlighted that most of the environmental problems having a global impact today have been caused primarily by environmentally unsound and poorly coordinated industrialization, consumption, and disposal patterns in the developed countries. *Development and Environment*, GA Res. 2849(XXVI), UN GAOR, 26th Sess., UN Doc. A/RES/2849 (1971).

86 See U.N. Doc. A/CONF.151/26, Vol. IV, at paragraph 16 for the U.S. reservation on principle 7. The United States noted, 'The United States understands and accepts that principle 7 highlights the special leadership role of the developed countries, based on our industrial development, our experience with environmental protection policies and actions, and our wealth, technical expertise and capabilities.'

87 It is based on this development that some writers have argued that the legal basis for the transfer of technology and financial resources from the industrial to developing countries is founded on entitlement not need. See Subrata Roy Chowdhury, 'Common but Differentiated Responsibility in International Environmental Law from Stockholm to Rio,' in Konrad Ginther et al, eds., *Sustainable Development and Good Governance* (Leiden: Brill, 1995) at 334.

88 Rio principle 7 reads: 'States shall cooperate in a spirit of global partnership to conserve, protect and restore the health and integrity of the Earth's ecosystem. In view of the different contributions to global environmental degradation, States have common but differentiated responsibilities. The developed countries acknowledge the responsibility that they bear in the international pursuit of sustainable development in view of the pressures their societies place on the global environment and of the technologies and financial resources they command.' 'Rio Declaration on Environment and Development,' in *Report of the United Nations Conference on Environment and*

Development, Rio de Janeiro, 3–14 June 1992, reprinted in (1992) 31 I.L.M. 874.

89 See *supra* note 8.

90 Decision 1/CP.1, The Berlin Mandate: Review of Adequacy of Articles 4, paragraph 2, sub-paragraph (a) and (b), of the Convention, including proposals related to a protocol and decisions on follow-up, in *Report of the Conference of Parties on its First Session*, FCCC/CP/1995/7/Add.1(1995); and Decision CP.8, The Delhi Ministerial Declaration, in *High-level Segment Attended by Ministers and Senior Officials, The Delhi Ministerial Declaration on Climate Change and Sustainable Development, Proposal by the President*, FCCC/CP/2002/L.6/Rev.1 (2002), are two of the many significant examples. All the recent political milestones in the lead up to Bali have also in some form or the other included references to the concept of common but differentiated responsibility. An example is the 'Sydney APEC Leader's Declaration on Climate Change, Energy Security and Clean Development,' 9 September 2007, online, http://www.state.gov/g/oes/rls/prsrl/2007/92037.htm

91 A term used by Vaughan Lowe in the context of the principle of sustainable development. See Vaughan Lowe, 'Sustainable Development and Unsustainable Arguments,' in Alan Boyle and David Freestone, eds., *International Law and Sustainable Development* (Oxford: Oxford University Press, 1999) at 19.

92 For an extended discussion of the legal status of the principle see Rajamani, *supra* note 8 at 158.

93 For a recent example see *supra* note 9.

94 UNFCCC Art. 4(7) (emphasis added).

95 Art. 5(5), *Montreal Protocol on Substances that Deplete the Ozone Layer*, 16 September 1987, reprinted in (1987) 26 I.L.M. 1541 [hereinafter *Montreal Protocol*].

96 Art. 20(4), *United Nations Convention on Biological Diversity*, 5 June 1992, reprinted in (1992) 31 I.L.M. 818.

97 Art. 13(4), *Stockholm Convention on Persistent Organic Pollutants*, 22 May 2001, reprinted in (2001) 40 I.L.M. 532.

98 In the context of the *Montreal Protocol*, the first of the multilateral environmental agreements to include such a clause, the erstwhile Indian minister of environment, Maneka Gandhi, said, '[f]inally, we reached a compromise that satisfied us. A clause was added saying that developing countries would fulfill their obligations only with transfer of technology and implementation of the financial mechanism.' See *supra* note 83 at 126. An important feature of Article 4(7) is the fact that the level of effective implementation required of developing countries is tied not to the level of

effective implementation of industrial countries' commitments under the UNFCCC more generally, but in particular to the level of effective implementation of their commitments relating to technology transfer and financial assistance. It follows that developing countries cannot claim, at least on the basis of this article, that industrial countries' leadership as evidenced by their fulfillment of their UNFCCC mitigation commitments is a precondition to developing countries' fulfillment of their commitments.

99 For an extended discussion of the linking clause see Lavanya Rajamani, 'The Nature, Promise and Limits of Differential Treatment in the Climate Regime' in *2005 Yearbook of International Environmental Law* (Oxford: Oxford University Press, 2007) at 81.
100 IPCC Working Group II Report, 'The Impacts of Climate Change,' 2007. For a flavour of the Indian reactions to this report see 'Forget Himalayan Glaciers,' *The Times of India*, 2 April 2007; and 'Fighting Warming in an Unequal World,' *The Times of India*, 4 April 2007.
101 Quoting Henry Miller, Stanford University, in Andrew C. Revkin, 'Poor Nations to Bear the Brunt as the World Warms' (*New York Times News Service*) *The Times of India*, 2 April 2007.
102 Of India's nearly 1.1 billion people, 34.7% live on less than US$1 a day. See *supra* note 4.
103 India's 700 million rural population depends directly on climate-sensitive sectors (agriculture, forests, and fisheries) and natural resources for their subsistence. See Jayant Sathaye et al., 'Climate Change, Sustainable Development and India: Global and National Concerns' (2006) 90 Current Science 3, 314 at 318.
104 See *supra*, note 9.
105 For anecdotal evidence see Roger Harrabin, 'How Climate Change Hits India's Poor,' *BBC News*, 1 February 2007, and Fu Jing, 'Climate Change Hits China's Poor Hardest,' *China Daily*, 26 September 2007.
106 See *supra* notes and ; also Joyashree Roy, 'A Review of Studies in the Context of South Asia with a Special Focus on India: Contribution to the Stern Review,' online, http://www.hm-treasury.gov.uk./media/5/0/roy.pdf, and Andrew Challinor et al., 'Indian Monsoon: Contribution to the Stern Review,' online, http://www.hm-treasury.gov.uk./media/3/4/Challinor_et_al.pdf
107 See *supra* note 103 at 318–19, and Challinor et al., *supra* note 106.
108 Ibid.
109 See *supra* note 16, Executive Summary at 6. See also 'Climate Change in South Asia: A Conversation with Sir Nicholas Stern,' 14 February 2007, online, http://web.worldbank.org.

110 See *supra* note 9.
111 See Erda Lin et al., 'China's National Assessment Report on Climate Change (II)' (2007) Advances in Climate Change Research 3 (Suppl.) 6.
112 *Supra* note 109. See also Roy and Challinor et al., *supra* note 106. Challinor et al. illustrate with the failure of the monsoon in July 2002 which resulted in a 3% drop in GDP.
113 See Roy *supra* note 106.
114 Erda Lin and Ji Zhou, 'Climate Change Impacts and Its Economics in China: Contribution to the Stern Review,' 28 August 2006, online, http://www.hm-treasury.gov.uk./media/8/1/ stern_review_china_impacts.pdf.
115 Ned Helme, Centre for Clean Air Policy, 'Creating Incentives to Reduce GHG Emissions Post-2012,' Presentation to the Dialogue on long term cooperative action to address climate change by enhancing implementation of the Convention, online, http://unfccc.int/files/meetings/dialogue/application/vnd.ms-powerpoint/061115_cop12_dial_helme.pps
116 See 'Developing Countries Fear on Emissions,' *Financial Times*, 25 October 2007.
117 The Dialogue process was initiated by the eleventh Conference of Parties to the UNFCCC held in Montreal in 2005. As part of the Dialogue process parties agreed to hold four workshops over the course of 2006 and 2007 and engage in an open and non-binding exchange of views on actions needed to respond to climate change. They also agreed not to open any negotiations leading to new commitments. See Decision 1/CP.11, Dialogue on long-term cooperative action to address climate change by enhancing implementation of the Convention, in FCCC/CP/2005/5/Add.1 at 3. Online, http://unfccc.int/resource/docs/2005/cop11/eng/05a01.pdf
118 Drawn from 'Scenario Note on the Fourth Dialogue Workshop, Note by the Co-facilitators, Dialogue Working Paper 6 (2007).' Online, http://unfccc.int/files/meetings/dialogue/application/pdf/final_scenario_note_wp6.pdf.
119 Some see this term as signifying climatic co-benefits of development actions, and others as development co-benefits of climate actions. Indian negotiators, for instance, argue for the need to identify and exploit development actions that have climatic co-benefits. Ajay Mathur, IGES Conference, New Delhi, 29–30 August 2007.
120 The Ad Hoc Working Group's work on further commitments for Annex I Parties was initiated by the first meeting of parties to the Kyoto Protocol in Montreal in 2005. See Decision 1/CMP.1, Consideration of commitments for subsequent periods for parties included in Annex I to the convention

under Article 3, paragraph 9, of the *Kyoto Protocol*, in FCCC/KP/CMP/ 2005/8/Add.1 at 3.
121 UNFCCC, 'Synthesis of Information relevant to the determination of mitigation potential and the identification of possible ranges of emissions reduction objectives of Annex I Parties,' Technical Paper, FCCC/TP/2007/ 1 (27 July 2007). Online, http://unfccc.int/resource/docs/2007/tp/01.pdf
122 Chair's summary, G8 Summit, Heiligendamm, 8 June 2007, at 2. Online, http://www.g-8.de/Content/EN/Artikel/_g8-summit/anlagen/chairs-summary,templateId=raw,property=publicationFile.pdf/chairs-summary
123 Presidency Conclusions, Brussels European Council, 8/9 March 2007. Online, http://www.consilium.europa.eu/ueDocs/cms_Data/docs/pressData/en/ ec/93135.pdf
124 Shinzo Abe, 'Invitation to Cool Earth 50: 3 Proposals, 3 Principles,' Speeches and Statements by Prime Minister, 24 May 2007. Online, http://www.kantei.go.jp/foreign/abespeech/2007/05/24speech_e.html
125 Major Economies Meeting on Energy Security and Climate Change, Washington DC, 27–8 September 2007. Further details online, http://www.state.gov/g/oes/climate/mem/
126 IPCC, 2007. Susan Solomon et al., eds., *Climate Change 2007: The Physical Science Basis. Contribution of Working Group I to the Fourth Assessment Report of the Intergovernmental Panel on Climate Change* (Cambridge: Cambridge University Press, 2007).
127 Ibid.
128 *See supra* note 16.

6 Comment – Across the Divide: The Clash of Cultures in Post-Kyoto Negotiations

STEVEN BERNSTEIN

The papers in this section suggest that the divisions between different regions and groupings in climate negotiations may be, at least in part, cultural in character. If this characterization is correct, the challenges facing future negotiations are more serious than analyses based solely on state interests suggest. Finding a way forward under such circumstances will depend as much on creative thinking and learning across the cultural divide as satisfying any financial, competitiveness, and fairness concerns that are most often the focus of climate negotiators and analysts alike. This comment focuses on two sets of issues in this regard: First, what does this cultural divide portend for the future of the global climate regime? Second, what are the implications of this divide for Canada and what can Canada learn from the experiences of these different regions and groupings.

Three Worlds of Climate Governance

Attempts to build multilateral institutions have long struggled with how to bridge the North-South divide. Environmental negotiations have been no different. The concept of 'sustainable development,' currently the dominant framing of nearly all global environmental problems, was explicitly designed as a way to overcome that divide. Built into the concept is the idea that environmental concerns and development goals ought to be considered as part of a whole, neither simply traded off nor dealt with independently of one another. The trick has always been to design practical policies to overcome this apparent trade-off. In practice, successful negotiations have been built around principles such as 'common but different responsibilities and capabilities,' discussed in some depth in the chapters in this section, as well as mechanisms for financing, phase-in periods for developing country commitments, and attempts to create investment and technology transfer in the hope that developing economies can leapfrog out of the most polluting trajectories. The problem with the sustainable development framing, however, is that it may underplay the need for tough regulatory action when time frames are shorter and problems require more fundamental interventions in the economy. It can also feed into a traditional North-South dynamic that denies the need for trade-offs.

Rajamani's chapter on India and China illustrates how difficult it is even for rapidly developing economies, which in many respects may already have great-power status, to step outside of the North-South culture of negotiations that has shaped their identities in multilateral forums for so long. In this 'developing world' context, the discourse of justice and equity has framed their understanding of their relationship to global governance and order, as well as been useful as a rhetorical strategy. The numerical advantage of developing countries in UN forums, including the climate change negotiations, has also worked to their advantage, providing little incentive to step out of the collective 'developing countries' identity that has served their interests. What some might consider an ironic support for strong global management in such negotiations – given that developing countries traditionally also jealously guard their own sovereignty and 'right' to develop – is understandable in a context in which the expectation is that developed countries take the lead in any commitments.

The North-South bargains that have come out of such processes have worked for other, simpler, environmental problems largely because deals could be struck that were low cost with mutual benefits obvious to most of the parties. Climate change is proving more intractable because it goes much more directly to the motor of development – energy consumption, industrial processes, and modern consumerism. These characteristics of climate change militate against a deal based purely on economic incentives or the traditional currencies of North-South bargains.

On top of this older divide, a new clash has emerged, primarily between the United States and Europe. This transatlantic rift is not a fundamental rift about values, economics, or fairness, but rather a divide over visions of global order.

For the EU, as Brunnée and Levin put it, 'there is an important normative dimension to European climate policy. The member states of the

EU have long been comfortable with a supra-national approach to law and policy making.' The gradual expansion of the EU and interest in 'constitutionalization' that they note is fully consistent with the behaviour and apparent understandings of EU member governments in international forums, including on climate change. A regime with strong governance and regulatory and legal capacity fits very well with a vision of global order consistent with the European experience. The noted 'internalization' of the goals, values, and principles of the UNFCCC is, again, understandable in this context because they are quite consistent with the EU's projection of its own values in its foreign and global policies.

Add to this overall culture of global governance the path dependencies being created by the early entry into regional and global carbon markets by the European Emission Trading Scheme. These sunk costs create a strong incentive for the continuity of a strong regulatory structure based on the Kyoto mechanisms and support for the compliance regime. This context also makes understandable the EU's apparent willingness, noted by Brunnée and Levin, to make further commitments even without a new post-Kyoto agreement in place.

In contrast, the United States – and increasingly Canada – views climate change governance through the lens of traditional problems of international cooperation and reciprocity.[1] This lens has assumed even greater importance under the current Bush administration, which has exaggerated the traditional U.S. reluctance to bind itself to international commitments. The ideological influence of the 'new sovereigntists,' though perhaps waning as the Bush presidency winds down, has turned this reluctance into an almost religious belief.[2]

But while it has been common to characterize the larger U.S. foreign policy orientation as a move to unilateralism, that is a misdiagnosis. Rather, there is a continuity that runs underneath the more traditional multilateral policies of previous U.S. administrations, whereby reciprocal obligation has guided its vision of international treaty making. As John Ikenberry has argued, the United States has supported a rules-based institutionalized international order in the post–World War II era in exchange for the 'acquiescence and compliant participation' of weaker states.[3] While such an order's legitimacy must be based on shared values and collective intentions, it operates through reciprocal bargains and mutual influence rather than any move toward a constitutional order.

If this analysis is correct, the current reticence in the United States towards the *Kyoto Protocol* may be at least partly independent of different views on the urgency of addressing climate change or the particular interests, motivations, or understandings of the Bush administration. Thus, even if the Democrats win the next election, the way in which the next administration will want to address the global problem is likely to differ from the European vision. A most likely scenario is that the new administration will re-engage along the lines of policies promoted at home rather than focusing on a ratcheting-up of global commitments. Still, there is also good reason to believe that the U.S. foreign policy culture is not allergic to a rule-based order in general terms; thus, the U.S. tendency to unilateralism and rejectionism under the Bush administration, as John Ruggie has argued, is 'largely discretionary, not structural.'[4] There is thus considerable room for progress in global negotiations.

The bad news is that the cultural character of these divides makes bargaining across them difficult. The good news is that even cultures may not be permanent, and some may be less rigid than they seem. Sometimes a shock can shake up a culture – although one hopes a climate shock is not necessary. More hopefully, cultures may also slowly change owing to tensions between their norms and their external environment that reveal potential self-destructive tendencies and disjunctures with lived experience. The chapters in this section suggest some hints of change in this regard.

Rajamani's contribution, for example, powerfully shows the negative implications of the current North-South bargaining culture, especially in the context of a problem with the potential for severe negative effects directly on developing countries. As she puts it in the case of India and China, 'While the equitable rhetoric may serve these countries well in international forums, lack of serious domestic action will hamper the ability of the international community to tackle climate change. Besides, climate change will have significant impacts – economic, social and environmental – in both these countries.' A growing recognition seems to be emerging that the current orientation to the climate regime is simply not 'sagacious.' To the degree such wisdom prevails, there is some opening for common ground between the United States and major developing country emitters.

The rift is already much smaller between developing countries and the EU, which, it seems, has internalized norms of developed countries 'taking the lead' and differentiated responsibilities. Still, even the EU will demand some movement on commitments over the medium to longer term. Current EU proposals, as noted by Brunnée and Levin, call for reductions in emissions following 2020 by developing countries, when their emissions are projected to overtake developed-country emissions.

Hunter's and Rajamani's contributions make clear that neither the United States nor major developing-country emitters will move significantly in terms of binding commitments without the other, though it remains unclear how much differentiation will be acceptable in the short or medium term. At the same time, there is little question that domestic pressure in the United States, combined with institutional pressure from the benefits of the clean development mechanism, will ratchet up. To the degree developing countries want to continue to have the comfort and political leverage offered by the UN framework, as well as the advantages of a robust clean development mechanism (which requires an institutional and legal foundation), the demand for reciprocity is likely to be more keenly felt. What form such reciprocity will take is an open question. Taking another page from the European example, a multi-speed, multi-track approach may be one solution, though that too would require a greater consensus on underlying goals and principles.

Implications for Canada

Canada's traditional support for a multilateral, rules-based international order would seem to present an opportunity for it to help bridge some of these divides. However, its recent bridge-building rhetoric notwithstanding, Canada's current shift towards the rejectionist camp undermines its ability to do so. The problem here is not specifically its scepticism over Kyoto, but rather its duplicitous refusal to either withdraw from the treaty or meet its commitments. It will be hard-pressed to make the case for a commitment to a rules-based means to address climate change with such a stance toward international obligations.

Returning to the themes of this volume, a more globally integrated climate policy could counteract this perception of Canadian policy, and offer new opportunities for engagement and even leadership (see the introduction to this volume). At a minimum, Rajamani's suggestion – that in the short term gaining the trust of India and China through confidence-building measures is absolutely essential – provides an opening for a middle power like Canada to play a constructive role. Canada can potentially achieve much in this regard with a modest commitment of resources to cooperative ventures and diplomatic initiatives with developing countries. Seen in the best light, the Asia-Pacific partnership does provide an institutional setting to pursue such a policy in the short term. Much depends on how serious the government is in bridging these various divides, which has been the stated policy of the current and previous governments on the climate change issue since the earliest days of global negotiations.

The U.S. and European internal policy experiences also offer valuable lessons for Canada. Both have experiments in emissions trading that Canada can learn from and ought to watch closely. In addition, if Hunter's analysis is correct, Canada risks quickly losing any advantage of being a first mover on a more active climate policy. At a minimum, any hope of leadership, as even the U.S. experience has shown, will require more engagement with international processes, whether inside or outside the Kyoto trajectory.

Notes

1 This understanding is also evident in the U.S.-based academic and policy analysis of the climate regime. A typical example of this understanding is the work of David Victor, an influential critic of the *Kyoto Protocol*. For example, see D.G. Victor, 'Toward Effective International Cooperation on Climate Change: Numbers, Interests and Institutions' (2006) *Global Environmental Politics* 6(3) 90.
2 See, for example, J.R. Bolton, 'Should We Take Global Governance Seriously?' (2000) *Chicago Journal of International Law* 1(2) 205; J. Rabkin, 'Is EU Policy Eroding the Sovereignty of Non-Member States?' *Chicago Journal of International Law* (2000) 1(2) 273; and J. Rabkin, *Law Without Nations? Why Constitutional Government Requires Sovereign States* (Princeton, NJ: Princeton University Press, 2005).
3 G.J. Ikenberry, 'Is American Multilateralism in Decline? (2003) *Perspectives on Politics* 1(3) 533 at 541.
4 J.G. Ruggie, 'Global Markets and Global Governance: The Prospects for Convergence,' in S. Bernstein and L. Pauly, eds., *Global Liberalism and Political Order: Toward a New Grand Compromise?* (Albany, NY: SUNY Press, 2007) 23 at 41.

PART THREE

Global Regime Building – Parameters and Imperatives for Canada

7 The Global Regime: Current Status of and *Quo Vadis* for Kyoto

MATTHEW J. HOFFMANN

1. Introduction

The global governance of climate change is in flux. The conventionally understood mode of global environmental governance – universal, interstate, multilateral negotiations quintessentially represented by the 1997 *Kyoto Protocol* – has been dangerously stymied in climate change by the yawning gulfs that exist between the negotiating positions of major states. The stalemate is such that many observers have given up hope that a process centred on the *Kyoto Protocol* or the *United Nations Framework Convention on Climate Change* (UNFCCC) can produce effective governance of climate change in time to avoid catastrophe (if it can produce effective governance mechanisms at all).[1] Are we to despair? In this chapter I consider this question and assess governance prospects from within the official UNFCCC/Kyoto process and the possibilities emerging outside the official negotiations.

2. Conventional Climate Governance: The Good, the Bad, and the Unknown

Good news in the realm of climate change has been in short supply since the international community brought the *Kyoto Protocol* into force in 2005. Reports of accelerating warming and ahead-of-schedule loss of Arctic sea ice have accompanied serious questions about the effectiveness of the UN-sponsored governance process. And while there has been some good news, unfortunately each bit of it seems to fall in the category of 'at least ...' This does not bode well for the state of climate governance through the UNFCCC/Kyoto process.

The *Kyoto Protocol*, while routinely criticized, did achieve some modest accomplishments. It requires moderate emission reductions from Northern states, an average of five per cent below 1990 emission levels. It includes flexible implementation measures aimed at keeping the costs of emissions reduction relatively low. Finally, it contains an initial foray into the delicate issue of the North–South impasse with the clean development mechanism.[2] At the very least, in 2005 the international community did agree to address the problem and adopted the 2001 *Marrakesh Accords* to implement the *Kyoto Protocol*.[3]

The year 2005 also saw hope for the future of the Kyoto process as the parties to the protocol agreed to the first steps of negotiating an agreement for a post-Kyoto commitment period. This process proceeds on two tracks – a protocol track without the United States and a UNFCCC track that includes the United States. While not a triumph by any means, the Montreal meetings in 2005 at least laid the groundwork for moving forward from within the UN-sponsored process. The negotiations in Nairobi in 2006 made no progress in fleshing out what a second commitment period might look like, but neither did they take a step backward. Indeed, the international community appears to have moved into a period of expectant waiting, either for a major push at upcoming negotiations (Bali 2007, Copenhagen 2008), or for a change in the U.S. presidential administration in 2009.

Other bits of good news remain more hopeful signs than concrete results. The recently concluded round of ozone depletion negotiations produced an agreement to accelerate the phase-out of hydrochlorofluorocarbons (HCFCs) – a potent greenhouse gas.[4] Driven largely by climate change concerns, this agreement bodes well, according to some, for the attempts to push forward in the upcoming climate negotiations. In addition, the work of the Intergovernmental Panel on Climate Change (IPCC), following the well-publicized launchings of its latest science, impacts, and policy options reports in the spring of 2007, was honored with the Nobel Peace Prize (shared with former U.S. vice-president Al Gore).[5] While the negotiations have slowed or stalled, at least the scientific process catalysed by the UN has worked as it should. There is now overwhelming consensus about the nature of the problem, the need to do something about the problem, and a menu of possible policy options. This consensus has created a sense of urgency around the December 2007 negotiations in Bali, evidenced by UN Secretary General Ban KiMoon's placement of climate change high on the international agenda.

Unfortunately this sense of urgency has not translated into effective action via the UN-sponsored process. The litany of problems is familiar and does not require detailed reiteration. Yes, the *Kyoto Protocol* is in force, but it took enormous compromises to get enough states to ratify it, compromises that weakened a treaty already light on significant emissions reductions. Further, of the two largest emitters in absolute terms, one is a party with no emission reduction requirement (China) and the other has withdrawn (the United States). Most states are having trouble meeting their targets, some have abandoned them explicitly (Canada), and those who might achieve their targets have done so mostly because of economic downturn (e.g. the EU's absorption of Eastern Europe) or fortuitously timed energy choices (e.g. the UK's switch from coal to natural gas as the main fuel for electrical generation), rather than climate-specific measures. Southern countries that, as the United States constantly reminds anyone who will listen, account for an increasing share of emissions, have no emissions reduction requirements and so far refuse to discuss them for the future.[6] Given the increasingly dire scientific reports about the pace of warming and the seemingly intractable nature of the problems facing the UNFCCC/Kyoto process, it is no wonder that the world is approaching the Bali meetings with both high expectations (last chance for saving the Kyoto process?) and a bit of fatalism concerning the prospects for addressing climate change within this framework. It is also no surprise that calls from observers to fundamentally overhaul or even abandon the Kyoto process are growing.

How precarious is the status of the UN climate regime? Is it the only game in town and thus must be the focus of the international community's efforts? What else is there to do? In the media, amongst observers, and even in broad sections of populations across the globe, there appears to be momentum building to do something and do it quickly. 'There is an increasing expectation – within the process and beyond it – that a shift in gear, or change in direction, is required if adequate progress is to be made.'[7] The important concern thus becomes how this expectation will be met and whether it can be met within a process where something as small as 'the agreement by all countries to continue discussing how to move forward' is seen as a successful negotiating outcome in the face of a climate crisis.[8]

3. Paths Forward for the Climate Regime

If the international policy goal is to keep global greenhouse gas emissions below dangerous thresholds, the existing process centred on the *Kyoto Protocol* and UNFCCC may no longer be viable. Time is short according to most scientific accounts (one to two decades) and the institutional tools currently in use seem ill suited to promoting quick and effective action. My focus here is thus on the political prospects for progress – how rules aimed at reducing greenhouse gas emissions may come about within the *Kyoto* process and outside of it. I am leaving for others the question of technical tools. Clearly, reducing greenhouse gases in our atmosphere has a significant technical aspect. Questions of the sufficiency of off-the-shelf technology, technological inertia or innovation, capital stock turnover, and patterns of technological investment are all crucial,[9] but my concern is the political dynamics that will (or will not) create the rules that shape the development and deployment of policy and technology.

This section is thus saturated with conjecture as I assess a number of possibilities for the climate regime. That the climate regime is in flux is obvious. The fact that 'since the negotiation of the *Kyoto Protocol*, more than fifty proposals for an international climate policy regime have been published' does not speak well of the confidence or esteem in which the Kyoto process is held.[10] In this context, I first examine possibilities of moving forward with the UNFCCC/Kyoto process. The legitimacy and inertia built into the UN-sponsored multilateral process ensures that the protocol will not be abandoned lightly and may even continue as the centrepiece of the global response to climate change. It will not go forward effectively unchanged, however. Second, I consider the emergence of climate governance experiments outside the UN-sponsored process, providing a glimpse into the institutional innovations that are springing up in response to the sluggish pace of the global regime. The question is whether these exciting developments can also be effective.

3.1. *Scenario One: Moving Forward with the UNFCCC/Kyoto Process*

While the calls to abandon the UNFCCC/Kyoto process are numerous and shrill, the necessity of doing so is far from obvious. Given that 'several veterans of the process seemed to be in a state of mild euphoria as the meeting [Montreal Meeting of the Parties in 2005] ended,' it may be too soon to declare the UNFCCC/Kyoto process dead.[11] At a practical level, it is the only framework for addressing climate change that encompasses all states. At a deeper level, this mode of governance – universal, state-centric negotiations toward a binding treaty – is still the dominant way that the international community approaches transnational issues (environmental or otherwise). Leaving aside the question of the necessity of a transition, switching to a radically new institutional structure may be difficult, especially in the compressed timeframe imposed by accelerating warming trends.

Indeed the Kyoto process is a quintessential example of a global governance model based on the primacy or legitimacy of universal, state-centric, multilateral mechanisms for the creation, maintenance, and alteration of rules – with all the good and bad that comes with it. The *Kyoto Protocol* is a universal agreement – negotiated, signed, and ratified by the vast majority of states in the world. By engaging and binding states, the Kyoto process has perhaps the best chance of producing enforceable rules to govern the transnational climate problem. As difficult as it is to reach agreement amongst states on a complex issue like climate change, there are few actors (if any) on the world political stage that have the authority to implement and enforce necessary changes in economies and societies across the globe. In addition, universality and consensual decision-making procedures are not inherently problematic. As we are told time and again, climate change is not a problem that can be solved by any single nation – global cooperation is necessary. A universal, state-centric model of climate governance thus has the advantage of widespread buy-in once decisions are reached, and has allowed a more prominent place for equity considerations. Southern states have been full participants in the negotiations from the outset and development concerns have always been at or near the top of the agenda.[12]

Unfortunately, Kyoto is also a model of the drawbacks of universal, state-centric global governance. First, universal state-centric participation does not guarantee progressive environmental agreements.[13] On the contrary, such universality has an at-best ambiguous effect on environmental improvement. Certainly, universality increases the chances for an equitable response to climate change, but it has also facilitated political stalemate by multiplying the interests represented at the negotiating table.[14] In addition, state centrism itself has potential drawbacks. While it is desirable to pursue global rule-making with authoritative actors who can enforce agreements, when approaching a con-troversial and complex issue like climate change a single comprehensive approach may not be optimal. This approach is subject to lowest-common-denominator pitfalls and to being held hostage to powerful veto states. States do provide legitimate authority and perhaps the most efficient means of enforcing global rules, but a state-centric approach can also be a straightjacket that hampers effective governance mechanisms.

The *Kyoto Protocol* is thus flawed, but a powerful combination of legitimacy and inertia are anchoring this kind of approach – universal or mega-multilateralism. A potential way forward is to use its legitimate status and reform the process from within. This process has already begun. At the 2005 negotiations in Montreal, the parties to the *Kyoto Protocol* and the UNFCCC decided upon a two-track path towards future commitments. This was done largely in response to the U.S. refusal to ratify the protocol, but it does set an important precedent for thinking about reform. One track within the Kyoto process is looking specifically at ways to enhance emissions reductions commitments in a post-2012 period for Northern states (Annex I parties). At the Montreal meeting in 2005, an ad hoc working group was deployed to begin these negotiations with the goal of ensuring that there would be no gap in commitments after the Kyoto requirements end in 2012.[15] This 'new cycle' of negotiations has started slowly with little of substance accomplished in 2006.[16] The second track within the UNFCCC, of which the United States is still a party, addresses more general provisions, continuing the commitment to dialogue on a broad climate-related agenda.[17]

Beyond the two-track approach, there are growing numbers of proposals for reforming the Kyoto process (along with the numerous calls for its abandonment) that can be broadly considered to advocate either substantive reforms or structural reforms. Substantive reform proposals call for gradual transformation within the Kyoto process. In other words, there could be a change in what gets negotiated within the treaty, not how the treaty is negotiated or its ultimate form. These kinds of proposals are usually aimed at getting the United States back into the process. Bang et al., for example, explore how a move toward binding targets for Southern countries, a switch to intensity rather than absolute emissions limits, or a modification of the compliance system could bring the United States back into the fold.[18] Christoff advocates a shift in focus away from the United States and towards a 'stronger culture of compliance' within the Kyoto process.[19] Unfortunately, few observers are sanguine about the prospects that substantive reform will lead to either a re-engagement of the United States (many are just waiting to see what happens after the current administration is replaced in 2009) or an effective agreement. Bang et al. dismiss the prospect of U.S. re-engagement, arguing that until 'the Kyoto process has been abandoned, a *global* climate regime with U.S. participation is not likely to emerge.'[20]

More radical than substantive reform, the essence of structural reform entails a change in one or two key institutional features of the *Kyoto* process: first, the number and categorization of states around the table, and second, the type of agreement under consideration. The current process is founded on the universal participation of states along with differentiated responsibilities for different categories of states. Its goal is a binding treaty (a renewal of the *Kyoto Protocol* for the next commitment period) that specifies emissions reductions. Haas worries that this process is a 'dysfunctional blueprint' for climate governance and the widespread acceptance of this concern has led to calls for sweeping change.[21]

The various structural reform proposals may be tantamount to abandonment of Kyoto in important ways because they seek to change the essence of the climate regime. For example, the *Kyoto Protocol* is *not* a technology development agreement and to change it into one would make it a larger version of the Asia-Pacific Partnership on Clean Development and Climate (APP).[22] It is *not* a small-club approach to negotiation. David Victor and others are now calling for this type of regime building, harkening back to James Sebenius's advice from 20 years ago.[23] This may be a good proposal, environmentally, but a shift away from universal participation is a fundamental renunciation of the Kyoto process. Finally, the *Kyoto Protocol* is *neither* a voluntary arrangement *nor* a policy-based treaty.[24] A reform that transitions away from binding emissions reductions, therefore, is also a repudiation of the Kyoto process. In this sense, then, structural reform of the Kyoto process can only tenuously be considered change from within. These proposals are fundamental challenges to the Kyoto process that seek to reconfigure what counts as a multilateral approach to climate governance.

There is a growing consensus that reform or at least a change in the direction of the UNFCCC/Kyoto process is necessary. Demand for change, however, does not guarantee actual change. Because of powerful inertial forces and states that are invested in the current process, fundamental reforms will be difficult to initiate. Breakthroughs in the upcoming negotiations are certainly possible, but governance efforts outside the UNFCCC/Kyoto process may be ascending to prominence.

3.2. *Scenario Two: Governance Experiments*

Aren't we a little too self-righteous to pretend that all strategy is here in the toolbox of Kyoto, where there are only numerical target, timeline, some flexible mechanisms and detailed punishment plan? Shouldn't we be a little more humble to the awesome might of nature and human action and start exploring many more tools and strategies on top of the Kyoto's tool box?

Japanese Presentation at Nairobi, COP12/MOP2, 2006[25]

While the *Kyoto Protocol* represents the apex of the universal state-centric model of global governance, it may have simultaneously triggered the decline of this governance model's dominance. For some the UNFCCC/Kyoto process has become irrevocably stalled. Depledge soberly observes that the Kyoto process 'has not only got "stuck" but is digging itself into ever deeper holes of rancorous relationships, stagnating issues, and stifling debates.'[26] This ossifying regime cannot produce the rules and innovations necessary to meet the climate challenge.[27]

Fortunately the UNFCCC/Kyoto process does not exhaust the extant approaches to climate governance. Stalemate at the interstate level has been met with the emergence of governance experiments at other levels. Like-minded groups and individuals have begun to fill the governance gap in climate change with new experiments. Communities, united by agreement about how to approach climate change, have emerged, challenging the reliance on state-centric governance mechanisms. This rule-setting at multiple levels is not as comprehensive as universal multilateral treaty-making, but these experiments are worthy of inspection for both how they instantiate new ideas of what counts as governance and how they do or do not comprise an effective global response to climate change.

3.3. *Governance through Uncoordinated Individual Actions: Carbon Offsetting*

At the most elemental level of political behavior, carbon-offsetting organizations, both for-profit and non-profit, have emerged to cater to individuals (and organizations like this conference) who want to reduce their carbon impact without necessarily changing their lifestyles. These organizations allow individuals or organizations to calculate or estimate their personal carbon emissions and pay to support projects and/ or businesses that pledge to reduce carbon dioxide emissions by the requested amount. This type of activity is not immediately recognizable as global governance in a traditional sense, yet rules are clearly emerging, not only in terms of how people should act in response to climate change, but also how offsetting organizations should act.

The former concerns the emergence of a voluntaristic, personal-responsibility approach to the response to climate change. Individuals, in an uncoordinated but intentional manner, follow rules about how to emit carbon dioxide or how to ameliorate their own emissions. This type of governance mechanism does not facilitate the emergence of global limits to carbon emissions and is not subject to enforcement, but it does at least move towards the notion that reductions (or even carbon neutrality) are appropriate goals that can be met through voluntary actions. Lest we reject such a governance model out of hand as inadequate, it is noteworthy that there are now at least 47 companies or organizations in this burgeoning industry.[28] In addition, while we lack good data on the amount of emission reductions from this industry, most organizations claim thousands of members and hundreds of businesses as clients, with at least one organization (Terrapass) claiming to have offset 500 million pounds of carbon dioxide emissions.[29]

This movement has also spawned meta-governance discussions.[30] As carbon offsetting grows in popularity, the need to monitor and govern offsetting organizations grows concomitantly – there is a demand for verification of the types and accuracy of the projects used to offset emissions. Global environmental governance studies are not unfamiliar with this dynamic – there is a lively literature on certification programs for sustainable forestry and other kinds of sustainable production.[31] Effective governance through individual offsetting behaviour is, in many ways, dependent on the development of standards amongst offsetting organizations.

Though this is a clear market-based mechanism of governance – the market will determine the cost of offsets and the appropriate or successful manner of offsetting – we should not naively consider that it is independent of the workings of other governance efforts. Specifically, the carbon-offset market is dependent on a stable understanding of the price of carbon dioxide – a value determined in part by governance efforts at other levels.

3.4. *Private Authority and Corporate Governance Mechanisms:*
     The Insurance Industry

Private authority as an alternative to state-centric governance is currently *en vogue* in the international relations literature and policy world.[32] Suggestions of corporate governance mechanisms have been viewed cautiously in climate change, yet for a variety of strategic and altruistic reasons, some industries have begun to push for climate governance and pursue rule setting of their own.[33] The insurance industry, perhaps the most aware of the potential costs of climate change, has long been a source of hope for the environmental movement.[34] Yet, while the insurance industry has not had the kind of impact on climate governance that some would have liked to see, the activities of this industry do point to an emergent source of governance – corporate rule-making.

In the 1990s, environmentalists had grand expectations that the insurance industry, as a powerful bloc of institutional investors, would change their investment patterns and move the global economy towards renewable energy, away from fossil fuels.[35] The hope was that the insurance industry would see self-interest in 'aggressively investing in renewables to mitigate global warming.'[36] An altered investment pattern by a major set of players in global finance could go a long way to reorganizing the economy along more climate friendly lines. To the dismay of the environmental community, this has not yet come to pass. This is not to say that the insurance industry has neglected climate change. On the contrary, it has been at the forefront of climate governance in a number of explicit and implicit ways; albeit not always in a manner approved of by environmentalists.

Explicitly, members of the insurance industry have teamed up with the United Nations Environment Program (UNEP) and formed an Insurance Working Group in 2006 under the auspices of the UNEP Finance Initiative (there is also a climate working group within this initiative). The objectives are relatively ambiguous; they focus mainly on integrating climate change adaptation measures with larger goals of sustainable development.[37] Beyond the UNEP linkage, the insurance industry has also experimented with carbon dioxide intensity baselines as a means for providing investors with a new view of risk (i.e. higher baselines translate to higher risk).[38] Of course, the effectiveness of these parameters is dependent upon the emergence of a set of governance mechanisms (global, multilateral, or otherwise) that limit the price of carbon by reducing demand.

Perhaps more importantly, the insurance industry constructs rules governing the global response to climate change in how it sets premiums and in its attempts to commodify the risks of catastrophe. Raising premiums in areas expected to bear the brunt of climate change (in the global north) and/or removing insurance coverage from certain areas or activities, serves to alter how people and eventually countries act in the face of climate change. It is a subtle, but powerful set of rules for adapting to climate change. Similarly, by commodifying the risk of catastrophe in the form of catastrophe bonds, the insurance industry is shaping how the problem of climate change and its management is understood by investors, countries, and the whole global economy.[39]

3.5. *Governance through Local Organization: Carbon-Rationing Action Groups (CRAGs)*[40]

Though the Kyoto process is bogged down, the provisions of the agreement have served as a model for political organization at other levels, specifically the notion that carbon dioxide emissions should be capped at reduced levels for certain groups. Kyoto caps the emissions of Annex I parties at reduced levels. CRAGs translate this idea to groups of individuals. In a remarkable example of self-organization, carbon-rationing action groups have sprung up in the UK and these groups exemplify the adaptive possibilities in climate governance. CRAGs are groups of people that are unwilling to wait for multilateral governance and have organized into groups where the individual members pledge to reduce their carbon dioxide emissions to specified levels, paying into a community fund if they miss their targets.

The emergence of these groups is very much a reaction to the stalemate at the multilateral level. A member of one CRAG in Glasgow noted that he is a member 'for many reasons but perhaps most importantly because it allows me to do at a local scale what I think our governments should be doing at a global scale.'[41] These local, self-governing groups decide as a community how much carbon dioxide each person is allowed to emit, how to calculate emissions, and how members compensate for exceeding limits. While this movement is currently taking place on a small scale (23 established CRAGs so far) and in only one country (United Kingdom), it is still an interesting governance experiment for a number of reasons. First, there is no reason that this could not be a transnational movement – there is nothing stopping groups of individuals anywhere from organizing similar communities. Second, CRAGs are clearly designed as steering mechanisms, setting up rules for responding to climate change that hold within the individual communities. Finally, the effect is not necessarily limited to the individual communities – the CRAGs also represent a growing sentiment that carbon emissions must be curtailed and that local communities are (or could be) authoritative actors in accomplishing it.

3.6. Globalized Subnational Governments: Municipalities, States, and Provinces

A similar principle is at work at a more established level of political organization – cities. The Cities for Climate Protection Program coordinates the actions of hundreds of municipalities (674 in 30 countries)[42] that pledge to work towards climate change mitigation through five steps: determine a baseline emissions inventory, decide upon an emissions reduction target, develop an action plan, implement policies to operationalize the plan, and monitor the results of the plan.[43] The cities that join tend to be significantly more progressive on emissions targets than the Kyoto targets given to their home states. For instance, Copenhagen has pledged a 35 per cent reduction as opposed to Denmark's 21 per cent reduction from 1990 levels.[44] Salt Lake City has agreed to uphold Kyoto targets at the U.S. level even after the United States withdrew from the agreement.[45]

This network of municipalities contributes to a global response to climate change in and of itself, given its transnational nature and the fact that the network represents 15 per cent of global carbon dioxide emissions.[46] As Betsill and Bulkeley argue, the cities program 'has created its own arena of governance through the development of norms and rules for compliance with the goals and targets of the network.'[47] Beyond this direct influence on governance – the direct steering of large populations across the globe – the cities network also has an impact on governance efforts at other levels. Because cities are embedded in larger governmental structures, their efforts to promote climate protection contribute to climate politics at the national and multilateral level.[48] This subtler influence is not just encompassed by traditional lobbying and is, instead, an attempt 'to reframe an issue which is usually considered in global terms within practices and institutions which are circumscribed as local.'[49] Thus, beyond challenging received notions of which actors are legitimately authoritative, the cities program also questions the conventional wisdom of where the global response to climate change should be centred.[50]

Similar to the cities program, but at a higher level of government, activist governors in the United States, and in potential partnership with provincial leaders in Canada, have begun working to establish carbon markets that are simultaneously subnational and transnational. The dynamics are nearly identical to the cities program and the CRAGs in that groups of like-minded political actors (individuals, cities, states and provinces) agree amongst themselves to address climate change in a specific way – mainly by choosing mandated reductions for members of the groups with flexible (market-based) mechanisms for achieving the reductions along with varying forms of enforcement:

> A new report from CIBC World Markets, the wholesale and corporate banking arm of the Canadian Imperial Bank of Commerce (CIBC), forecasts that all jurisdictions in Canada and the US will have carbon dioxide regulations in place by the end of the decade to address global warming concerns. The report predicts that every province and state in North America will follow the lead of California and implement not only a CO_2 emissions cap but also an emissions trading system that will allow larger polluters to buy emissions credits from other firms whose emissions are less than what is allowed under the cap.[51]

What is emerging is a patchwork of progressive action on climate change self-organizing at multiple levels of government and political community. These measures lack the kind of comprehensive approach that multilateral treaty-making begets, but they do serve to catalyse potentially climate-friendly action at multiple levels, provide a range of activities for reducing emissions, and question the received understanding of multilateral approaches as the sole legitimate form of global environmental governance.

3.7. *Multilateral Competition for Kyoto: The Asia-Pacific Partnership*

The stalemate in the Kyoto process has not stifled all multilateral approaches to climate governance. One response has been the emergence of a competing multilateral approach – the Asia-Pacific Partnership on Clean Development and Climate. Catalysed by the United States and Australia, the most vocal Kyoto opponents, the APP brings together six countries (China, India, Australia, South Korea, Japan, and the United States) in a market based, voluntary governance arrangement designed to respond to global climate change through technological diffusion. This new pact, which Canada has recently indicated it would formally join,[52] approaches climate change in a market-based manner. One of its members claims that 'the Asia-Pacific Partnership on Clean Development and Climate (AP6) is a ground-breaking climate change approach bringing together key developed and developing countries on practical, pro-growth, technology-driven efforts.'[53] While the members of APP have taken pains to say that this initiative is not a challenge to Kyoto, the contrast is striking and it certainly represents a governance experiment that challenges the dominance of the universal, state-centric, multilateral treaty-making model of governance that is exemplified by the *Kyoto Protocol*.

By putting tangible material resources behind the APP, the members of the pact are legitimating a governance experiment that addresses climate change as a side effect of the pursuit of technological innovation and economic development. While many observers remain sceptical of the sincerity of the program, there are over 90 proposed projects in eight sectors.[54] By taking this route – proposing an alternative institutional structure – the United States and its partners are proposing a governance mechanism that stresses voluntary action by states, a significant departure from the foundation that Kyoto is built upon.

Further, while the Asia-Pacific partnership is ostensibly open to other states (and thus in principle universal), in practice it is a sub-group approach to the global response to climate change – like-minded states are grouping together and pursuing governance mechanisms that fit their understanding of climate change and their policy priorities. In essence this development could challenge universal state-centric approaches and harkens back to suggestions on the appropriate structure of a climate-change regime made before universal participation became the dominant frame – governance through the limited participation of key states.[55]

3.8. Let 100 Flowers Bloom?

In one sense, this sampling of climate-governance experiments should be cause for optimism. There are diverse actors taking climate change very seriously and attempting to implement rules and policies to address the problem. Yet, while exciting, the emergence of these experiments is also foreboding because it signals the dissatisfaction with and ineffectiveness of conventional multilateral approaches – the standard tool in global environmental governance. Further, we simply do not know whether these experiments will add up to, or evolve into, an effective response to climate change. As Bulkeley and Moser argue, 'the motivations for different actors to get involved [in climate governance] differ considerably and as such are not all driven by a common ethical sentiment or material goal.'[56]

4. Conclusion: Predictions, Hopes, and Fears

The instability in the global response to climate change is palpable and the way ahead is murky. This is a troubling conclusion to reach, given the urgency of this problem. In this concluding discussion, let me venture two predictions for the global response to climate change, along with the hopes and fears that accompany them.

4.1. *Predictions*

First, I predict that the experimentation that has recently emerged will continue and expand. This pattern will dominate in the near future of the global response to climate change. This does not mean an end to multilateral processes, but it does mean that multilateralism will cede pride of place to a diverse mix of approaches. The global response to climate change is heading for significant decentralization in the identity of important players (from states to multiple political actors), diversification in the mode of response (from dominant focus on treaty-based, binding emissions reductions to a menu of options dependent on the community doing the responding), and the transition to a 'networked' response (from a hierarchy of global negotiations leading to national policy to a more horizontal structure with nested polities and responses at multiple and linked scales).[57] This is a brave new world with all sorts of climate-governance experiments in it.

Second, and very much related to the first prediction, I foresee that the conventional mega-multilateral approach to climate change will fade from prominence in the near future. It will not disappear entirely and I substantially agree with Christoff when he argues that 'bilateral or limited regional agreements cannot successfully replace' the Kyoto process.[58] Yet, for all its advantages, a multilateral response built on binding treaty arrangements, binding emissions reductions, and universal participation with differentiated responsibilities faces challenges that will likely be insurmountable in the near term when crucial measures must be taken. A multilateral and global response is necessary for climate change. I am just not convinced that they will look the same way in the future as they have to date or that the multilateral response will continue to be the global response.

4.2. *Hopes and Fears*

The preceding paragraph was a painful one to write for someone who has followed the Kyoto process and hopes that the world will come to its collective senses in addressing climate change. It is infuriating to predict the demise of a governance mechanism into which so many have put so much effort and hope. It is infuriating to consider that such a prediction could provide succour to those who argue against universal negotiations aimed at binding emissions reductions as a way to delay significant action on climate change. Yet, I find that experimentation is ascendant in the global response to climate change and that the mega-multilateralism represented by the UNFCCC/Kyoto model is ebbing. Such a potential transition is both exciting and worrisome.

My fears, admittedly, arise from the uncertain nature of experimentation and from nostalgia for mega-multilateralism. There is comfort in a single authoritative response that encompasses the globe and achieves binding treaties. A proliferation of experimental governance responses leads to uncertainty. If everyone is responsible for the response to climate change does that imply that no one is? The calls for smaller, like-minded groups responding to climate change at multiple scales through a variety of tools make a great deal of sense. There is, after all, plenty to do in responding to climate change. But will the whole be at least as much as the sum of the parts, and will the sum of efforts be enough?[59] If the age of effective mega-multilateralism in global environmental governance is over (assuming it ever existed in more than our imaginations), I fear that we have yet to reach the age of effective multi-scale global environmental governance. I fear that, without global coordination, the overall effectiveness of the global response to climate change is both open to question and difficult to measure.

But uncertainty and instability present opportunities as well as challenges. Indeed, if the Kyoto process is as ossified and ineffective as its critics claim, an uncertain response made up of governance experiments across the world can hardly do worse. In their examination of complex adaptation in organizations, Smith and Stacey find hope that an adaptive response like we are currently observing can lead to innovative change:

> Change becomes a possibility only when the informal system seeks to undermine the formal, but no one can guarantee that change will in fact happen, nor can anyone guarantee that the outcome will necessarily be

good in some sense ... The processes we would expect to observe as key in bringing about innovative change would be networking, politics of an alternating cooperative and competitive sort, learning communities of practice swapping stories that embody their learning and redundancy taking the form of overlapping processes and activities.[60]

The governance experiments in the global response to climate change are relatively new, but even at this stage we can see hope for innovative change. None of the experiments exists in isolation. There are clear and growing communities of practice. There is emerging redundancy that provides resilience. All of these efforts are undermining what is a currently stalemated formal global response embodied in multilateral negotiations. In so doing, these efforts provide hope that, either through their combined efforts or through innovations that lead to a new global response, the world will respond effectively to the climate crisis.

Quo vadis? Into a great unknown of experimental governance responses. Binding multilateral emission reduction treaties will and should continue to be negotiated. However, they will exist, sometimes competitively and sometimes cooperatively, in conjunction with responses at multiple scales driven by diverse ideas about what the global response to climate change should be. Perhaps the best role for a universal global forum like the UNFCCC is to accept experimentation and work to coordinate it, helping communities find and exploit synergies.[61] Perhaps the best global approach to climate change in this context is a multilateral forum that accompanies and manages local, national, and regional responses rather than one that supersedes them.

Notes

1 See *United Nations Framework Convention on Climate Change*, U.N. Doc. A/AC.237/18 (PartII)/Add.1, reprinted in (1992) 31 I.L.M. 849 [hereinafter UNFCCC]; and *Kyoto Protocol to the United Nations Framework Convention on Climate Change*, 10 December 1997, U.N. Doc. FCCC/CP/1997/L.7/add. 1, reprinted in (1998) 37 I.L.M. 22 [hereinafter the *Kyoto Protocol*].

2 See Article 12, *Kyoto Protocol, supra* note 1. The clean development mechanism allows developed-country parties to the *Kyoto Protocol* to invest in emissions reduction projects in developing-country parties, which do not have emissions reduction commitments under the protocol. The

developed-country party can count emission reductions toward its Kyoto commitment; the developing country receives technology transfer and assistance with its transition to a low-emissions development path.

3 See 'Summary of the Eleventh Conference of the Parties to the UN Framework Convention on Climate Change and First Conference of the Parties Serving as the First Meeting of the Parties to the Kyoto Protocol' (2005) 12 Earth Negotiations Bulletin 1 [hereinafter COP-11/MOP-1].

4 See 'Summary of the Nineteenth Meeting of the Parties to the Montreal Protocol on Substances that Deplete the Ozone Layer' (2007) 19 Earth Negotiations Bulletin 1.

5 See The Norwegian Nobel Committee, 'The Nobel Peace Prize for 2007,' Press Release (12 October 2007) online, http://nobelprize.org/nobel_prizes/peace/laureates/2007/press.html.

6 Frank Jotzo, 'Developing Countries and the Future of the Kyoto Protocol' (2005) 17 Global Change, Peace & Security 77 at 86.

7 'Summary of the Twelfth Conference of the Parties to the UN Framework Convention on Climate Change and Second Meeting of the Parties to the Kyoto Protocol' (2006) 12 Earth Negotiations Bulletin 1 at 19 [hereinafter COP-12/MOP-2].

8 Lisa Schipper and Emily Boyd, 'UNFCCC COP 11 and COP/MOP 1: At Last, Some Hope?' (2006) 15 Journal of Environment and Development 75 at 76.

9 See generally, Steinar Andresen and Jon Birger Skjærseth, 'Science and Technology: From Agenda Setting to Implementation,' in Daniel Bodansky, Jutta Brunnée, and Ellen Hey, eds., *Oxford Handbook of International Environmental Law* (Oxford: Oxford University Press, 2007) at 182.

10 Alex Michaelowa, 'Principles of Climate Policy after 2012' (2006) 41 Intereconomics 60 at 61.

11 See COP-11/MOP-1, *supra*, note 3 at 20.

12 M.J. Hoffmann, *Ozone Depletion and Climate Change: Constructing a Global Response* (Albany: SUNY Press, 2005).

13 Ibid. at chs. 1, 5.

14 Ibid. at chs. 2, 5.

15 See COP-11/MOP-1, *supra* note 3; COP-12/MOP-2, *supra* note 7. See also 'Twenty-Fourth Sessions of the Subsidiary Bodies of the UNFCCC and First Session of the Ad Hoc Working Group under the Kyoto Protocol' (2006) 12 Earth Negotiations Bulletin 1.

16 See COP-11/MOP-1, *supra* note 3; COP-12/MOP-2, *supra* note 7.

17 See COP-11/MOP-1, *supra* note 3; COP-12/MOP-2, *supra* note 7.

18 Guri Bang et al., 'The United States and International Climate Cooperation:

International "Pull" versus Domestic "Push"' (2007) 35 Energy Policy 1282.
19 Peter Christoff, 'Post-Kyoto? Post-Bush? Towards an Effective "Climate Coalition of the Willing"' (2006) 82 International Affairs 831 at 832.
20 Bang et al, *supra* note 18 at 1290.
21 Peter Haas, 'Post-Kyoto Governance Prospects,' Paper presented at (2007) Human Dimensions of Global Environmental Change Conference at 7 (on file with author).
22 See, e.g., Aynsley Kellow, 'A New Process for Negotiating Multilateral Environmental Agreements? The Asia-Pacific Climate Partnership beyond Kyoto' (2006) 60 Australian Journal of International Affairs 287.
23 David Victor, 'Toward Effective International Cooperation on Climate Change: Numbers, Interests and Institutions' (2006) 6 Global Environmental Politics 90. See also James Sebenius, 'Designing Negotiations toward a New Regime: The Case of Global Warming' (1991) 15 International Security 110.
24 Joanna Lewis and Elliot Diringer, 'Policy-based Commitments in a Post-2012 Climate Framework,' Pew Center on Global Climate Change Working Paper (May 2007); online, http://www.pewclimate.org/policy_center/international_policy/.
25 Work by Annex 1 Parties on the scientific basis for determining their further commitments, including on scenarios for stabilization of atmospheric concentrations of GHG and on the implications of these scenarios, First in-session workshop of the AWG on Further Commitment for Annex I Parties under the Kyoto Protocol (6–14 November 2006) online, http://unfccc.int/files/meetings/cop_12/in-session_workshops/application/pdf/061107_4_awg_japan_1.pdf.
26 Joanna Depledge, 'The Opposite of Learning: Ossification in the Climate Change Regime' (2006) 6 Global Environmental Politics 1 at 1.
27 Ibid. at 3.
28 Anja Kollmuss and Benjamin Bowell, 'Voluntary Offsets for Air-Travel Carbon Emissions: Evaluations and Recommendations of Voluntary Offset Companies,' Tufts Climate Initiative (2006) online, http://www.tufts.edu/tie/tci/pdf/TCI_Carbon_Offsets_Paper_Jan31.p df.
29 'What Is Terrapass?' Online, http://www.terrapass.com/about/index.html, accessed 30 October 2007.
30 See *supra* note 26.
31 See, e.g., B. Cashore et al., *Governing through Markets: Forest Certification and the Emergence of Non-state Authority* (New Haven: Yale University Press, 2004).

32 See, e.g., John G. Ruggie, 'Global_governance.net: the Global Compact as Learning Network' (2001) 7 Global Governance 371; Benedicte Bull, Morten Bøås, and Desmond McNeill, 'Private Sector Influence in the Multilateral System: A Changing Structure of World Governance?' (2004) 10 Global Governance 481; V. Haufler, *A Public Role for the Private Sector: Industry Self-Regulation in a Global Economy* (Washington, DC: Carnegie Endowment for International Peace, 2001).

33 Beyond the insurance industry, the Chicago Carbon Exchange is a voluntary arrangement for businesses to cap and trade carbon emissions with members whose total emissions exceed those of Great Britain. See online, http://www.chicagoclimatex.com/about/, and http://www.liveneutral.org/who_we_are.

34 Jeremy Leggett, 'Climate Change and the Insurance Agency: Solidarity among the Risk Community,' (Greenpeace, 1993). See also Matthew Paterson, 'Risky Business: Insurance Companies in Global Warming Politics' (2001) 1 Global Environmental Politics 18; and Sverker C. Jagers and Johannes Stripple, 'Climate Governance beyond the State' (2003) 9 Global Governance 385.

35 See Paterson, *supra*, note 34, and Jagers and Stripple, *supra* note 34.

36 See Paterson, *supra* note 34 at 19.

37 See United Nations Environment Program (UNEP) Finance Initiative. Online, http://www.unepfi.org/work_streams/insurance/index.html.

38 See Jagers and Stripple, *supra* note 34.

39 Ibid.

40 Carbon Rationing Action Group, 'What is Carbon Rationing?' online, http://www.carbonrationing.org.uk/what.

41 Carbon Rationing Action Group, 'Who are we?' online, http://www.carbonrationing.org.uk/glasgow/threads/who-are-we.

42 International Council for Local Environmental Initiatives (ICLEI), 'Cities for Climate Protection (CCP) Participants,' online, http://www.iclei.org/index.php?id=809.

43 ICLEI, 'How It Works,' online, http://www.iclei.org/index.php?id=810.

44 Associated Press, 'European Cities Tackling Climate Change' (14 October 2007) online, http://www.ctv.ca/servlet/ArticleNews/story/CTVNews/20071014/clim ate_change_071013, accessed 30 October 2007.

45 Salt Lake City Corporation, 'On the Eve of Olympics, Salt Lake City Joins World in Support of Kyoto Protocol' (3 February 2002) online, http://www.slcgov.com/mayor/pressreleases/kyoto%20protocol.htm.

46 ICLEI, 'About CCP,' online, http://www.iclei.org/index.php?id=811.

47 Michelle Betsill and Harriet Bulkeley, 'Cities and the Multilevel Governance of Global Climate Change' (2006) 12 Global Governance 141 at 151
48 Ibid.
49 Harriet Bulkeley, 'Reconfiguring Environmental Governance: Towards a Politics of Scales and Networks' (2005) 24 Political Geography 875 at 893.
50 Ibid.
51 Green Car Congress, 'Forecast: US and Canada Will Regulate CO_2 by End of Decade' (10 January 2007) online, http://www.greencarcongress.com/2007/01/forecast_us_and.html#more.
52 See CBC News, 'Kyoto Alternative – What Is This New Asia-Pacific Partnership All About?' (27 September 2007) online, http://www.cbc.ca/news/background/kyoto/asia-pacific-partnership.html.
53 Government of Australia, Department of Foreign Affairs and Trade, 'Asia Pacific Partnership on Clean Development and Climate,' online, http://www.dfat.gov.au/environment/climate/ap6/.
54 To this point the funding is 'in principle' but Australia has pledged AUS$59 million and the United States is proposing $52 million for the fiscal 2007 budget. Beyond direct dollars, the material incentive is in opening up investment opportunities. Ibid.
55 See Hoffmann, *supra* note 12 and Sebenius, *supra* note 23.
56 Harriet Bulkeley and Suzanne C. Moser, 'Responding to Climate Change: Governance and Social Action beyond Kyoto' (2007) 7 Global Environmental Politics 1 at 3.
57 For more on networked governance structures see Bulkeley, *supra* note 49.
58 See Christoff, *supra* note 19 at 850.
59 An example of this concern is the carbon market. So much of the response to climate change through market measures relies on a stable price for carbon. A concatenation of multi-scalar responses may not provide the stability of a globally integrated response and the global carbon market that would accompany it.
60 Michaela Smith and Ralph Stacey, 'Governance and Cooperative Networks: An Adaptive Systems Perspective' (1997) 54 Technological Forecasting and Social Change 79 at 85.
61 My thoughts on this have been influenced by the insights of Ngaire Woods who is examining parallel dynamics in the context of the International Monetary Fund and the World Bank. See Ngaire Woods, 'Current Revolutions in Development Finance,' Presentation at the University of Toronto (19 October 2007).

8 Grandfathering, Carbon Intensity, Historical Responsibility, or Contract/ Converge?

J. TIMMONS ROBERTS AND BRADLEY C. PARKS

Introduction

Who is responsible for climate change? What are the different ways of accounting for responsibility? And who is making progress towards resolving the problem?[1] This brief article reviews four yardsticks that have been proposed for measuring carbon emissions responsibility across countries during fifteen years of international negotiations. Each method reflects a different set of principled beliefs and identifies a different set of nations as most responsible. Politicians have used these yardsticks to defend their positions on what they believe to be 'fair' and 'just.' And, not surprisingly, these measures have lead to scientific uncertainties and rancorous debate.[2]

First, we describe the range of principles that have been proposed to assign responsibility for carbon emissions. We then briefly explain why none of these rules and principles (by themselves) will likely produce a fairness consensus, and proceed to discuss some of the more promising hybrid proposals currently being considered. While recognizing the need to move away from particularistic notions of justice, we argue that the 'negotiated justice' literature obscures the more central question of whether and to what extent an agreement must favour rich or poor nations.[3]

Four central – and radically different – methods of differentiation have been proposed during the fifteen or so years of climate negotiations, each with crucial assumptions about and implications for climate stabilization, social justice, political expediency, and who will bear the greatest burden of change if it is accepted as the basis for a climate treaty.[4] These are grandfathering; carbon intensity; historical responsibility; and contraction and convergence to a global per capita norm.

The *Kyoto Protocol* as it was negotiated in 1997 was based on *grandfathering*: the notion that nations should reduce their emissions incrementally from a baseline year, in this case, 1990.[5] Large emitters, therefore, had their high discharges of greenhouse gases grandfathered in, with relatively minor adjustments averaging 5.2 per cent, for the foreseeable future. The *carbon intensity* approach, introduced by the World Resources Institute and favoured by the second Bush administration starting in 2002, calls for voluntary efficiency changes to drive improvement of emissions. In this approach, the goal is to have strong economic growth with as few carbon emissions as possible. Both of these proposals have the modest effect of departing incrementally from the current status quo without radical requirements on powerful nations.

On the other side of the spectrum are two proposals that strongly favour developing countries: *historical responsibility* and per capita *contraction and convergence*. India, China, and much of the developing world favour a per capita approach, in which each person on earth is given an equal right to the ability of the atmosphere to absorb carbon. Under the per capita proposal, countries whose per capita consumption of fossil fuels is significantly lower than the world average would be given significant room to grow and emit. Most per capita plans would allow them to trade their extra carbon emission credits for the capital they need for development, while allowing them to trade extra credits for the capital they need for development. By comparison, nations with highly fossil-energy-intensive economies would face sharp requirements to cut their consumption of fuels.

Brazil also introduced a proposal in 1997 that would take into account the amount of damage done by nations in the past to the atmosphere's ability to absorb more greenhouse gases. This *historical responsibility* approach puts the onus on nations that put greenhouse gases in the atmosphere in past decades to reduce their emissions quickly, most notably Britain and the United States. Some developing countries have supported this approach and demanded that some indemnification be paid for the so-called carbon debt. We take up each alternative approach below, examining its roots and implications, the principles upon which it is based, and how it has been approached or ignored in global climate negotiations over the past dozen years.

Grandfathering

The treaty that emerged from the back-room bargaining at Kyoto in 1997 was based on the concept of 'grandfathering' – that the world's wealthier nations would make efforts to reduce their carbon emissions relative to a baseline year, in this case, 1990.[6] After a series of drawn-out negotiations, individual reductions targets were agreed upon among rich nations, mostly 6–7 per cent below the 1990 baseline by 2010.[7] The approach was decided upon for Kyoto because of its political expediency.

For more than a decade of climate negotiations, similar arguments have been made in response to calls for sharp and immediate cuts in emission levels. Many countries contend that national circumstances and economic hardships affect their ability to make immediate and deep reductions. At Kyoto and the many meetings since, the United States, Russia, and several other high-emissions nations bargained hard for minor changes to the current state of affairs. The U.S. Senate and current U.S. administration have underscored that even a Kyoto-type treaty would unfairly damage the nation's economy and send jobs overseas. On the Senate floor in the fall of 2003, dozens of senators argued the McCain-Lieberman Climate Stewardship Act would place terrible limits on U.S. economic growth and create terrible suffering among different groups of Americans.[8]

The principle of grandfathering has not been applied to developing countries. However, were it to resurface (as focal points often do)[9] most experts agree it would have the effect of punishing late-developers. As Aslam explains, 'current emissions of developing countries ... are very low compared with those of industrialized countries, but are rising rapidly. This places developing countries at a severe disadvantage when it comes to negotiating emission control targets that are based on a grandfathering system.'[10]

Figures 8.1 and 8.2 illustrates the startling extent of inequality in total emissions.[11] According to the latest figures at the time of writing, the United States emitted the largest proportion of carbon of any nation, with China closing in rapidly (figure 8.1). Since the 1990 baselines, emissions have in fact increased in most nations, and in some nations substantially. Besides the ex-Soviet Union republics, whose economies largely collapsed after the transition to capitalism, virtually no countries in the developed world are on track to meet even their modest Kyoto goals. Wealthier nations who accepted Kyoto targets were listed in Annex I of the treaty, and the rest, who were expected to take up limits only in future rounds of the treaty, were classified as 'non-Annex I.'

While many argue that the grandfathering approach is amoral and baldly based on political power in the international system, it does rep-

Figure 8.1. Thousands of metric tons of total emissions of CO_2 from fossil fuels: selected nations, 2004.

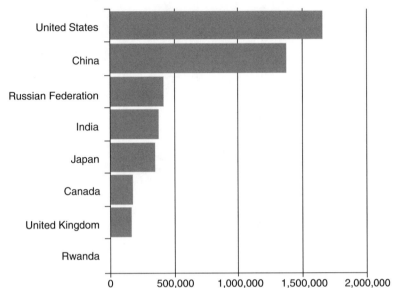

Source: Carbon Dioxide Information Analysis Center, *Trends: A Compendium of Data on Global Change*.

resent at least three understandings of justice. Entitlement theories of justice, both in their libertarian and Marxist forms, hold that individuals are entitled to what they have produced.[12] As such, every nation possesses a common law (inherent) right to emit carbon dioxide. Grandfathering also exemplifies the justice principle of proportional equality – that nations are unequal and should therefore be treated unequally. While developing countries were not required to commit to scheduled reductions of emissions in the first round of negotiations, the decision to use 1990 as a baseline year is an implicit recognition of these two principles. Finally, grandfathering represents the pragmatic principle that if we can solve the problem we are closer to justice than if we insist upon a utopian plan which makes no progress. Cecilia Albin argues that in spite of the fact that international environmental agreements regularly institutionalize fairness norms – for example, the 'polluter pays,' 'no harm,' and 'shared, but differentiated responsibility' principles – their success is first and foremost dependent upon their

Figure 8.2. Per capita emissions of CO_2 in metric tons of carbon per person: selected nations, 2004.

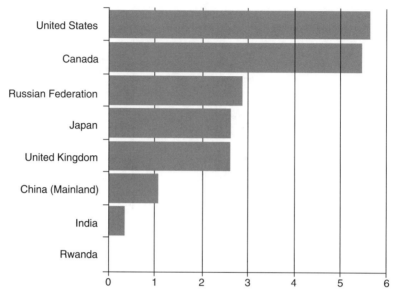

Source: Carbon Dioxide Information Analysis Center, *Trends: A Compendium of Data on Global Change*.

ability to yield joint gains for all parties.[13] She offers the example of Sweden and Finland, which, despite being victimized by the air pollution of neighbouring Baltic countries, did not insist that the 'polluter pays' principle be strictly enforced. Quite the opposite. They financed a large foreign aid initiative to help the less developed countries responsible for the air pollution adopt more environmentally friendly technologies.[14]

Carbon Intensity

Faced with pressure to sign the Kyoto treaty, President George W. Bush promised during his 2000 campaign to do so. However after entering office, his position shifted and he withdrew the United States from the treaty entirely. U.S. national security advisor Condoleezza Rice told EU members in the spring of 2001 that the Kyoto treaty to address climate change was 'dead' without U.S. participation, since the treaty requires that countries responsible for 55 per cent of the total amount of emissions from the world's wealthy nations ratify it. A firestorm of reaction from Europe and environmentalists in the United States forced the Bush administration to provide an alternative plan to address the problem.

At the science centre of the National Oceanic and Atmospheric Administration in Maryland on 14 February 2002, President Bush stated flatly that 'As president of the United States, charged with safeguarding the welfare of the American people and American workers, I will not commit our nation to an unsound international treaty that will throw millions of our citizens out of work.' Rather, he proposed a 'New Approach on Global Climate Change' in response to the treaty, and provided a new benchmark by which the United States offered to measure its own progress on the issue. He 'committed America to an aggressive new strategy to cut greenhouse gas intensity by 18% over the next 10 years.'[15] The White House press releases argued that 'the President's yardstick – greenhouse gas intensity – is a better way to measure progress without hurting growth.' It continued, 'A goal expressed in terms of declining greenhouse gas intensity, measuring greenhouse gas emissions relative to economic activity, quantifies our effort to reduce emissions through conservation, adoption of cleaner, more efficient, and emission-reducing technologies, and sequestration. At the same time, an intensity goal accommodates economic growth ...'[16]

The carbon intensity approach is an outgrowth of Bentham's utilitarian theory of justice, which states that mutually advantageous and cost-effective solutions are also just solutions. Since everyone is worse off in the absence of aggregate net benefits, utilitarians argue that inefficient solutions are also unjust.[17] The fair solution, with respect to greenhouse gas emission reductions, would therefore be to stabilize the climate as cost-effectively as possible, while maximizing global economic growth. Since developing nations currently offer the most cost-effective opportunities to reduce greenhouse gas emissions, the international effort to stabilize greenhouse gas emissions would predominantly focus on the developing world.[18]

On the positive side, the carbon intensity approach forces the international community to think about designing solutions that will allow growth to occur while minimizing impact on the global climate. A number of analysts have also suggested that the carbon intensity approach creates greater opportunities for developing country buy-in, since it does not impose a hard cap on their total emissions (hard caps are often viewed as 'caps on development').[19] An added advantage to this approach is that industrialized nations tend to do better in intensity terms, since their infrastructure is typically much better than that of poorer nations.[20] As such, a carbon intensity approach could promote early action, which, according to Baumert et al., is important because 'many developing countries believe that the industrialized countries lack credibility on the issue of international cooperation to curb greenhouse gas emissions, having done little to address a problem largely of their own making.'[21]

On the downside, the proposals made by the Bush administration place no real restrictions on the future emissions of the United States (since most analysts see the nation's efficiency as improving on its own by at least 18 per cent) and are widely perceived as a repudiation of earlier commitments. The Bush administration's plan also does nothing about the existing stock of emissions and makes no effort to include 'exported emissions' caused by the offshoring of U.S. industries to poorer nations. In addition, the carbon intensity approach has become a tool of political manipulation. The United States used this approach strategically at COP-8 and COP-9 in an effort to torpedo the Kyoto Protocol and delay post-2012 talks. U.S. negotiator Harlan Watson urged Western nations at the New Delhi negotiations to 'recognize that it would be unfair – indeed, counterproductive – to condemn developing nations to slow growth or no growth by insisting that they take on impractical and unrealistic greenhouse gas targets.'[22] The following year at the Milan negotiations, U.S. undersecretary of state Paula Dobriansky tried to forge an unusual coalition with China and the G77 by rejecting the need for developing countries to undertake scheduled commitments to reduce emissions.[23]

Per Capita

India, China, and the G77 (actually a group of about 133 nations) have developed and advocated a series of proposals for accounting for carbon dioxide and other greenhouse gases on the basis of a simple, egalitarian principle. The idea is that every human on earth has equal rights to the global atmosphere, and therefore allocations of how much each can pollute should be done on a per capita basis.[24] France, Switzerland, and the European Union have all endorsed this proposal in one form or another.[25] Climate policy expert Michael Grubb calls it 'the most politically prominent contender for any specific global formula for long-term allocations with increasing numbers of adherents in both developed and developing countries.'[26]

Per capita proposals place rich nations at a sharp disadvantage, since most of them already far exceed the stabilization target (roughly one metric ton of carbon equivalent per capita). Poor nations, by comparison, stand to gain considerably from a per capita allocation of carbon entitlements because their existing levels of income and industrialization place them well below the one metric ton threshold.

Environmentally sustainable per-capita proposals typically require that a global 'emissions budget' first be specified. The scientific consensus is that to avoid the worst effects of climate change, we need to stabilize the concentrations of carbon dioxide around or below 450 parts per million. However, others suggest that 350 and 550 parts per million are more appropriate targets. In any case, these proposals suggest drastic reductions for the world's richest nations, and commitments very soon for the poorer ones to reduce growth rates of their emissions and eventually stop and reverse them.

Under most per capita proposals, including the contraction and convergence model proposed by the Global Commons Institute, once the size of the emissions budget is specified, every global citizen is allocated an equal entitlement to the atmosphere. Rich countries, whose relatively small populations have already used a disproportionate amount of their atmospheric space, must 'contract' their annual carbon budget to a level of roughly one metric ton of carbon equivalent over the next century. Poor nations, whose citizens have thus far occupied very little atmospheric space, are allowed to increase their emissions for some time and eventually 'converge' with rich nations. Developing countries willing to restrict their emissions growth below their allowance have the opportunity to trade those allowances in exchange for funding or technical assistance through the Clean Development Mechanism, Joint Implementation, and other emissions trading mechanisms.

The key question surrounding the per capita approach is its political feasibility. Egalitarian principles played a prominent role in UN Convention on Law of the Sea negotiations.[27] However, many analysts consider the application of egalitarian principles to climate policy politically explosive and economically inefficient. Grubb and his colleagues describe one very telling interaction between rich and poor nations at the Kyoto negotiations that lasted late into the evening.[28] At 3 o'clock in the morning, amidst heated debate over global emissions trading, China, India, and the Africa Group of Nations expressed their support for a per capita allocation of global atmospheric property rights. In response, Chairman Raul Estrada and a representative of the U.S. delegation stated bluntly that the contraction and convergence proposal

Table 8.1. Total and per capita CO_2 emissions and comparison ratios: selected nations, 2004

Nation	Total CO_2 emissions from fossil-fuels (thousand metric tons)	Per capita CO_2 emissions (metric tons of carbon)	Avg. emissions per citizen compared to Rwandan	Number of avg. people to emit as much as one avg. Canadian
Canada	174,401	5.46	273	1
China	1,366,554	1.05	52.5	5
India	366,301	0.34	17	16
Japan	343,117	2.69	134.5	2
Russia	415,951	2.89	144.5	2
Rwanda	156	0.02	1	273
United Kingdom	160,179	2.67	133.5	2
United States	1,650,020	5.61	280.5	1

Source: Carbon Dioxide Information Analysis Center, *Trends: A Compendium of Data on Global Change*.

was a political non-starter and negotiations were immediately brought to a close.

It is important for readers to understand just how far apart the people of the world are in per capita terms. Twenty per cent of the world's population in the high income countries is responsible for 63 per cent of the emissions, while the bottom 20 per cent of the world's population is only releasing three per cent.[29] According to our calculations with 2004 CDIAC figures, the average Canadian citizen dumps as much greenhouse gas into the atmosphere as five Chinese citizens, sixteen citizens of India, and 273 Rwandans.[30] In 183 nations, people emit on average less than half the emissions Americans do. In 130 nations, it would take at least five citizens to generate as much carbon dioxide from fossil fuel burning as one U.S. citizen does. In 90 nations it would take over ten citizens to generate as much as one American. And in 30 of those nations, it would take over *one hundred*.[31]

Historical Responsibility

The 'polluter pays' principle has been central to domestic and international environmental law for more than thirty years.[32] In the mid-1990s, Brazilian scientists and government experts developed a sophisticated proposal to address climate change based on this principle. They argued that a country's greenhouse gas reductions should depend on its relative contribution to the global temperature rise.[33] The reasoning behind the 'historical responsibility' proposal is that carbon dioxide burned now stays in the atmosphere for 100 to 120 years. Therefore, it is important to account not only for future emissions, but for all of the damage done in earlier years. The political implications are obvious: since virtually all the carbon emitted since 1945 is still in the atmosphere and early industrializers are almost exclusively responsible for that damage, rich nations would be required to make deep and immediate cuts.[34] Early estimates suggest that by 2010 Britain would have to reduce emissions by 66 per cent, the United States by 23 per cent, and Japan 8 per cent.[35]

Given their tiny contribution to the existing stock of carbon emissions, it is not surprising that developing countries have been strong advocates of the historical responsibility approach. At their 2000 South Summit in Havana, the G77 submitted the following statement as part of a larger manifesto: 'We believe that the prevailing modes of production and consumption in the industrialized world are unsustainable and should be changed for they threaten the very survival of the planet ... We advocate a solution for the serious global, regional, and local environmental problems facing humanity, *based on the recognition of the North's ecological debt* and the principle of common but differentiated responsibilities of the developed and developing countries.'[36]

However, the historical responsibility proposal has failed to gain much traction in the policy community. To be broadly acceptable to people around the world, proposals for addressing climate change generally need to be easy to understand, and making the historical responsibility principle operational requires fairly complex methods of calculation.[37] One of the more straightforward proposals for estimating the historical responsibility of nations is to sum their cumulative emissions from some date in the past. The 2000 *Special Report on Emissions Scenarios* of the IPCC found that such summed emissions 'supply a reasonable "proxy" for the relative contribution to global warming' of different nations, if 'limited to a few decades.'[38] For the purposes of illustration, we have taken a similar approach, summing all industrial emissions of carbon dioxide from 1950–2004 for each nation, to create indicators of a nation's total historical responsibility and another such value of cumulative emissions per capita.[39]

Table 8.2 indicates that when emissions since 1950 are summed, the

Table 8.2. Carbon emissions intensity and historical responsibility, 2004.

Country	Emissions intensity[1]	Per capita historical responsibility[2]
Canada	0.18	221.91
China	0.70	22.16
India	0.53	8.54
Japan	0.07	95.49
Russia	0.71	121.61[3]
Rwanda	0.10	0.65[4]
United Kingdom	0.08	157.36
United States	0.14	280.39

Source: Carbon Dioxide Information Analysis Center (CDIAC), *Trends: A Compendium of Data on Global Change.*

Notes
1 Emissions intensity is in kg of carbon emitted per unit of GDP in 2004 U.S. dollars.
2 Historical responsibility is the sum of per capita emissions from 1950 to 2004.
3 Russian values unknown before 1992. The figure here is based on an average of 2 tons per capita per year from 1950–1991.
4 Figures before 1962 for Rwanda are for Rwanda-Burundi.

gap between rich and poor nations is much higher and not narrowing nor going away any time soon.[40] The summed emissions from the high-income nations amount to nearly twice the tons of carbon of the middle income nations, and four times the cumulative emissions of the majority who live in the poorest nations. This is a highly contentious issue, but one which we believe we have to consider if we are to address inequality and climate change. The polluter pays argument is that high-emitting nations, even if they did not know the danger of their behaviour, still benefited from it and should be held responsible. This logic holds in many national laws, including Superfund and other pollution laws in the United States. Table 8.2 demonstrates very clearly that rich nations are hundreds of times more responsible for global warming than are poor nations.

Moving toward 'Hybrid Justice'

To recap, there are four very different approaches to measuring national responsibility for greenhouse gas emissions: the Kyoto grandfathering approach, which relies on entitlement principles of justice; the carbon intensity approach, which rests on utilitarian principles of justice; the historical responsibility approach, which operationalizes the 'polluter-pays' principle; and the egalitarian per capita approach. Discussions of 'climate justice' – taking place within the Brazilian government, among influential Indian NGOs, between developing country negotiators at environmental conferences, in the U.S. Congress, the European Parliament, the Canadian Parliament, and at G77 summits – could therefore be placed along a hypothetical 'principled-beliefs spectrum.' What our analysis shows is that poor nations and rich nations hold almost diametrically opposed views of climate justice, largely for configurational reasons having to do with their position in the global hierarchy of economic and political power. Such inequality unfortunately makes it very unlikely that a North–South fairness consensus will spontaneously emerge on the basis of one of these four principles described above. Therefore, it appears that a moral compromise, or 'negotiated justice' settlement may be necessary. Both scholars and negotiators increasingly recognize that strict adherence to particularistic notions of justice is a perfect recipe for a stalemate.[41] So it may be that an optimal mix of efficiency and equity principles will enable rich and poor nations to overcome previous barriers to cooperation.[42]

However, there are no hard and fast distinctions between efficient and equitable proposals in the burden-sharing debate. If an equitable policy principle lacks implementability – that is, if it fails to yield joint gains – it ends up being inefficient.[43] Moral ambiguity has also forced analysts to slur this efficiency-equity distinction. Should burden-sharing proposals substantively favour those with the greatest need, the least responsibility, or the greatest ability to pay?

Nevertheless, a number of proposals representing moral compromise have emerged in recent years. Bartsch and Müller propose a 'preference score' method, which combines the grandfathering and per capita approach through a voting system.[44] Their proposal allows each nation – weighted by their population – to choose the methodology that they prefer. Each global citizen's 'vote' is then used to calculate national carbon emission allowances. According to their preliminary model, under this proposal, roughly three-quarters of the global emissions budget would be based on the per capita approach and one-quarter on grandfathering.[45] Others have focused on more politically feasible per capita proposals that provide for 'national circumstances', or allowance factors, like geography, climate, energy supply, and domestic economic structure, as well as 'soft landing scenarios.'[46]

The Pew Center for Global Climate Change has developed a hybrid proposal that assigns responsibility based on past and present emissions, carbon intensity, and countries' ability to pay (i.e. its per capita GDP).[47] It separates the world into three groups: those that 'must act now,' those that 'could act now,' and those that 'should act now, but differently.'[48] The 'Triptych' proposal, designed by scholars at the University of Utrecht (and already used to differentiate commitments among EU countries), 'accounts for differences in national circumstances such as population size and growth, standard of living, economic structure and fuel mix in power generation.'[49] Its novel contribution is that it divides each country's economy into three sectors: energy-intensive industry, power generation, and the so-called domestic sector (transport, light industry, agriculture, and commercial sector).[50] It applies the carbon intensity approach to the energy-intensive sector, 'decarbonization targets' to the power generation sector, and a per capita approach to the 'domestic' sectors. Similarly, the multi-sector convergence approach, developed by two research institutes in Northern Europe (ECN and CICERO), treats sectors differentially and integrates per capita, carbon intensity, and ability to pay (GDP per capita) approaches.[51]

EcoEquity.org has also created a 'Greenhouse Development Rights' framework as a reference to evaluate proposals for the post-2012 period.[52] They argue that countries below a "global middle class" income of US$9,000 per capita should be assured that they will not be asked to make binding limits until they approach that level, while countries above that level should be responsible for rapid reductions of emissions and payments to assist those below the line in improving their social and economic status while adjusting to a less carbon-intensive path of development.

We believe these hybrid proposals are among the most promising solutions to break the North–South stalemate. However, simply asserting that a 'negotiated justice' settlement is necessary avoids the more central question of whether and to what extent an agreement must favour rich or poor nations.[53] As we have argued elsewhere, the greatest barriers to meaningful North–South cooperation are not differences in principled understandings of 'what is fair.' Rather, divergent principled beliefs are a consequence of more fundamental root causes: incongruent worldviews and causal beliefs, persistent global inequality, and an enduring deficit in North–South trust.[54] Therefore, along with developing a workable and fair 'hybrid justice' proposal, we believe policy makers must redouble their efforts to allay the fears and suspicions of developing countries; rebuild conditions of generalized trust, forge long-term, constructive partnerships with developing countries across multiple issue areas; and create greater 'policy space' for governments to pursue their own development strategies.[55]

Attention is finally beginning to be paid to sectors and pathways of development. Several South Asian authors working on the Fourth Assessment Report of the IPCC have recently argued that 'development pathways ... societies choose today may be as important, possibly even more important, as the climate measures they take.'[56] Initial discussions of this issue also began at the COP-10 in Buenos Aires in 2004. Our own research suggests that some development pathways insulate countries from economic volatility more than others, cause less local environmental damage, and give more options to planners; others are much more difficult to change.[57]

More in-depth analysis of development pathways under globalization is needed, but one can imagine a sophisticated hybrid proposal for assigning national carbon-dioxide emission quotas based initially on economic profiling of the consumption and production of nations. This would require that a future treaty be developed from the physical science of what the atmosphere can likely handle, principled decisions about which approach is fairest, and the practical social science of how different types of nations will meet their allowed emissions. This picture of national responsibilities for the world's emissions requires more than a static accounting of tons of carbon emitted in each nation. Rather, the rapid shifting of the energy- and natural resource-intensive stages of production to developing nations requires responsibility to be tied to the total carbon 'footprint' of products where they are consumed. Climate justice will require complex physical and social science calculations and many normative decisions about how to assess responsibility. Simply put, brute bargaining strength will never lead us to a workable climate treaty, in neither the sense of atmospheric stability nor the political or social sense.

Notes

The views expressed in this paper are those of the authors and do not necessarily represent those of their employers.

1 To be more precise, we are here investigating only carbon dioxide emis-

sions from fossil fuels and from cement manufacturing and waste-gas flaring. The latter two categories account for only about 3 per cent of industrial CO2 emissions, and carbon dioxide is currently estimated to cause over 70 per cent of greenhouse warming. Commercial and residential sources are included, but emissions from changes in land-use (mostly deforestation, which accounts for about an additional 18–25 per cent of carbon releases) are not included, since these are quite uncertain scientifically: see G. Marland, T.A. Boden, and R.J. Andres, 'Global, Regional, and National Fossil Fuel CO_2 Emissions,' in Carbon Dioxide Information Analysis Centre (CDIAC), *Trends: A Compendium of Data on Global Change* (Oak Ridge: Oak Ridge National Library, 1999, 2000, and 2003) at 6–11. See the sections below for more on this issue.

2 While much of this is the result of the normal process of an emerging field of science, some of this debate about the validity of different measures and estimates has been fuelled by those concerned about their nations or industries being saddled with tough requirements under any particular system of measurement. After the science is fed into the agreement-making system, the real political implications of who would win and who would lose drive more controversy. See David Levy and Daniel Egan, 'Capital Contests: National and Transnational Channels of Corporate Influence on the Climate Change Negotiations' (1998) 26 Politics and Society 337; David Levy and Ans Kolk, 'Strategic Responses to Global Climate Change: Conflicting Pressures on Multinationals in the Oil Industry' (2002) 4 Business and Politics 275; R. Gelbspan, *Boiling Point: How Politicians, Big Oil and Coal, Journalists, and Activists Have Fuelled a Climate Crisis – And What We Can Do to Avert Disaster* (New York: Basic Books, 2004).

3 See, for example, I. William Zartman et al., 'Negotiation as a Search for Justice' (1996) 1 International Negotiation 79; Adam Rose et al., 'International Equity and Differentiation in Global Warming Policy' (1998) 12 Environmental and Resource Economics 25; Frank Biermann, 'Justice in the Greenhouse: Perspectives from International Law,' in F. Toth, ed., *Fair Weather? Equity Concerns in Climate Change* (London: Earthscan, 1999) 160; Ellen Wiegandt, 'Climate Change, Equity, and International Negotiations,' in U. Leterbacher and D. Sprinz, eds., *International Relations and Global Climate Change* (Cambridge, MA: MIT Press, 2001) 127. See particularly Odile Blanshard et al., 'Combining Efficiency with Equity: A Pragmatic Approach,' in I. Kaul, P. Conceicao, K. Le Goulven, and R.U. Mendoza, eds., *Providing Global Public Goods: Managing Globalization* (New York: Oxford University Press, 2003) 280 at 286, who argue that, 'any future burden-sharing agreement involving developing countries will probably be based on a complex differentiation scheme combining different basic rules.'

4 Many other options exist these are but the four most frequently suggested.
5 *Kyoto Protocol to the United Nations Framework Convention on Climate Change*, 10 December 1997, U.N. Doc. FCCC/CP/1997/L.7/add. 1, reprinted in (1998) 37 I.L.M. 22, [hereinafter the *Kyoto Protocol*].
6 The year 1990 was chosen because climate science became well known then, with the first assessment report of the Intergovernmental Panel on Climate Change, and by being seven years in the past it would reward and not punish early-adopters of efficiency measures.
7 Because national emissions can vary greatly depending on economic conditions in any year, the target date of 2010 was expanded into a five year average of 2008–2012.
8 See Audrey Leath, 'Senate Votes against Mandatory Greenhouse Gas Emissions Controls' (24 November 2003) 152 *FYI: The AIP Bulletin of Science Policy News*, American Institute for Physics, online: http://www.aip.org/fyi/2003/152.html.
9 J. Goldstein and R.O. Keohane, eds., *Ideas and Foreign Policy. Beliefs, Institutions, and Political Change* (Ithaca: Cornell University Press, 1993).
10 Malik A. Aslam, 'Equal per Capita Entitlements: A Key to Global Participation on Climate Change?' in K.A. Baumert, ed., *Building on the Kyoto Protocol: Options for Protecting the Climate* (Washington, DC: World Resources Institute, 2002) 175 at 176.
11 Figure 8.1 reflects how of all the 190 billion tons of carbon being released into the atmosphere in 2000, 120 billion come from the 24 high-income nations at the top of the world's class structure. The 115 low-middle- and low-income nations only emitted a total of 45 billion tons. Fifty-three middle- and high-middle income nations emit only 24 billion tons. See CDIAC 2003, *supra* note 1.
12 Benito Müller, 'Varieties of Distributive Justice in Climate Change: An Editorial Comment' (2001) 48 Climatic Change 273. Cecilia Albin, 'Negotiating International Cooperation: Global Public Goods and Fairness' (2003) 29 Review of International Studies 365.
13 Albin, *supra* note 12; C. Albin, *Justice and Fairness in International Negotiation* (Cambridge: Cambridge University Press, 2001).
14 See Albin, *supra* note 12 at 373.
15 The White House, 'Fact Sheet: President Bush Announces Clear Skies & Global Climate Change Initiative' (White House, 2002) online, http://www.whitehouse.gov/news/releases/2002/02/20020214–5.html.
16 The White House, 'Global Climate Change Policy Book' (White House, 2002) online, http://www.whitehouse.gov/news/releases/2002/02/climatechange.html.
17 D. Gauthier, *Morals by Agreement* (Oxford: Clarendon Press, 1986).

18 Robert Stavins, 'Can an Effective Global Climate Treaty be Based upon Sound Science, Rational Economics, and Pragmatic Politics?' (2004) Discussion Paper, Washington, DC: Resources for the Future 8.
19 See Yang G. Kim and Kevin A. Baumert, 'Reducing Uncertainty through Dual-Intensity Targets,' in K.A. Baumert, ed., *Building on the Kyoto Protocol: Options for Protecting the Climate* (Washington, DC: World Resources Institute, 2002) 109; Joanna Depledge, 'Continuing Kyoto: Extending Absolute Emission Caps to Developing Countries,' in ibid. 31.
20 See J. Timmons Roberts and Peter Grimes, 'Carbon Intensity and Economic Development 1962–1991: A Brief Exploration of the Environmental Kuznets Curve' (1997) 25 World Development 181; Roberts et al., 'Social Roots of Global Environmental Change: A World-Systems Analysis of Carbon Dioxide Emissions' (2003) 9 Journal of World-Systems Research 277. Figure 2 shows how the most carbon intense nations in terms of CO_2/GDP are the poorest, and the most efficient are the most wealthy. This very suggestive pattern allows many scholars to project that efficiency improvements will sweep around the world with prosperity, a point we will return to shortly.
21 Kevin A. Baumert, James F. Perkaus, and Nancy Kete, 'Great Expectations: Can International Emissions Trading Deliver an Equitable Climate Regime?' (2003) 3 Climate Policy 137.
22 'Harlan L. Watson Senior Climate Negotiator and Special Representative and Head of the U.S. Delegation – Remarks to the Eighth Session of the Conference of Parties (COP-8) to the UN Framework Convention on Climate Change' (New Delhi: United Nations Framework Convention on Climate Change, 2002).
23 Paula Dobriansky, 'Only New Technology Can Halt Climate Change,' *Financial Times* (1 December 2003).
24 Two groups have been promoting the idea of a per capita framework for years. The Global Commons Institute, led by Aubrey Meyer, has been detailing a 'contraction and convergence' approach which makes tough demands for reductions on the global North, but allows a transition period and lots of tradable permits to emit greenhouse gases in the short term transition period. The other group, with perhaps more clout because of their location in New Delhi, India, is the Centre for Science and the Environment, led by Anil Agarwal and Sunita Narain. More recently, a third group called EcoEquity, has been forcefully arguing for a per capita climate accord, saying it would be efficient, fast, equitable, and global. They argue that nations such as China will never agree to unequal limits on emissions, and so 'climate equity, far from being a "preference," is essential to ecological sustainability.' See T. Athanasiou and P. Baer, *Dead Heat: Global Justice*

and Climate Change (New York: Seven Stories Press, 2002) 74 at 75. To be sure, it is difficult to imagine any rapid convergence of nations where one is consuming twice the fuel and emitting twice the carbon of another. This is why tables 8.1 and 8.2 and figures 8.1 and 8.2 reveal such a desperate picture: so many of the world's people emit so little, while very few emit so much.

25 Other rich countries (e.g. Japan, Norway, Iceland, Poland) would reportedly accept the per capita principle if it were integrated into a larger approach (i.e. a multi-sectoral, menu approach). See Baumert et al., *supra* note 19.

26 M. Grubb, C. Vrolijk, and D. Brack, *The Kyoto Protocol: A Guide and Assessment* (London: Royal Institute of International Affairs, 1999) 270.

27 See Baumert et al., *supra* note 19.

28 See Grubb et al., *supra* note 26.

29 See Figure 5.3, 'Appendix – Carbon profiles: Countries' Annual per Capita Carbon Dioxide Emissions from Fossil Fuel Combustion and Cement Production, 1950–1995' (Population Action International, 1998) online, http://www.populationaction.org/why_pop/carbon/carbon_pdf.htm.

30 Data for the current analysis come from the World Bank and from the Carbon Dioxide Information and Analysis Center.

31 This article and nearly all the analysis and discussion of emissions inequality focuses on inequality *between* nations. However it is important to acknowledge and suggest future research on inequality of emissions *within* nations. We currently lack much data on intra-country variation in carbon emissions, especially in the poor nations, but Loren Lutzenheizer's 1996 analysis shows how U.S. citizens with incomes over $75,000 emitted nearly four times the amount of carbon as those whose income is under $10,000. See Loren Lutzenheizer, 'Riding in Style' (1996) Presentation at the American Sociological Association Annual Meeting, 16 August, New York. We lack analysis on this inequality within other nations, but if the average American and Canadian emits 273 times that of the average Rwandan, it is clear that 10,000–100,000 or more poor residents of these nations probably emit as much as one millionaire in the United States or Canada. In most poor nations there are wealthy elites who emit at levels near or exceeding those of the average citizen of wealthy nations like the United States. So it can be said with confidence that the world's richest people cause emissions *thousands of times* that of the world's poorest.

32 The 'polluter pays' principle was endorsed by all OECD countries in 1974, see Organization for Economic Co-operation and Development (OECD), *Recommendation on the Implementation of the Polluter-Pays Principle* (Paris: OECD, 1974) at 223.

33 E.L. La Rovere, L. Valente de Macedo, and K.A. Baumert, 'The Brazilian

Proposal on Relative Responsibility for Global Warming,' in K. Baumert, ed., *Building on the Kyoto Protocol: Options for Protecting the Climate* (Washington, DC: World Resources Institute, 2002) 157, 158. Since the late-1990s, the Brazilian proposal has been significantly revised with improved understanding of how carbon is absorbed and released by the oceans, land, and plants.

34 See CDIAC 1999, *supra* note 1.
35 See La Rovere et al., *supra* note 35.
36 Group of 77, *Declaration of the Group of 77 South Summit held in Havana from 10 to 14 April 2000* at para. 44; online, http://www.g77.org/summit/Declaration_G77Summit.htm (emphasis added). On the South's perception of responsibility for anthropogenic climate change, see Marc Williams, 'The Group of 77 and Global Environmental Politics' (1997) 7 Global Environmental Change 295; Adil Najam, 'International Environmental Negotiations: A Strategy for the South' (1995) 7 International Environmental Affairs 249; A. Agarwal and S. Narain, *Global Warming: A Case of Environmental Colonialism* (New Delhi: Centre for Science and Environment, 1991); J. Parikh, *Consumption Patterns: The Driving Force of Environmental Stress* (New Delhi: Indira Ghandi Institute of Development Research, 1992); Joyeeta Gupta, 'India and Climate Change Policy: Between Diplomatic Defensiveness and Industrial Transformation' (2001) 12 Energy and Environment 217; Kristian Tangen, Gørild Heggelund, and Jørund, Buen, 'China's Climate Change Positions: At a Turning Point?' (2001) 12 Energy and Environment 237; K. Von Moltke and A. Rahman, 'External Perspectives on Climate Change: A View from the United States and the Third World,' in T. O'Riordan and J. Jager, eds., *Politics of Climate Change: A European Perspective* (London: Routledge, 1996) 330.
37 See Baumert et al., *supra* note 19.
38 See La Rovere et al., *supra* note 35 at 168.
39 Readers should note that more nations are missing for this analysis: we have 150 cases instead of the 170 to 190 for the other measures.
40 This is probably legitimate since virtually all the carbon emitted since 1945 is still in the atmosphere. Carbon dioxide has an atmospheric lifetime of 120 years, methane 12 years. See CDIAC 1999, *supra* note 1.
41 Adil Sagar and Ambuj Najam, 'Avoiding a COP-Out: Moving towards Systematic Decision-Making under the Climate Convention' (1998) 39 Climatic Change iii; see Müller, *supra* note 12; Blanchard et al., *supra* note 3; Biermann, *supra* note 3; Wiegandt, *supra* note 3; Zartman et al., *supra* note 3; Rose et al., *supra* note 3; Albin, *supra* notes 13 and 14.
42 See Stavins, *supra* note 18; Joseph E. Aldy, Scott Barrett, and Robert N.

Stavins, 'Thirteen Plus One: A Comparison of Global Climate Policy Architectures' (2003) 3 Climate Policy 373. See also Baumert et al., *supra* note 19 and Baumert et al., *supra* note 21. See particularly Blanchard et al., *supra* note 3 at 286, who state, '[A]ny future burden-sharing agreement involving developing countries will probably be based on a complex differentiation scheme combining different basic rules.'

43 Scott Barrett and Robert Stavins, 'Increasing Participation and Compliance in International Climate Change Agreements' (2003) 3 International Environmental Agreements: Politics, Law, and Economics 349. Take for instance the precarious position in which small island states find themselves.

44 U. Bartsch and B. Müller, *Fossil Fuels in a Changing Climate: Impacts of the Kyoto Protocol and Developing Country Participation* (Oxford: Oxford University Press, 2000).

45 Critics charge even this approach freezes North-South inequalities and is therefore unlikely to build consensus. See Alex Evans, *Fresh Air? Options for the Future Architecture of International Climate Change Policy* (London: New Economics Foundation, 2002).

46 Sujata Gupta and Preety M. Bhandari, 'An Effective Allocation Criterion for CO_2 Emissions' (1999) 27 Energy Policy 727; Blanchard et al., *supra* note 3; Baumert et al., *supra* note 21 at 190; A. Agarwal, S. Narain, and A. Sharma, eds., *Green Politics: Global Environmental Negotiation-1: Green Politics* (New Delhi: Centre for Science and Environment, 1999); J.R. Ybema, J.J. Battjes, J.C. Jansen, and F. Ormel, *Burden Differentiation: GHG Emissions, Undercurrents and Mitigation Costs* (Oslo: Center for International Climate and Environmental Research (CICERO), 2000); Lasse Ringus, Asbjorn Torvanger, and Arild Underdal, 'Burden Sharing and Fairness Principles in International Climate Policy' (2002) 2 International Environmental Agreements: Politics, Law and Economics 1; Asbjorn Torvanger and Lasse Ringius, 'Criteria for Evaluation of Burden-Sharing Rules in International Climate Policy' (2002) 2 International Environmental Agreements: Politics, Law and Economics 221; Asbjorn Torvanger and Odd Godal, 'An Evaluation of Pre-Kyoto Differentiation Proposals for National Greenhouse Gas Abatement Targets' (2004) 4 International Environmental Agreements: Politics, Law and Economics 65.

47 E. Claussen and L. McNeilly, *Equity and Global Climate Change: The Complex Elements of Fairness* (Arlington: Pew Center on Climate Change, 1998).

48 Not surprisingly, developed countries tend to fall under the former category, and developing countries under the latter.

49 Helen Groenenberg, Dian Phylipsen, and Kornelis Blok, 'Differentiating Commitments World Wide: Global Differentiation of GHG Emissions

Reductions Based on the Triptych Approach – A Preliminary Assessment' (2001) 29 Energy Policy 1007.
50 See Groenenberg et al., ibid.; Evans, *supra* note 47.
51 See J.P.M. Sijm et al., *The Multi-Sector Convergente Approach of Burden Sharing: An Analysis of its Cost Implications* (Oslo: CICERO, 2000); see also Ybema et al., *supra* note 48.
52 Paul Baer, Thomas Athanasiou, and Sivan Kartha, *The Right to Development in a Climate Constrained World: The Greenhouse Development Rights Framework*. Report of EcoEquity and the Stockholm Environment Institute. 24 September 2007. Online, www.ecoequity.org/docs/TheGDRsFramework.pdf.
53 See B. Müller, *Justice in Global Warming Negotiations: How to Obtain a Procedurally Fair Compromise* (Oxford: Oxford Institute for Energy Studies, 1999); Biermann, *supra* note 3; Wiegandt, *supra* note 3; Zartman et al., *supra* note 3; Rose et al., *supra* note 3. See also Blanchard et al., *supra* note 3 at 286, for example, who argue that 'any future burden-sharing agreement involving developing countries will probably be based on a complex differentiation scheme combining different basic rules.'
54 J. Timmons Roberts and Bradley C. Parks. *A Climate of Injustice: Global Inequality, North-South Politics, and Climate Policy* (Cambridge, MA: MIT Press, 2007).
55 Roberts and Parks, ibid.
56 Adil Najam, Saleemul Huq, and Youba Sokona, 'Climate Negotiations beyond Kyoto: Developing Countries Concerns and Interests' (2003) 3(3) Climate Policy 221.
57 Roberts and Parks, *supra* note 55.

9 Global Carbon Trading and Climate Change Mitigation in Canada: Options for the Use of the Kyoto Mechanisms

MEINHARD DOELLE

1. Introduction

Since the negotiation of the *Kyoto Protocol* in 1997,[1] carbon trading has increasingly captured the attention of policy makers, politicians, business leaders, and academics worldwide. Carbon or greenhouse gas (GHG) emission trading can take many forms. It usually involves the allocation and trading of emission permits. It can also involve the identification and certification of GHG emission reductions. Whatever units are used in a scheme of emission reductions and permit allocations, the essence of carbon trading is the ability to trade credits. The ability to trade emission rights and reduction credits is expected to provide financial incentives to achieve emission reductions.[2]

The basic aim of any form of carbon trading is to minimize the cost of emission reductions by utilizing the market to identify where the reductions can be achieved at the lowest cost. Carbon trading is based on the belief that the market is in a better position than government to identify the most cost-effective way to achieve a given level of emission reductions. It assumes that the lowest cost should be the primary factor in deciding where emission reductions take place. In turn, it discourages participants from distinguishing among the emission reduction options included in the trading system on any basis other than price.

Alternatives to carbon trading include other broad financial mechanisms, such as carbon taxes. The essential characteristic of a carbon tax is that it allows for the price of emissions to be controlled directly, while the emission reductions achieved will depend on a variety of factors. By contrast, under a carbon trading system, the quantity of emissions are controlled directly, not the price. The alternatives to broad financial mechanisms are directed measures. Such measures aim to control where and how emission reductions are achieved. They require law and policy makers to determine in which sector or industry emission reductions are to be achieved rather than allowing the market to make this determination based on price.[3] Directed measures can serve either as alternatives or as complements to broad financial mechanisms. Examples of directed measures include emissions or efficiency standards for buildings, vehicles, and appliances, and renewable energy portfolio standards for electricity suppliers.

When treated as alternatives, the choice between broad financial mechanisms and directed measures is fundamentally about whether all reductions are to be treated as equal or whether certain reductions are to be preferred over others. Carbon trading and taxes tend to treat emissions from a range of sources in the same way, whereas directed measures tend to treat each source individually. A basic policy choice therefore is whether to be guided by the price of carbon alone or whether distinctions should be made based on factors such as the permanence of the emission reductions, the contribution made to a low-emissions development path, and other collateral benefits and costs associated with the choice that are not reflected in the price of carbon.

There are two basic design options for carbon trading systems. One is the establishment of a firm limit or cap on emissions. This cap is then combined with an ability to sell emission units in case of emissions below the limit and to buy units to make up for emissions above that limit. The end result is a cap-and-trade system. The other involves the issuance of credits for emissions below a baseline, but without a requirement to buy units to compensate for emissions above the baseline. With the proper incentives, this kind of system can result in emission reduction efforts without a binding target. Combinations of the two are also possible.

The ultimate question I pose in this contribution is what role the Kyoto mechanisms can and should play in Canada's climate change policy. It is assumed for the purposes of this chapter that climate change is a serious global threat, that a global response is needed to address climate change, and that Canada has a self-interest and a moral obligation to take a leadership role in mitigating climate change. These assumptions lead to the conclusion that it is in Canada's interest to meet its international obligations under the *Kyoto Protocol*. While this is at odds with current federal policy, it is nevertheless assumed for the purposes of this chapter that Canada's policy goal is to meet its Kyoto target through some combination of domestic action and access to the Kyoto mechanisms, on the path to much deeper reductions.[4]

The appropriate role of the Kyoto mechanisms in Canada's climate-change policy is explored in three steps. First, a brief overview of the current rules under which the mechanisms operate is provided. This overview is followed by a discussion of the range of possible and likely future scenarios for the international climate change regime with respect to the Kyoto mechanisms. Against this backdrop, three possible approaches to the use of the mechanisms by Canada are explored: no use, unrestricted access to the mechanisms, and controlled use. The implications of these basic choices are then considered in light of Canada's prospects for compliance through domestic action alone.

2. Global Carbon Trading and the Kyoto Mechanisms

The Kyoto mechanisms, consisting of emissions trading (ET), the clean development mechanism (CDM), and joint implementation (JI), use the cap-and-trade system and create emission reduction credits. The emissions trading system under Kyoto is essentially a cap-and-trade system.[5] The CDM generates credits for emission reductions below the business-as-usual baseline in developing nations, without imposing any obligation on developing countries to reduce emissions. JI is a hybrid between the two. These three Kyoto mechanisms, and the role they play in the international climate change regime, are reviewed in this section.

2.1. *Emissions Trading (ET)*[6]

The emissions trading system under the *Kyoto Protocol* uses emissions reduction targets for developed countries as the basis for a cap-and-trade system. These targets are implemented through an allocation of allowed emissions for the first commitment period called assigned amount units (AAUs). The emissions trading system establishes rules for trading these AAUs as well as the trading of credits generated from the other two mechanisms. ET sets the rules for trading units that are created through developed-country targets, and those accredited through the CDM and JI, thereby introducing a trading system that allows developed countries to buy and sell permission to emit greenhouse gases during a given commitment period.[7]

Kyoto allows more or less unrestricted trading of AAUs and credits from CDM and JI projects, with a few largely symbolic measures to address competing concerns about compliance, liability, and the environmental integrity of the trading system.[8] Parties are free to trade credits generated from the use of the other two mechanisms in addition to being able to trade AAUs. Trading can take place between parties or private entities. There is a limit imposed on 'banking' credits, in other words, saving credits for future commitment periods.[9] Banking has become particularly important because a significant surplus of AAUs is expected to be available for the first commitment period, from 2008 to 2012. This is due to two factors: the sharp reduction in emissions in former Soviet Union states resulting from the collapse of their economies in the early 1990s, and the absence of the United States as a major buyer. Without banking, the price of carbon globally would likely collapse for the first commitment period.

2.2. *The Clean Development Mechanism (CDM)*[10]

The CDM was a relatively late addition to the *Kyoto Protocol*.[11] Its objective is twofold. It is designed to give developed nations a release valve if domestic action becomes too expensive. At the same time, the CDM can provide developing countries with much-needed development assistance in the form of technology transfer and economic activity that can help to place them on a low-emissions development path. In the process, CDM will engage developing countries in climate change mitigation without imposing emissions reduction targets. The CDM is currently the only mechanism available under Kyoto to encourage investment in mitigation efforts in developing nations. Support for the CDM has the potential to be an effective way to engage developing nations in mitigation in a manner consistent with the principle of common but differentiated responsibilities.

The CDM reflects the view that if reductions can be achieved more cost-effectively in a developing nation without a reduction target,[12] that country should be able to join forces with a developed nation to achieve the latter's reduction target.[13] The reductions count toward the target of the developed country in return for providing the financial and technical assistance needed to effect reductions in the developing country that otherwise would not be achieved.

Emission reductions are calculated relative to a counterfactual baseline of what would have happened without the CDM support. Provisions regarding baselines for CDM are contained in the *Marrakesh Accords*.[14] Only reductions below the baseline are to be accredited. The baseline is therefore critical to the quantity of credits issued for a CDM project. A related issue in the accreditation of CDM projects is the requirement of additionality, which essentially asks CDM project proponents to demonstrate that the emission reductions achieved are additional to what would have happened without CDM support. Additionality is about demonstrating that the CDM accreditation is responsible for the reductions below the baseline.

A key issue for the CDM was how to ensure approved projects are sustainable. The *Marrakesh Accords* deal with this issue by giving developing nations who host CDM projects the final say on whether a project assists it in achieving sustainable development.[15] There is generally no international oversight to ensure CDM projects actually move the host nation to a sustainable low-emissions development path. One exception to this is that developed nations agreed to refrain from using reductions achieved from nuclear projects.[16]

Participation in a CDM project is voluntary for both the host country and the prospective purchaser of CDM credits.[17] If a host party chooses to participate, it is required to have a designated national authority responsible for CDM projects.[18] Preliminary conservative estimates of CDM credits likely to be available by the end of the first commitment period are in the range of 2,000 to 5,000 megatons.[19]

2.3. *Joint Implementation (JI)*[20]

JI combines elements from ET and the CDM and is directed mainly at economies in transition. It consists of two tracks. One resembles ET, and the other is project-based and is based on the CDM. The two tracks resulted from a concern among developed nations interested in JI activities that some of the states with economies in transition would have capacity problems in implementing the eligibility requirements (e.g. monitoring and reporting requirements) for emissions trading. This resulted in one JI track that required these eligibility requirements to be met, and another track that can operate without them.

The main difference between the two tracks therefore relates to the establishment of the baseline and verification of the additionality of emissions reduction units to be issued to the funding party. For track one, the emissions trading track, it is up to the host country to ensure the credits sold are in line with the reductions achieved from the project. Track two is available where the host country has not met monitoring and reporting eligibility requirements (such as annual emissions reporting), and is therefore treated more like a developing nation without an emissions reduction target.

It is too early to make accurate predictions about the amount of CDM and JI credits likely to be available for the first commitment period. JI, however, is expected to make only a relatively modest contribution to the carbon market.

2.4. *Choice of Carbon Credits for Kyoto Compliance*

The Kyoto mechanisms offer Canada a range of compliance alternatives to domestic action. Canada's emissions are more than 30 per cent above its Kyoto target, with the start of the commitment period only months away, and only five years left to reduce domestic emissions to comply. To achieve compliance under these circumstances, some reliance on carbon credits from the Kyoto mechanisms seems inevitable. In this section, the choice of available international credits is considered from a Canadian perspective, based on the previously stated assumption that Canada's policy goal is to meet its Kyoto commitments. The implications of choosing among AAUs, clean development credits, and JI credits are identified, as well as the implications of choices within each category of credits. Factors taken into account include the likely cost of the credits, the contribution the purchase will make to GHG emission reductions, and the impact on low-emissions sustainable development in Canada as well as globally.

The first category considered is AAUs. As they reflect the limited allowance of emissions in developed countries, one might expect that the purchase of these credits would reflect actual emission reductions beyond what was required under Kyoto in another developed nation. While this is generally true, there are a few issues to note in considering this category of credits.

Firstly, AAUs can be sold and purchased before the end of the commitment period, at a time when it may not be clear whether the credit is actually a surplus credit for the selling party, or whether the sale will place the seller in a non-compliance situation. Secondly, a number of parties have lower emissions for reasons other than climate change policy. Most notably, emissions in the former Soviet Union states have dropped dramatically as a result of the collapse of economies in these countries. Many of these countries therefore have emission credits to sell for emissions reductions that were part of the business-as-usual scenario.

The bottom line for AAUs is that they are a source of credible carbon credits, subject to whether the credits are available as a result of emissions reduction efforts or through unrelated circumstances. The significance of these issues for a potential buyer will depend to some extent on whether one takes a short- or long-term view of carbon trading under the international climate change regime. If trading is to be an important part of the regime over the long term, a key feature of the Kyoto trading system is the ability to bank credits. Banking refers to the ability to hold units eligible for compliance within the first commitment period for use in a future commitment period. Given the unrestricted ability to bank excess assigned-amount credits, it is reasonable to take the position that the quality of credits purchased is immaterial, as the elimination of any credit will force additional action by the seller in the future, regardless of how the selling state acquired the credit. Such an approach would favour the purchase of the lowest-cost credits available.[21]

An alternative view would be to focus on supporting the most meaningful emission reductions through the purchase of AAUs. This perspective would lead to an assessment of which developed countries have made the most effective efforts to move toward a low-emissions development path. Such credits are likely to be more expensive and limited. Furthermore, liberal rules for the recycling and banking of credits may undermine any effort to support more meaningful reduction efforts through the selective purchasing of high-quality credits. The main reason for this is that these liberal trading rules allow any party to purchase lower-quality credits, use them for compliance, and then either sell assigned-amount units or keep them for future commitment periods. In either case, it is difficult to foresee the possibility of low-quality credits being abandoned without being used to meet emissions reduction obligations.

Similar issues arise with respect to credits generated through JI. The emissions trading track of JI would essentially allow parties that have excess emission credits, whether from a generous assigned-amount allocation, general emissions reduction efforts, or from specific projects, to sell those credits. JI credits from business as usual will tend to demand the lowest price. Emissions reduction credits more clearly linked to projects that are assisting economies in transition to choose a low-emissions development path will likely demand a higher price.

With respect to the CDM, the starting point is that it does offer the potential for achieving the dual function of meeting mitigation obligations and assisting developing nations in their sustainable-development efforts. For the CDM, there is a clear distinction between credits generated from a given project and any allocation of credits to developed countries with targets under the *Kyoto Protocol*. The reason is that CDM projects have to be located in developing party states with no emissions reduction targets and therefore no assigned-amount unit allocation. Nevertheless, there are still potential choices to be made in the purchase of CDM credits. There are considerable differences among CDM projects in terms of technology used, the baseline used to quantify reductions, the basis for establishing additionality, and the contribution the project can be expected to make to a low-emissions development path in the host nation.

Lowest-cost CDM credits will likely involve credits from projects that are questionable from a sustainability perspective.[22] As with credits generated by the other two mechanisms, CDM credits generated through projects that make a more decisive contribution to long-term solutions are likely to be more limited and to demand a higher price. What cannot be lost in the discussion about the relative quality of CDM credits from different projects is the secondary purpose of the CDM: to assist developing countries with mitigation and sustainable development.

Canada will have to choose whether, to what extent, and how to make use of international carbon credits for compliance with the *Kyoto Protocol*. Given Canada's record to date on domestic emissions reductions, and the short time left to make reductions at home, it seems clear that some reliance on international credits will be required for compliance with the first commitment period. Looking at the three basic options and the range of credits available within each, one policy decision to be made is whether to purchase the lowest-cost credits available, or whether to make purchasing decisions based on which credits were generated with long-term solutions to climate change in mind. If the latter approach is chosen, criteria for selection of appropriate credits will have to be considered (see section 3 below).[23] Furthermore, Canada will have to decide whether and how to limit access to international carbon credits to ensure some level of domestic action. Finally, Canada will have to decide to what extent it wants to integrate domestic mitigation with international carbon trading over the long term.

3. The Future of the Kyoto Mechanisms

In the previous section, the range of credits available for the first commitment period was discussed. The use of international credits as part of a domestic climate change mitigation strategy may also be influenced by the expected quality and quantity of credits available in the future. In this section, therefore, three options for the role of the Kyoto mechanisms in the future of the climate change regime beyond 2012 are explored. They are the status quo, the elimination of carbon trading from the international regime, and proposals for major changes to carbon trading.

What form the future climate change regime will take and the role of carbon trading within it are difficult to predict, especially in light of significant uncertainties about the role and position of the United States and major developing countries such as China in a post-2012 regime. Predictions about the future of carbon trading in the international regime are nevertheless essential for a coherent domestic compliance strategy. This section therefore concludes with some observations about the future of carbon trading within the international regime.

One possible outcome of the negotiations on the post-2012 climate-change regime is that the three Kyoto mechanisms will remain essentially unchanged. Under this scenario, developed countries would likely adopt tougher targets, and some developing countries may accept binding targets. Some minor adjustments would likely be made to improve the functioning of the mechanisms based on the experience with the first commitment period. The basic infrastructure, however, would remain in place. Countries with binding targets would have access to AAUs and credits from JI and the CDM to supplement domestic emissions reduction efforts.[24]

A second possible outcome for a post-2012 approach would be the elimination of carbon trading from the international climate change regime. Under this scenario, there would be no opportunity to buy and sell emissions credits to meet emissions reduction obligations. Rather, each party would focus exclusively on domestic emissions reduction efforts to meet its commitments. This scenario would also involve the elimination of JI and the CDM. Any assistance from more- to less-developed parties would have to be provided directly rather than through the trading of emissions reduction credits. It is conceivable that parties may become less dependent upon the mechanisms if *Kyoto Protocol* experience allows parties to better estimate the cost of domestic effort to reduce emissions to certain levels. The more significant the emissions reduction commitments taken on, the less likely this scenario becomes.

The third scenario involves major adjustments to the three mechanisms. One possible adjustment involves the elimination of one or more of the mechanisms. It is conceivable that the future regime would not require JI, because many of the economies in transition will have recovered. ET and the CDM could also each stand alone, but it is harder to conceive of circumstances under which these mechanisms would be eliminated. A second type of adjustment is a fundamental change to one of the mechanisms. An example would be the shift to a 'sectoral' CDM, which would expand the CDM from individual projects to whole sectors. For example, the CDM could be applied to public transportation in a developing nation rather than a single project to encourage public transit in one city. It could involve a shift to renewable energy for the entire electricity sector rather than for individual wind farms. A third form of adjustment would involve the introduction of a new mechanism, such as action targets or credits for sustainable-development policies and measures.[25]

Having considered a range of options for the future of carbon trading as part of the international regime, it is at least possible to narrow down these options. The elimination of carbon trading from the international regime must be considered unlikely for a number of reasons. The abandonment of carbon trading is only conceivable in combination with a move away from binding emissions reduction commitments. The abandonment of binding targets for states with such targets for the first commitment period, in turn, is unlikely for a number of reasons.

First, few countries met the aspirational target of returning to 1990 levels by 2000 as established in the *UN Framework Convention on Climate Change* (UNFCCC) in 1992.[26] Given this experience, it is unlikely that an aspirational target alone would be considered by developing states to be a sufficient signal of commitment from developed states. Binding targets are seen by developing nations as essential in light of the principle of common but differentiated responsibility adopted under the UNFCCC. Furthermore, the *Kyoto Protocol* has created a reasonable expectation from developing states that developed states will lead on climate-change mitigation by accepting and meeting binding emissions reduction targets. Finally, developed states and industries have invested heavily in building the infrastructure and expertise needed for emissions trading based on binding targets. It is reasonable to conclude, therefore, that some combination of binding targets and carbon trading will remain part of the climate change regime.

As long as there are binding commitments for significant emissions reductions, parties will look for some kind of release valve to allow states that are struggling to make domestic reductions a reasonable alternative to meet their obligations in a manner that still contributes to the goal of global emissions reductions. Carbon trading remains the most likely release valve to ensure that states can accept binding commitments for significant emissions reductions for the medium and long term.

It is likely that the United States will re-engage through some form of integration between its domestic emissions reduction efforts and international carbon trading.[27] With tough targets coming for the medium and long term, parties will be looking for ways to limit the risk. ET and the CDM offer such opportunities. Developed-country parties will be prepared to accept higher levels of emissions reductions for 2020 if they have access to the Kyoto mechanisms. There are signs of this in recent negotiations on future emissions reduction obligations under the *Kyoto Protocol*.[28] In light of increasing domestic and international pressure on developed states to accept significant and binding emissions reductions for the medium and long term, reduced access to international credits must be considered unlikely.[29]

4. Use of the Kyoto Mechanisms Domestically

Having considered the Kyoto mechanisms as currently designed and the likely future of carbon trading under the international climate change regime, I now turn to the role of carbon trading in a domestic emissions reduction strategy that is integrated with Canada's international obligations. Three options are briefly reviewed, followed by some recommendations for a coherent domestic policy on the use of the Kyoto mechanisms. The three options are: full integration of international carbon trading with domestic mitigation efforts; no use of international credits; and the controlled use of international credits.

Full integration of domestic emissions reduction and international carbon credits would essentially provide any domestic entity responsible for reducing emissions full access to international credits recognized under the *Kyoto Protocol* to meet its domestic obligations. This would mean unlimited access to credits created under JI and the CDM. It would also mean access to the AAUs of other parties. Access could be through emissions trading systems in other states or regions, such as the trading system established in the EU. It could also take the form of direct purchases of credits from CDM or JI host nations, or nations willing to sell AAUs.

Under this approach there would be no government control over how much of Canada's target is met through domestic action and how much is met through the purchase of international credits. There would similarly be no control over the source of international credits, the choice of mechanisms, or how mechanisms are used. A fully integrated approach would allow the energy sector free access to international credits eligible under the *Kyoto Protocol*. This, in turn, would allow each energy producer to choose whether to produce energy in a manner consistent with the target, or whether to carry on with business as usual and purchase credits to offset emissions above the emissions limit set per unit of energy produced.

The second option involves the rejection of international credits and a focus on domestic action. If fully implemented, it would eliminate international carbon trading as an option not only for domestic private entities with emissions reduction obligations, but also for the Canadian government. Given the significant uncertainties involved with GHG emissions reductions, this approach inevitably has a significant amount of risk associated with it. With sufficient lead time and planning, this approach can, however, result in an early start to domestic action to effectively move toward a low-emissions development path. The steps necessary to ensure that Canada can meet its first-commitment-period target through meaningful domestic action alone have not been taken.

Due to the absence of meaningful domestic action toward a low-emissions development path to date, the rejection of international credits at this late stage would force Canada to choose between two undesirable options. One would be potentially very expensive domestic action to achieve compliance in the short time left. The other would be to face the consequences of being perhaps the only country to fail to comply with its obligations under the *Kyoto Protocol* at a time when climate change is increasingly seen as one of the greatest threats to the global community. To do so would place Canada in the uncomfortable position of being subjected to the non-compliance procedure under the *Kyoto Protocol*. More importantly, perhaps, it would undermine Canada's moral basis for pushing other nations to meet the much tougher future targets needed to avoid the most dangerous consequences of climate change.

The third option involves the controlled use of and access to international carbon credits. Control over access can be approached in many different ways. One approach under this option would be to restrict access to the federal government. The government would determine some time before the final accounting of our emissions and credits for the first commitment period how much of the *Kyoto Protocol* target will be met through domestic action and would take steps to secure sufficient international credits to make up the difference.[30] An effective way to implement this approach would be in combination with a domestic financial mechanism such as a carbon tax. An alternative might be a domestic emissions reduction target for various sectors and industries in combination with a penalty high enough to allow the federal government to purchase the international credits necessary to make up the gap from a source that meets predetermined criteria.

Another approach would be to allow domestic entities with emissions reduction obligations access to international credits, but to limit this access. One form of control could be to limit the percentage or share of emissions reductions a domestic entity can acquire through the purchase of international credits. Another form of control would be to specify the types of credit that can be purchased. This could involve a choice among AAUs, credits from the CDM, and credits from JI. The control could also be specific within one type of international credit. For example, the federal government could require that CDM credits be from projects involving certain technologies. It could also specify that AAUs from former Soviet Union states are not eligible for compliance with domestic emissions reduction targets.

Having reviewed the nature of the Kyoto mechanisms and the basic options for their use domestically in Canada now and in the future, I will consider what a coherent domestic strategy for the use of international carbon credits might look like. The following are starting points in light of the discussion above:

- Canada will have to make some use of carbon trading to ensure it will meet its first commitment period target under the *Kyoto Protocol*.
- There are credits available from each of the three Kyoto mechanisms.
- Canada can control the type and amount of international credits to be used for domestic compliance.
- Not all international credits represent actual past or current emission reductions.
- Credits can differ greatly in their contribution to a low-emissions sustainable development path.
- There is considerable uncertainty about the future of the climate-change regime.

If banking and carbon trading are part of the international regime over the long term, the source and characteristics of the credits purchased for compliance with first-commitment-period obligations arguably do not matter. This position is based on the assumption that eliminating any credit from use today will require an additional credit to be generated in the future to ensure compliance with future targets. Assuming more or less unlimited ability to bank credits once they are generated, which is currently the case for assigned-amount units, credits are not likely to be eliminated until they have been used for compliance. The logical conclusion of taking this position would be that there is no need to control the quality of international credits to be made eligible for domestic compliance, only the quantity.[31]

Whether or not it is important to control the quality of international credits used by Canada for compliance depends, therefore, on predictions about the future. If credits will be part of the system over the long term, it is arguably unnecessary to impose controls on the quality of credits for domestic compliance. Instead, the focus would be on controlling the quantity of international credits used, and Canada would have to take a strong position internationally to improve the quality of international credits generated in the future.[32] To be clear, the control of the quality of international credits for the future would still be critical to meeting the dual objective of the international regime, which is to prevent dangerous climate change and to encourage low-emissions sustainable development. In fact, ensuring that credits motivate parties to implement integrated solutions will be critical for an effective climate change regime.

The goal internationally, regardless of decisions about their domestic use, should be to focus credits generated through mechanisms such as the CDM and JI on initiatives that move host countries toward a low-emissions development path. This requires cooperation between the international community and host countries. In the energy sector, for example, it means a focus on technologies that are sustainable, that minimize energy demand, and that allow nations to meet their energy needs in the long term. It also means eliminating from CDM eligibility projects that do little to move participating states to a low-emissions sustainable development path. All this requires a rethinking of the CDM rules to ensure that the focus shifts to the key challenges of sustainable, low-emissions energy and transportation, and generally more efficient use of energy. For ET, the problem with the quality of credits results from the generous allocation of AAUs to some parties. More careful and appropriate allocation in the future will resolve current concerns about the quality of AAUs.

If we knew that trading would be part of the climate change regime in the long term, and that the quality of international credits would improve after 2012, Canada might reasonably choose not to impose controls on the quality of credits to be used domestically to achieve compliance with our first-commitment-period target. This is perhaps best illustrated using an example. Under the conditions described, what would be the effect of Canada purchasing Hot Air from an economy in transition? It would reduce the AAU's that state would be able to bank for future commitment periods, forcing that state to make domestic emission reductions in the future it would otherwise not have to make. The effect of the purchase would therefore be future emission reductions in the selling state. Having said this, there may still be political or optical reasons to refrain from the purchase of Hot Air AAU's if Canada's target can be met without them.

Given the current uncertainties around the future direction of the climate change regime, however, it would be sensible to delay decisions about the type of credits to be eligible for domestic use until the rules for the second commitment period have been negotiated. It is currently expected that the rules for post-2012 will be in place by the end of 2009. Depending on the outcome, and the availability and cost of the various types of credits, more- or less-stringent quality requirements may be warranted.

This leaves the question of the quantity of credits to be permitted to meet Canada's first-commitment-period obligations. Should there be any quantitative restrictions on access to international credits? If so, how are decisions to be made on which domestic emissions are to be offset rather than reduced? Taking a long term view, it is difficult to avoid the conclusion that some control over the quantity of international credits used to meet Canada's emissions reduction obligations will be necessary.

A key objective of the *Kyoto Protocol* and the UNFCCC is for developed countries to lead the way in demonstrating that a high quality of life is compatible with low emissions.[33] What is needed in Canada to achieve this goal is a transition to a low-emissions development path, particularly in the transportation and energy sectors. Free access to international credits would mean giving up control over whether, how fast, and how precisely this transition occurs in Canada. It would also place us at risk of funding other nations in their efforts to make this transition while being left behind ourselves. In fact, the experience from the European emissions trading system so far has been that the combination of overly generous free allocation of emissions permits and too much access to CDM and JI credits has flooded the European carbon market, thereby discouraging the investment needed to transition the electricity sector in Europe to a lower-emissions path.[34]

Unrestricted access to international credits is clearly not in Canada's interest. At the same time, some access is necessary to meet our international obligations and uphold the integrity of the international regime designed to ensure our efforts are supported with equitable contributions from other nations. How, then, should decisions be made about the extent to which international credits should be permitted to replace domestic emissions reductions? Two principles are offered to guide decisions about access to international credits.

First, decisions about access to international credits should be based on the principle that such credits not take away from domestic efforts to make the transition to a low-emissions sustainable development path. To be clear, if the barriers to domestic implementation of such measures are economic, this is not a reason to offer a release valve in the form of access to international credits. Other ways have to be found to ensure the implementation of these measures. Second, international credits should be freely available to make up the gap between what can economically and technically be achieved domestically within the time frames imposed through the international regime and what is required under Canada's international emissions reduction obligations. In other words, there is no reason to pay an economic price to force domestic reductions that do not move Canada to the low-emissions development path.

These two principles set the boundaries for the use of international credits. They identify when international credits should be prohibited, and when open access should be allowed. They do not deal with the middle ground, where the choice is between international credits and domestic emissions reductions that do not contribute to the low-emissions development path, but that are technically and economically feasible. An example might be the retrofit of a coal-burning power plant that is not part of the low-emissions energy future of a region, or a forest management project that offsets emissions but does not contribute to a low-emissions development path. These choices cannot and should not be made primarily based on climate change considerations. Rather, they have to be made in light of a full range of social, environmental, and economic considerations. In other words, the decision involves an overall consideration of the long-term utility of the domestic-reduction option and a long-term view of the international regime.

Unlimited access to international credits should be reserved for industries or sectors that are considered to be an essential part of the Canadian economy into the future, but that are not in a position to make meaningful[35] reductions domestically within the time frame of Canada's international emissions reduction obligations. Access should be strictly controlled for industries or sectors that are able to make some meaningful reductions domestically, but not enough to make an equitable contribution to Canada's international emissions reduction obligations. Access should be denied to any sector or industry that is able to achieve emissions reductions that represent an equitable contribution to Canada's international emissions reduction obligations.

There are multiple challenges in implementing this approach. An equitable contribution to Canada's international obligations has to be identified for each sector or industry.[36] To determine what portion of that overall contribution should be met through domestic efforts, it would be necessary to identify sectoral or industry-specific emissions reduction efforts that lower emissions and promote sustainable development. To bridge the gap between the overall contribution of a sector and the domestic contribution identified, there would then need to be some opportunity to fund emissions reductions outside the sector. Options include the purchase of domestic credits from other sectors, the purchase of international credits, or some form of payment to allow government to purchase the credits.

The implementation of this approach might involve a combination of directed measures, a domestic trading system, and some financial mechanism. Expected domestic actions have to be translated into the quantity of domestic emission reductions to be expected within the various time frames set through international commitments. Sector issue tables set up in Canada following the signing of the *Kyoto Protocol* carried out much of this analysis; however, the results are likely too dated to serve as the basis for implementing this approach today.

For the first commitment period, time is running out to overcome these challenges. An alternative approach in the short term would be a combination of some directed measures on issues such as renewable energy, energy efficiency, conservation, public transportation – an emissions reduction target for each sector that represents that sector's equitable contribution to the first-commitment-period target, and the option to pay into a domestic compliance fund at a rate that is sufficiently high to encourage domestic action and at the same time allow the government to purchase green international credits to make up any shortfall. Over time, the flexibility on how much of the sector targets have to be met through domestic action should be reduced in sectors where domestic action is critical to move Canada towards a low-emissions sustainable development path.

In other words, in the short term, the best way forward may be to implement laws and policies to encourage maximum effort in key areas, and put financial mechanisms in place that encourage domestic action for the rest, but to only draw a firm line between domestic action and access to international credits in the key areas identified. This would mean characterising domestic action according to whether it is part of the low-emissions sustainable development path, and mandating measures in that area first and foremost. Any measure that clearly does not fit this category could be regulated through cap and trade, a carbon tax, a domestic compliance fund, or other financial mechanisms that would serve the dual purpose of encouraging domestic action while generating the funds necessary to purchase credits to make up the gap in the end. Over time, the broad financial mechanisms should be continuously supplemented and refined through directed measures as opportunities to move Canada towards a low-emissions development path are identified.

5. Conclusion

In the short term, access to international carbon credits will be an essential component of any coherent, globally integrated domestic climate-change mitigation strategy. Given the quantity of international credits available, access to the global carbon market will ensure that Canada meets its international obligations and upholds the integrity of the international regime – the best hope for a global solution to the global climate challenge. The proposed approach will allow Canada to focus on measures needed to get to deep cuts over the long term while upholding the integrity of the international regime at the lowest cost. This approach has the best chance to serving the dual purpose of making progress at home and setting the stage for global action. As serious efforts are made to reduce emissions at home and to place Canada on a low-emissions path to sustainable development, the dominant reason for nevertheless encouraging the use of international credits is that it can be an effective way to assist developing nations in choosing a low-emissions development path in a manner that formally recognizes Canada's contribution to this effort.

As we get to long-term targets in the 70–80 per cent range of reductions below 1990 for developed nations, access to mechanisms could very well become more important than the target. In combination with access to a well-functioning CDM, and possibly other mechanisms to fund reductions in developing states to offset domestic emissions, developed nations could eventually get close to 100 per cent emissions reduction targets.[37] It is quite conceivable that within a decade or two the international debate will shift from what developed-country targets should be, to what combination of domestic reduction and offset through Kyoto mechanism–type carbon trading will lead to 100 per cent offsets in developed nations.[38] In other words, the debate will be about access to carbon trading and generating credits in developing countries more than about overall targets. There are some early signs of this shift in the results of the preparatory meeting for Bali held in Vienna in August, 2007.[39]

Notes

1 *Kyoto Protocol to the United Nations Framework Convention on Climate Change*, 10 December 1997, U.N. Doc. FCCC/CP/1997/L.7/add. 1, reprinted in (1998) 37 I.L.M. 22 [hereinafter *Kyoto Protocol*].
2 See J. Robinson et al, *Climate Change Law: Emissions Trading in the EU and the UK* (London: Cameron May, 2007).
3 Such measures can also direct which technologies have to be utilized to achieve reductions.
4 See M. Doelle, *From Hot Air to Action? Climate Change, Compliance and the Future of International Environmental Law* (Toronto: Carswell, 2005).
5 ET, which involves trading of emissions allowances, allows emissions reduction credits from the other two mechanisms into the trading system.
6 See A. Bachelder, 'Using Credit Trading to Reduce Greenhouse Gas Emissions' (2000) 9 J. Envtl. Law and Practice 281. For a general discussion, see also R.B. Stewart, J.L. Connaughton, and L.C. Foxhall, 'Designing an International Greenhouse Gas Emissions Trading System' (2001) 15 Nat. Resources & Env't 160.
7 Most of the key rules for the use of the Kyoto mechanisms were negotiated in Marrakesh and are included among the following decisions: Conference of the Parties, United Nations Framework Convention on Climate Change, *Report of the Conference of the Parties on its Seventh Session*, 29 October–10 November 2001, U.N. Doc. FCCC/CP/2001/13/Add.1 (Decisions 1/CP.7 –

14/CP.7), FCCC/CP/2001/13/Add.2 (Decisions 15/CP.7 - 19/CP.7), FCCC/CP/2001/13/Add.3 (Decisions 20/CP.7 – 24/CP.7), FCCC/CP/2001/13/Add.4 (Decisions 25/CP.7 – 39/CP.7 & Resolution 1/CP.7 – 2/CP.7), online: http://unfccc.int/2860.php [hereinafter *Marrakesh Accords*].
8 See Doelle, *From Hot Air to Action*, *supra* note 4 at 28–40. See also, F. Yamin and J. Depledge, *The International Climate Change Regime: A Guide to Rules, Institutions and Procedures* (Cambridge: Cambridge University Press, 2005).
9 The limit for CDM credits is 2.5% of a party's assigned amount. There is no limit on banking AAUs.
10 For a general discussion on the CDM, see D.V. Wright, *The Clean Development Mechanism: Climate Change Equity and the South-North Divide* (Berlin: VDM Verlag, 2007). See also R. Stewart et al., *The Clean Development Mechanism: Building International Public-Private Partnerships Under the Kyoto Protocol*, UN Doc. UNCTAD/GD5/GF5B/Misc.7 (2000) at 9; William L. Thomas, D. Basurto, and G. Taylor, 'Creating a Favorable Climate for CDM Investment in North America' (2001) 15 Nt. Resources & Env't 172.
11 See J. Werksman, 'The Clean Development Mechanism: Unwrapping the "*Kyoto* Surprise"' (1998) 7 Rev. Eur. Community & Int'l Envtl. L. 147.
12 It is assumed, therefore, that the developing country has no direct, short-term incentive to make the reductions. Certainly, no such direct incentive exists under the protocol itself. The question of what a developing country would do, however, in the absence of assistance under this mechanism is a very difficult question to answer. An answer would have to consider other motivations for taking action to reduce emissions, including, but not limited to, the possibility of a future target, or an effort to demonstrate to Annex I countries that their efforts are inadequate, and not based on best efforts.
13 Annex I parties to the protocol have emissions reduction targets, and thereby an incentive to support efforts to reduce emissions, if they can get credit for the reductions realized.
14 See *Marrakesh Accords*, *supra* note 7, Draft decision –/CMP.1 (Article 12) Annex, paras. 45–48.
15 Ibid.
16 The same principle applies to JI projects. Nuclear energy projects are therefore practically excluded from JI and the CDM.
17 Credits under Kyoto are generally presented in terms of tonnes of GHG emissions translated into CO_2 equivalent. The initial purchaser of the CDM credits could be a developed nation or a private entity.
18 See *Marrakesh Accords*, *supra* note 7, Draft decision –/CMP.1 Annex, Part F, paras. 28–34.

19 See online: http://cdm.unfccc.int/Statistics/index.html. See also A. Michaelowa, 'How Many CERs Will the CDM Produce by 2012,' Climate Strategies Discussion Paper CDM-2, September 2007, online: http://climatestrategies.org. See also World Bank, *State and Trends of the Carbon Market, 2007* (Washington, DC, May, 2007) online, http://carbonfinance.org/docs/Carbon_Trends_2007-_FINAL_-_May_2.pdf.
20 See *Marrakesh Accords, supra* note 7. For a more detailed discussion of JI, see, e.g., A.G. Hanafi, 'Joint Implementation: Legal and Institutional Issues for an Effective International Program to Combat Climate Change' (1998) 22 Harv. Envtl. L. Rev. 441.
21 This approach assumes that these credits will be used eventually, that the banking of credits does not affect future targets, and that carbon trading remains part of the international climate change regime.
22 For a more detailed discussion, see M. Doelle, 'The Cat Came Back, or the Nine Lives of the *Kyoto Protocol*' (2006) 16 J.E.L.P. 261 at 269.
23 See section 3 below.
24 For a discussion of the future of the CDM, see C. Streck and T.B. Chagas, 'The Future of the CDM in a Post-*Kyoto* World' (2007) 1 Carbon & Climate L. Rev. 53.
25 See, e.g., K.A. Baumert et al., eds., *Building on the Kyoto Protocol: Options for Protecting the Climate* (Washington, DC: World Resources Institute, 2002) online, http://pdf.wri.org/opc_full.pdf.
26 *United Nations Framework Convention on Climate Change*, U.N. Doc. A/AC.237/18 (PartII)/Add.1, reprinted in (1992) 31 I.L.M. 849 [hereinafter UNFCCC].
27 See D.M. Driesen, 'The Changing Climate for United States Law' (2007) 1 Carbon & Climate L. Rev. 35. See also M.B. Gerrard, ed., *Global Climate Change and US Law* (Chicago: American Bar Association, 2007).
28 Report of the Ad Hoc Working Group on Further Commitments for Annex I Parties under the *Kyoto Protocol* on the first part of its fourth session, held at Vienna from 27 to 31 August 2007, U.N. Doc. FCCC/KP/AWG/2007/4.
29 The most realistic alternative to international carbon trading might be an internationally controlled compliance fund. Such a fund has been rejected as a compliance tool in the past by parties to the *UNFCCC*. See J. Brunnée, 'A Fine Balance: Facilitation and Enforcement in the Design of a Compliance Regime for the *Kyoto Protocol*' (2000) 13 Tul. Envtl L.J. 223. See also Doelle, *From Hot Air to Action, supra* note 4 at 60.
30 The final opportunity to purchase credits for compliance with the first commitment period will be during the so-called 'true up' period, likely in 2015. See Doelle, *From Hot Air to Action, supra* note 4 at 138.

31 This conclusion does not take away from the critical need to ensure that, for future commitment periods, the rules for the allocation, certification, and trading of credits are improved to ensure that the dual objectives of GHG emissions reduction and sustainable development are met.
32 The other assumption implicit here is that the purchase of credits does not affect future targets. For example, it is assumed here that Russia's target for post 2012 does not depend on how much of its excess AAUs it sells during the first commitment period. A principled based allocation of future targets would ensure this assumption is valid.
33 UNFCCC, *supra* note 26, Article 3.
34 See Robinson et al, *supra* note 2 at 173.
35 'Meaningful' here means that they make a contribution to a low-emissions sustainable development path for Canada.
36 The question of what is equitable is actually an issue that can be separated from the analysis of the emission reductions to be achieved in various sectors. This is accomplished by separating where the reductions should take place from who contributes to achieving the reductions.
37 Improvements to the CDM might include a shift to a more limited range of technologies more clearly part of the solution, and a focus on projects that offer integrated solutions to host nations. Other measures with promise to engage developing nations in climate change mitigation include sustainable development policies and measures and action targets.
38 For a discussion of equitable principles that point to developed nations carrying most of the responsibility for mitigation, see H. Ott et al, *South-North Dialogue on Equity in the Greenhouse*, May 2004, Wuppertal Institute, Germany, online: www.wupperinst.org/download/1085proposal.pdf.
39 What would have to be addressed separately is how also to engage developing nations directly in mitigation, particularly those that have the capacity and responsibility to do so. There are a number of ways to bring developing nations on-board gradually with mitigation, such as aspirational targets, intensity targets, dual targets, action targets, and sustainable-development policies and measures.

PART FOUR

Domestic Policy Tools – The Right Mix

10 Renewable Energy under the Kyoto Protocol: The Case for Mixing Instruments

DAVID M. DRIESEN

This paper argues that effective climate policy may require a mixture of policy mechanisms to encourage technological development necessary to facilitate an eventual fossil fuel phase out. This idea contrasts with the view that broad global environmental benefit trading[1] offers a climate change panacea. The trading as panacea view suggests that the *Kyoto Protocol*'s trading mechanisms assure adequate attention to renewable energy and that neither trading design nor technology policy measures are important as long as a broad and wide carbon market exists. This paper explains why this view is mistaken and puts forward the idea of taxing fossil fuels in order to pay for the introduction of more renewable energy, the use of other targeted programs to encourage renewable energy, and trading design principles to encourage technological progress.

This paper begins with an explanation of why renewable energy is important to efforts to address global climate change. The paper's second part presents data showing that global environmental benefit trading has not encouraged renewable energy as well as more targeted programs. This part of the paper also briefly presents some of the theoretical reasons to expect global trading to perform suboptimally in encouraging renewable energy.

The paper's final part discusses policy implications. It argues that the proper goal of policy should be to move toward an eventual phase-out of fossil fuels, rather than maximizing short-term cost minimization. It suggests a mix of policy tools and emissions trading design principles that can help move us toward this goal.

1. Renewable Energy's Importance in Addressing Global Climate Change

Practically all climate change experts recognize that meaningfully addressing climate change will require fundamental changes in the production and use of energy. Predictions of the amount of carbon reductions needed to avoid dangerous climate change tend to coalesce at around a 50 per cent global reduction by 2050, followed by even more drastic cuts.[2] Because developed countries like Canada have better technological capabilities than developing countries and bear historical responsibility for most of the problem, developed countries will need to produce substantially more than a 50 per cent reduction in national carbon emissions to make a worldwide cut of that magnitude possible. Since carbon dioxide, a product of fossil fuel combustion, accounts for about 80 per cent of the global warming potential of world greenhouse gas emissions, the overwhelming majority of these cuts must come from drastic reductions in the amount of fossil fuel burned.[3] Achieving anything less than this creates a high probability of very serious consequences, such as melting polar ice caps, the inundation of much of Bangladesh, and widespread drought in Africa.[4] Seen in this light, a successful climate change program will deliver large short-term reductions in a manner that sets the stage for making more serious cuts in the future.

The economics of energy suggest that strategies employed now to reduce carbon emissions must contribute to lowering the price and increasing the utility of renewable energy in order to make more drastic future cuts feasible. Economists frequently use top-down approaches to modelling the costs of addressing climate change, relying on basic macroeconomic data about the effects of previous rises in energy costs.[5] This approach, which implicitly assumes that our technological capabilities today resemble those of the 1970s, sometimes generates numbers so high that they raise serious questions about whether the world's governments would ever agree to make the reductions needed to avoid dangerous climate change.[6] Yet, some bottom-up models, which consider current technologies available to reduce greenhouse gas emissions,[7] predict that the drastic cuts scientists call for can be achieved at zero net cost, meaning no more cost than continuing with business as usual would generate.[8] They arrive at this conclusion by assuming that the continued refinement of existing technologies will generate cost savings in energy technology comparable to what we have seen in the past, especially for renewable energy.[9] This suggests that policies must support the deployment of renewable energy in the near term in order to make avoiding dangerous climate change feasible. Deployment of renewables creates 'learning by doing,' which will enable renewable energy providers to improve the utility and lower the cost of their products.[10]

Renewable energy has enormous advantages over competing forms of energy, when viewed from a long-term perspective. Some forms of renewable energy, such as solar, have zero fuel costs.[11] This implies that even with large initial capital investment, if the technology is durable, then the non-discounted long-term costs should be low. And renewable energy offers zero direct carbon emissions while simultaneously delivering significant reductions in particulate and ground-level ozone, which constitute serious health hazards in Canada and in many other places around the world. Furthermore, a transition to renewable energy promises to ameliorate acid rain and the ecological damage associated with burning fossil fuels. Thus, renewable energy would be worthwhile, even if it costs a lot more than 'end-of-the-pipe' approaches to greenhouse gas reductions because it simultaneously offers a host of incidental environmental benefits.

The history of renewable energy suggests that its employment will cost much less in the future than it does today, if appropriate policies support its development. We have seen large decreases in the cost of all forms of renewable energy as manufacturers learn how to more efficiently deploy renewable energy with increased production opportunities.[12] Indeed, wind power has become competitive with fossil fuels as a source of peak generating power. Generally, progress has been most dramatic in countries and regions with supportive policies.[13] It is reasonable to assume that with appropriate policies, costs can continue to fall.

By contrast, continued reliance on fossil fuels presents grave risks to the economy and our security, not just to the environment. Since fossil fuels are finite resources, their price will eventually rise and then they will run out. If we fail to implement policies that will stimulate development, refinement, and deployment of alternative technologies, these price increases and shortages will likely prove very disruptive.

To put it another way, we will stop using oil and coal as energy sources sooner or later, because they will cease to exist. The only question is whether we switch to alternatives before or after we commit the atmosphere to very dangerous global warming. It would seem prudent to begin the switch as soon as possible in light of the long residence times of greenhouse gases. Once emitted, these gases remain in the atmosphere trapping heat for a century or more, so that we cannot subsequently adjust to failures to make sufficiently ambitious cuts early on.

In emphasizing renewable energy's importance, I do not mean to deny the relevance of other technological options. Indeed, cogent analysis of how to avoid dangerous climate change envisions a mixture of technologies and strategies, as renewables may not solve the climate change problem alone.[14] Improved energy efficiency is an important part of making renewables, which are difficult to introduce at a large scale, more viable. But many other technological options have significant drawbacks that make renewable energy relatively attractive. Nuclear power poses the risk of accidents and creates security and waste disposal issues. Carbon sequestration may have potential, at least for addressing part of the problem, but it leaves many environmental problems associated with burning coal unaddressed and may not provide a long-term solution to climate change.[15] Advances in renewable energy then are important to our future and sustainable development, even though they are not the sole means we have of addressing climate change.

2. Trading, Targeted Incentives, and Renewable Energy

The literature on trading suggests that it encourages innovation, which might lead one to suspect that the trading of carbon credits will stimulate large increases in the production of renewable energy. So far, trading does not seem to have done so. China and India add significant amounts of coal powered-generating capacity to their power grids every year.[16] One might object that this does not represent a failure of trading, but rather the lack of caps on those countries. But that is precisely the point. Cap-and-trade programs' environmental improvements come from the caps, not from the trading. The trading simply provides a means of lowering the cost of meeting the environmentally valuable carbon reductions required by setting strict caps. Cap-and-trade programs must demand large carbon reductions from very significant sectors of the economy in order to provide a meaningful impetus for carbon reduction.

At first glance, the Clean Development Mechanism (CDM), the *Kyoto Protocol*'s trading program allowing developing countries to provide emission reduction credits to offset emissions in the developed world,

Figure 10.1. Distribution of total CDM credits issued through 18 July 2007

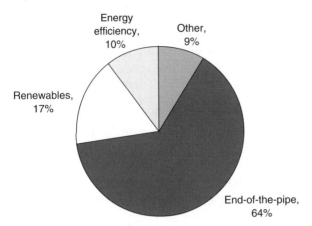

seems to provide a countervailing force. The majority of CDM projects involve renewable energy. But a more careful analysis makes the countervailing force vanish. One should evaluate CDM distribution by carbon credits, because this gives a picture of how much of the greenhouse gas reductions being provided come from renewable energy. An evaluation of the distribution of carbon credits reveals that renewable energy has provided approximately 17 per cent of the total credits, and energy efficiency, which is vital to renewable energy's long-term prospects, constitutes a paltry 10 per cent.[17] End-of-the-pipe control, as figure 10.1 shows, generates the lion's share of the credits.

The main reason that trading encourages end-of-the-pipe controls is that the market favours the cheapest means of meeting greenhouse gas reductions. End-of-the-pipe projects controlling emissions of potent greenhouse gases, such as HFC 23, cost less than most renewable energy projects on a dollar per carbon dioxide equivalent basis.[18] The data available so far reflect only the experience with the reductions encouraged by Phase I of the European Union's emissions trading scheme (ETS) and the newly proposed stricter targets in Phase II. The limits imposed on pollution sources create demand for credits, making the ETS the primary driver of private sector purchases. One would expect the role of renewable energy to increase somewhat as targets become stricter, as they will in Phase II of this two-part scheme.

However, competition among CDM projects will tend to encourage the cheapest emissions reductions. This means that the market does not systematically account for renewable energy's long-term and non-carbon advantages, that is, its capacity to contribute to long-term technological advances and its ability to limit conventional air pollution. Emissions trading markets are better vehicles for seizing low hanging fruit than for planting new fruit trees. Trading's maximization of short-term cost effectiveness conflicts with the goal of maximizing technological progress, collateral benefits from carbon motivated technological changes, and even long-term cost effectiveness.[19]

While the suggestion that trading does not optimally encourage valuable innovation may appear novel, it enjoys increasing support in the economics literature.[20] Furthermore, the most careful students of the acid rain trading program's technological impact have concluded that it encouraged innovation less effectively than the traditional performance standards that preceded it.[21]

By contrast with trading, more targeted energy policies have increased deployment of renewable energy significantly, creating a huge increase in wind generating capacity and more modest increases in deployment of other alternative energy sources.[22] Fiscal incentives have played a key role in encouraging these advances, but some quantity mechanisms have also played an important role. The most successful fiscal incentive appears to be the 'feed-in tariff,' which is widely used in Europe.[23] Countries employing feed-in tariffs require electricity producers to pay a fixed above-market price for all alternative energy produced. Countries experiencing the greatest growth in wind power have employed this system.[24] Many regions of the world employ renewable portfolio standards to encourage renewable energy, rather than feed-in tariffs. These programs require the purchase of a fixed quantity of renewable energy. Recently, these programs have taken the form of 'renewable certificate' programs, which allow trades of certificates representing renewable energy generation. While these limited trading programs offer some advantages in tracking compliance with renewable portfolio standards, it is not clear that the extra flexibility they provide is vital to achieving program goals.[25] Targeted measures aimed at supporting renewable energy, unlike broad global trading, have played a major role in encouraging the increases in production and decreases in price that we have seen in the renewable energy sector.

Trading programs and targeted incentives for renewables clash philosophically. A philosophy of market liberalism undergirds trading programs. Devotees of these programs want to limit governments' role to goal setting and leave technological choices to the free market.[26] By contrast, renewable programs stem from government decisions to favour a particular class of technologies. Governments around the world have decided that sustainable development goals – including energy security, long-term economic development, protection of the health of current generations from routine air pollution, and the creation of local employment – all favour renewable energy. From the standpoint of market liberalism such decisions smack of unjustified interference in the free market of emissions trading. One frequently hears the mantra, 'Let the market decide.' This slogan seems to reflect the notion that the only legitimate goal involved in choosing future energy technology is the short-term reduction of carbon at the lowest possible cost.

The analysis offered above, however, suggests that the market, even markets that have internalized some environmental costs associated with current carbon dioxide emissions, do not offer proper incentives for the long-term changes needed to address global warming and the economic well-being of future generations. Economists recognize that markets tend to under-invest in technological innovation. Those who advance technological development create positive economic spillovers – that is, benefits that do not generate revenues for the firm bearing the cost of technological development.[27] For example, an advance in the design of photovoltaic cells that may make them cheaper or more useful in cloudy climates may gain some additional revenue for the person making the advance. But if another firm looks at the design of these cells and uses the information gleaned from this examination to further advance the state of the art, the second firm, not the first, may gather the revenue. The contribution of the first firm's technology to the second technological advance constitutes a positive spillover from the first firm's investment. Companies can be reluctant to make investments that will substantially benefit competing firms. Furthermore, investments in new technologies are inherently risky and their benefits are difficult to predict.[28] This can make firms, especially firms that control lucrative conventional technologies, reluctant to invest in innovation, which is a major driver of economic prosperity. The collateral benefits of reduced conventional air pollution also constitute a positive spillover that carbon markets do not encourage firms to internalize. This means that even with optimal carbon targets, an emissions trading program does not provide optimal incentives for broadly rational technological choices unless supplemented with some corrective mechanism. This suggests that governments should supplement carbon markets with more targeted measures aimed at advancing technology.

3. A Mix of Policies

Instrument selection reflects implicit normative choices. For the most part, writers assume that the norm of minimizing the short-term cost of meeting a given carbon reduction goal should govern instrument selection, and therefore use a short efficiency criterion to guide instrument choice. In the climate change context, however, society should instead choose a goal of pushing the price differential between renewables and fossil fuels to a tipping point where renewable energy becomes the more economic option. This goal commends itself because of the significant costs associated with failing to avoid dangerous climate change.[29] If one wants to avoid those costs, then maximizing long-term technological change is more important than minimizing short-term costs. While we can expect some movement in this direction to occur through scarcity-induced increases in fossil fuel prices, this price-induced change will not move the world toward the drastic cuts we need to meaningfully address global warming in a timely manner. The question is: What can be done to make this change happen?

3.1. *New Economic Incentive Programs*

The goal of encouraging a shift from fossil fuels to renewables suggests that we should tax fossil fuels to fund renewable energy. Mikael Skou Andersen has pointed out that systems that employ this kind of approach, employing a negative economic incentive to fund a positive economic incentive, can effectively encourage innovation.[30] This approach sends a signal that private actors should invest in alternatives to fossil fuels, instead of building new coal-fired plants with years of useful life ahead of them that pump large amounts of carbon into the atmosphere.

This proposal also helps rectify the problem of fossil fuel subsidies artificially delaying the introduction of renewable energy by making fossil fuels too cheap relative to renewable energy.[31] Of course, that observation leads to the question, why not simply abolish fossil fuel subsidies? Abolishing fossil fuel subsidies would be a good idea, although we probably need to increase positive incentives for renewable energy as well.[32] But eliminating long established subsidies from tax codes and government budgets may prove even more politically difficult than enacting a bold new program to shift the incentives in the right direction.

China provides an example of deliberate government use of tax incentives to meet sustainable development goals. China has taxed CDM projects differentially, imposing high taxes on projects generating carbon credits through end-of-the-pipe controls of industrial gases while imposing low taxes on renewable energy.[33] Furthermore, China has announced an intention to fund renewable energy.[34] It is at least possible that some of the funds for the renewables will come from taxes on less desirable CDM projects. China, then, has created incentives that favour renewable energy, which makes sense for the climate and economic development. It would be even better policy to tax the generation of carbon, rather than carbon reductions, to fund renewable energy.

An Environmental Competition Statute, which also funds positive economic incentives with negative economic incentives, has the potential to avoid many of the governmental weaknesses that tend to interfere with the achievement of ambitious goals, like those needed to address global warming. Such a statute would simply authorize any entity lowering its carbon dioxide emissions to recoup the costs incurred in reducing its carbon output from a competitor of its choosing that has higher emissions, along with a pre-set premium.[35] It allows the capacity of the most environmentally progressive entities in an industry to establish benchmarks that other companies must adhere to if they wish to avoid paying competitors. It thus encourages a race to maximize environmental performance, comparable to the race to improve product quality that can occur because of market share concerns in highly competitive markets.

This approach helps avoid a problem with emissions trading and pollution taxes; namely, their dependence on tough government decision-making. Trading can only work when government officials set ambitious caps. Similarly, pollution taxes only provide significant reduction incentives if officials set reasonably high tax rates. The Environmental Competition Statute allows the capabilities of the most environmentally capable companies to drive programmatic achievements, relying on the government only to set up a law requiring these sorts of payments and establishing the profit margin that companies besting their competitors will be awarded through the premium. In this approach the limited bravery of government officials in setting caps or tax rates does not limit the programs' achievements.

3.2. Sectoral Programs

Countries can expand and strengthen the sectoral programs that have proven more successful than trading in stimulating renewables, such as renewable portfolio standards and feed-in tariffs. These programs can include demand-side management programs for electric utilities and energy efficiency standards for appliances and vehicles.[36]

3.3. Trading Design

While an environmental benefit trading program offers a suboptimal tool for stimulating relatively expensive renewable energy, renewable energy can play a role in such programs. A country can maximize that role by attending to the following design principles if it employs environmental benefit or emissions trading.

STRICT CAPS

Ambitious caps requiring large emission reductions will raise the cost of compliance sufficiently to make some renewable energy viable. While those purchasing credits will still prefer to avoid relatively expensive renewables in favour of lower cost conventional options, they may find some purchase of renewable credits unavoidable if they must purchase large volumes of credits in order to comply. And, of course, strict caps ameliorate climate change more effectively than lax caps, simply because they lower the amount of carbon warming the earth.

LIMITED OFFSETS

A country that wishes to make sure that its credit purchases make a long-term contribution to sustainable development can limit the classes of projects eligible for offset credits and the amount of offsets that are allowed. Indeed, a country could accept only renewable energy projects as offsets, thereby ruling out projects that contribute little to the long-term process of technological development to facilitate fossil fuel replacement.[37] The Regional Greenhouse Gas Initiative, a program to limit utility emissions in the northeastern states, provides quantitative and qualitative limits on offsets in order to protect the program's environmental integrity.[38]

LIMITED BREADTH

Large linked markets will tend to maximize cost savings, thereby dropping the cost of credits.[39] These cost drops reduce incentives for expensive innovation. This point is consistent with economic models of taxes, which link higher tax rates to greater innovation rates. Narrower programs stimulate valuable high-cost innovation better than broad ones, by limiting the availability of cheap conventional credits.

INPUT ALLOWANCES

A program limiting the total amount of fossil fuel consumed through tradable allowances in dirty inputs (such as coal and oil) will better encourage innovation than a broad program focused on end-of-the pipe allowances.[40] It will also prove much easier to monitor and enforce. Such a program can reach the transport sector, which has been left out of caps enacted so far, thereby allowing for a comprehensive economy-wide cap. Some of the bills pending in the United States Congress use a variant on this idea to create an economy-wide cap.[41]

4. Conclusion

A tension exists between maximizing short-term cost effectiveness and maximizing long-term investments needed to address global warming. We can creatively employ a mixture of policy tools to maximize incentives to shift away from fossil fuels to cleaner approaches.

Notes

The author would like to thank Melina Williams, Janet Moon, and Myriah Jaworski for research assistance.

1 I use the term 'environmental benefit' trading, rather than emissions trading, because the climate change regime leaves open the possibility of offsetting emissions with carbon sequestration credits, rather than reducing emissions. See D.M. Driesen, 'Free Lunch or a Cheap Fix? The Emissions Trading Idea and the Climate Change Convention' (1998) 26(1) B.C. Envtl. Aff. L. Rev. at 32–3 [Driesen, 'Free Lunch'], (describing how the concept of emissions trading became broadened to go beyond trading of emission reductions). This means that the term 'emissions trading' does not accurately describe the Kyoto trading system, at least when sequestration credits are used, since more than just emissions can be traded.
2 See A. Dessler and E. Parson, *Science and Politics of Global Climate Change: A*

214   David M. Driesen

Guide to the Debate (Cambridge: Cambridge University Press, 2006) at 155–8 (suggesting that avoiding a 3°C temperature rise may require a 40 per cent cut in global carbon dioxide emissions from 2010 levels by 2050 and more than a 60 per cent cut by 2100); J.E. Hansen, 'A Slippery Slope: How Much Global Warming Constitutes "Dangerous Anthropogenic Interference"?' (2005) 68 Climate Change 269 at 277. (stating that a 2°C temperature rise 'almost surely takes us well into the realm of dangerous' climate change); M. Meinshausen, 'What Does a 2°C Target Mean for Greenhouse Gas Concentrations? A Brief Analysis Based on Multi-Gas Emission Pathways and Several Climate Sensitivity Uncertainty Estimates,' in H.J. Schnellnhuber et al., eds., *Avoiding Dangerous Climate Change* (New York: Cambridge University Press, 2006) 265 at 269–70 (estimating that limiting a temperature rise to less than 2°C likely requires a 55 per cent reduction below 1990 emission levels by 2050).

3 See R.A. Posner, *Catastrophe: Risk and Response* (New York: Oxford University Press, 2004) at 15 (describing global warming as largely a product of fossil fuel combustion).

4 See R.L. Glicksman, 'Global Climate Change and the Risks to Coastal Areas from Hurricanes and Rising Seas: The Costs of Doing Nothing' (2006) 52 Loyola L. Rev. 1127 at 1135–42 (reviewing the science on ice melting and sea level rise); J.E. Hansen, 'Global Warming: Is There Still Time to Avoid Disastrous Human-Made Climate Change? I.e. Have We Passed a Tipping Point?' (Discussion at the National Academy of Sciences, Washington, DC, 23 April 2006) at 26–9, online, http://www.columbia.edu/~jeh1/nas_24april2006.pdf (providing maps of areas that would probably be under water if temperature increased by 3°C); see also James E. Neumann et al., 'Sea-Level Rise & Global Climate Change: A Review of Impacts to U.S. Coasts' (Paper prepared for the Pew Center on Global Climate Change, Arlington, February 2000) at 12, online, http://www.pewclimate.org/global-warming-in-depth/all_reports/sea_level_rise/index.cfm (providing maps of areas likely to experience flooding under three different scenarios).

5 R. Repetto and D. Austin, *The Costs of Climate Protection: A Guide for the Perplexed* 6 (Washington, DC: The World Resources Institute, 1997). See also T. Barker et al., Working Group III Contribution to the Intergovernmental Panel on Climate Change Fourth Assessment Report, *Climate Change 2007: Mitigation of Climate Change, Summary for Policymakers* (2007) 10. T. Barker et al., Working Group II Contribution to the Intergovernmental Panel on Climate Change Fourth Assessment Report, *Climate Change 2007: Mitigation of Climate Change, Summary for Policymakers* (2007) 10.

6 See R.J. Sutherland, '"No Cost" Efforts to Reduce Carbon Emissions in the U.S.: An Economic Perspective' (2000) 21(3) Energy J. 89 at 90–1 (comparing 'mainstream economic analysis' with an 'energy conservation paradigm'); but see T. Barker and P. Ekins, 'The Costs of Kyoto for the U.S. Economy'(2004) 25(3) Energy J. 53 at 53–4 (finding that some top-down models predict high costs, but that the view that addressing climate change will be very costly is 'not well founded').
7 Repetto and Austin, *supra* note 5 at 6. See also Barker et al., *supra* note 5 at 10; R.N. Stavins, J. Jaffe, and T. Schatzki, 'Too Good to Be True? An Examination of Three Economic Assessments of California Climate Change Policy' (Washington, DC: AEI-Brookings Joint Centre for Regulatory Studies, January 2007), online, http://ssrn.com/abstract-973836.
8 Sutherland, *supra* note 6 at 91.
9 Ibid.
10 See A.B. Jaffe et al., 'Technological Change and the Environment,' in K.G. Mäler and J.R. Vincent, eds., *Handbook of Environmental Economics*, vol. 1 (North Holland, 2003) 461 at 493 (discussing price drops due to learning by doing); Sabine Messner, 'Endogenized Technological Learning in an Energy Systems Model' (1997) 7 J. Evolutionary Econ. 291 at 293 (characterizing 'learning by doing' as 'among the best empirically corroborated phenomena characterizing technological change in industry'); L. Argote and D. Epple, 'Learning Curves in Manufacturing' (1990) 247 Science 920 (reporting learning curve effects documented in building *inter alia*, ships, aircrafts, petroleum products, oil refineries, and trucks).
11 M. Ilkan, E. Erdil, and F. Egelioglu, 'Renewable Energy Resources as an Alternative to Modify the Load Curve in Northern Cyprus' (2005) 30 Energy 555 at 565 ('similar to other renewable energy systems, Photovoltaic and [wind-electric conversion] systems are characterized by high capital, low operation and maintenance cost, and zero fuel cost').
12 J. Greenblatt et al., 'Baseload Wind Energy: Modeling Competition between Gas Turbines and Compressed Air Energy Storage for Supplemental Generation' (2007) 35 Energy Pol'y 1474 at 1474 (attributing a 30 per cent annual increase in installed wind capacity to a 'twofold drop in capital costs between 1992 and 2001' and 'government initiatives'); Commission of the European Communities, *The Share of Renewable Energy in the EU: Evaluation of the Effect of Legislative Instruments and Other Community Policies on the Development of the Contribution of Renewable Energy Resources in the EU and Proposals for Concrete Actions* 2004 (SEC) 547 at 19 [hereinafter 2004 Commission Energy Evaluation] (finding that wind costs have fallen by 50 per cent over the last 15 years); L. Barreto and S. Kypreos, 'Emissions Trading

and Technology Deployment in an Energy-Systems 'Bottom-up' Model with Technology Learning' (2004) 158 Eur. J. Operational Res. 243 at 246–8 (estimating an 80 per cent progress ratio for solar photovoltaics, representing the rate of cost decline per doubling of production).

13 See F.C. Menz and S. Vachon, 'The Effectiveness of Different Policy Regimes for Promoting Wind Power: Experiences from the States' (2006) 34 Energy Pol'y 1786 at 1794 (finding that renewable portfolio standards have stimulated increased production of wind power); 2004 Commission Energy Evaluation, *supra* note 12 at 19 (finding that wind power grew by 23 per cent in 2003, exceeding EU wind target). See also A. Ford et al., 'Stimulating Price Patterns for Tradable Green Certificates to Promote Electricity Generation from Wind' (2007) 35 Energy Pol'y 91 at 92 n. 4 (explaining that the Texas renewable portfolio standard produced the 'Texas Wind Rush,' the installation of 10 new wind projects in 2001 producing 930 megawatts of power); N.H. van der Linden et al., *Review of International Experience With Renewable Energy Obligations Support Mechanisms* (Netherlands: ECN Policy Studies, 2005) at 35, online, http://eetd.lbl.gov/ea/ems/reports/57666.pdf (suggesting that a number of policy instruments have contributed to increased renewable energy production in Sweden and that investment in wind power has slowed as the governmental support for land-based wind power is phased out); J.W. Moeller, 'Of Credits and Quotas: Federal Tax Incentives for Renewable Resources, State Renewable Portfolio Standards, and the Evolution of Proposals for a Federal Renewable Portfolio Standard' (2004) 15 Fordham Envtl. L.J. 69 at 73–7 (explaining that a federal requirement that electric utilities purchase power from renewable energy sources played a 'significant, though not precisely quantifiable, role' in expanding renewable power generation).

14 See S. Pacala and R. Socolow, 'Stabilization Wedges over the Next Fifty Years With Current Technologies' (2004) 305 Science 968. This analysis bases its estimate of renewable energy's contribution to limiting climate change on existing technologies. While innovation should enable the role to grow, this analysis suggests that the growth would have to be truly enormous to make it the sole solution to climate change.

15 See Working Group III, Intergovernmental Panel on Climate Change, *Special Report on Carbon Dioxide Capture and Storage* (New York: Cambridge University Press, 2005) at 24, 60–6 (noting that the potential for carbon capture corresponds to only 9–12 per cent of 2020 global carbon dioxide emissions and 21–45 per cent of 2050 emissions, that 38 per cent of carbon dioxide emissions are from dispersed sources that are generally not considered suitable for carbon capture technologies, that other environmental

costs and risks may arise from the storage of carbon dioxide such as impacts on marine ecosystems due to the slow seepage of carbon dioxide into the water, and that the amount of carbon that remains in storage once deposited varies with the type of storage system).

16 See K. Bradsher, 'China to Pass U.S. in 2009 in Emissions,' *The New York Times* (7 November 2006) C1 (discussing new coal-fired power plants in China, India, Germany, and Britain).

17 See J. Fenhann, UNEP Risoe CDM/JI Pipeline Analysis and Database, CDM Pipeline Overview (July 2007), online, http://cdmpipeline.org/publications/CDMpipeline.xls; see also Capacity Development for the CDM, Guidance to the CDM/JI Pipeline 1 (2006), online, http://cdmpipeline.org/publications/GuidanceCDMpipeline.pdf (explaining that data comes from UNFCC homepage located at http://cdm.unfccc.int/index.html, including Project Design Documents also available there). Joergen Fenhann, UNEP Risoe CDM/JI Pipeline Analysis and Database, *CDM Pipeline Overview* (July 2007) online, http://cdmpipeline.org/publications/GuidanceCDMpipeline.pdf.

18 See K. Capoor and P. Ambrosi, *State of the Carbon Market, 2006* (Washington, DC: The World Bank and IETA, 2006), online, http://carbonfinance.org/docs/StateoftheCarbonMarket2006.pdf (characterizing HFC projects as the 'lowest-cost options' and therefore becoming the 'first asset class to be systematically tapped globally'); X. Zhao and A. Michaelowa, 'CDM Potential for Rural Transition in China Case Study: Options in Yinzhou District, Zhejiang Province' (2006) 34 Energy Pol'y 1867 at 1876 (finding the initial cost of solar installation high, even though over the long-term it is cost competitive).

19 I have discussed the tension between maximizing short-term cost effectiveness and encouraging innovation before. For a detailed discussion of the theory underlying this tension, see D.M. Driesen, 'Design, Trading, and Innovation,' in J. Freeman and C. Kolstad, eds., *Moving to Markets in Environmental Protection: Lessons after 20 Years of Experience* (Oxford: Oxford University Press, 2007); D.M. Driesen, 'Does Emissions Trading Encourage Innovation?'(2003) 33 Envt'l L. Rep. (Envtl. L. Inst.) 10094.; Driesen, 'Free Lunch,' *supra* note 1; D.M. Driesen, 'Is Emissions Trading an Economic Incentive Program? Replacing the Command and Control/Economic Incentive Dichotomy' (1998) 55 Wash. & Lee L. Rev. 289.

20 See J.F. Bruneau, 'A Note on Permits, Standards, and Technological Innovation' (2004) 48 J. Envtl. Econ. & Mgmt. 1192; J.-P. Montero, 'Permits, Standards, and Technology Innovation' (2002) 44 J. Envtl. Econ. & Mgmt. 23; J.-P. Montero, 'Market Structure and Environmental Innovation' (2002) 5 J.

Applied Econ. 293 (trading, taxes, or traditional regulation can best encourage research and development when firms' products are strategic substitutes); D.A. Malueg, 'Emissions Credit Trading and the Incentive to Adopt New Pollution Abatement Technology' (1987) 16 J. Envtl. Econ. & Mgmt. 52 (pointing out that firms purchasing credits under trading face lesser incentives to innovate than they would under a traditional performance standard); W.A. Magat, 'Pollution Control and Technological Advance: A Dynamic Model of the Firm' (1978) 5 J. Envtl. Econ. & Mgmt. 95.

21 M.R. Taylor, E.L. Rubin, and D.A. Hounshell, 'Regulation as the Mother of Innovation: The Case of SO_2 Control' (2005) 27 Law & Pol'y 348 at 370 (finding that 'command and control' regulation created more innovation than the trading program and did not change the type of innovation found); D. Popp, 'Pollution Control Innovations and the Clean Air Act of 1990' (2003) 22 J. Pol'y Analysis & Mgmt. 641 (finding more innovation under 'command and control' but a different type of innovation under trading).

22 See, e.g., van der Linden, *supra* note 13 at 38 (suggesting that a number of policy instruments have contributed to increased renewable energy production in Sweden); Moeller, *supra* note 13 at 73–7 (explaining that a federal requirement that electric utilities purchase power from renewable energy sources played a 'significant role' in expanding renewable power generation); cf. I. Choi, 'Global Climate Change and the Use of Economic Approaches: The Ideal Design Features of Domestic Greenhouse Gas Emissions Trading and an Analysis of the European Union's CO_2 Emissions Trading Directive and the Climate Stewardship Act' (2005) 45 Nat. Resources J. 865 at 891 n. 86 (claiming that the acid rain program has discouraged the use of renewable energy, in spite of the establishment of reserve allowances to provide incentives to use it).

23 See M. Mendonça, *Feed-in Tariffs: Accelerating the Deployment of Renewable Energy* (London: Earthscan Publications Ltd., 2007) at 19 (claiming that numerous empirical studies have documented the superiority of feed-in tariffs as an instrument and listing countries using them).

24 M. Ringel, 'Fostering the Use of Renewable Energies in the European Union: The Race between Feed-in Tariffs and Green Certificates' (2006) 31 Renewable Energy 1 at 10.

25 See O. Langniss and R. Wiser, 'The Renewables Portfolio Standard in Texas: An Early Assessment' (2003) 31 Energy Pol'y 527 at 532 (stating that 'certificate trading may not be essential for the effective design of a state RPS, and little trading has yet taken place in the Texas market').

26 See B. Hansjürgens, ed., *Emissions Trading for Climate Policy: U.S. and Euro-*

pean Perspectives (Cambridge: Cambridge University Press, 2005) at 3 (stating that once government allocates allowances its 'action is limited to supervising the market, monitoring, and applying sanctions in the case of non-compliance').

27 See B. Frischman and M.A. Lemley, 'Spillovers' (2007) 107 Colum. L. Rev. 257 at 258–61.
28 Posner, *supra* note 3 at 123 (commenting that uncertainty lies at the 'core' of technological innovation, because 'scientific progress is unpredictable').
29 See N. Stern, *The Economics of Climate Change: The Stern Review* (Cambridge: Cambridge University Press, 2007) at xvi–xvii (finding that the costs of climate change are very high and justify strong measures to address climate change). Cf. F. Ackerman, 'Debating Climate Economics: The Stern Review vs. Its Critics' (2007) online, http://www.ase.tufts.edu/gdae/Pubs/rp/SternDebateReport.pdf (reviewing various criticisms of the *Stern Review* and offering some of his own).
30 M.S. Andersen, *Governance by Green Taxes: Making Pollution Prevention Pay* (Manchester: Manchester University Press, 1994) at 206–10 ('In general, earmarked taxes represent a second-best solution to user payment and must be regarded as especially appropriate for the production of environmental goods, such as environmental protection.').
31 See M.L. Hymel, 'The United States Experience with Energy-Based Tax Incentives: The Evidence Supporting Tax Incentives for Renewable Energy' (2006) 38 Loy. U. Chi. L.J. 43 at 73 (finding that the United States has invested three times as much money in fossil fuels as in renewables).
32 Ibid. at 80 (concluding '[the United States needs] to formulate a strategy to eliminate fossil fuel subsidies in favor of alternatives'); A.D. Owen, 'Environmental Externalities, Market Distortions, and the Economics of Renewable Energy Technologies' (2004) 25(3) Energy J. 127 at 155 (concluding that the removal of indirect and direct subsidies for power generation and the appropriate pricing of fossil fuels are essential policies for stimulating development of alternative energy technologies).
33 China's Office of National Coordination Committee on Climate Change, *Measures for Operation and Management of Clean Development Mechanism Projects in China* (2005) art. 24, online, http://cdm.ccchina.gov.cn/english/ then follow 'Measures for Operation and Management of Clean Development Mechanism Projects in China' hyperlink under 'Domestic Policy & Regulation' (providing that the Chinese government will take 65 per cent of the certified emissions reductions from HFC and PFC projects but only 2 per cent of the certified emissions reductions from renewable energy projects).

34 See BBC News, 'China Accelerates Construction of Renewable Energy Projects' (31 July 2006) (reporting that the Chinese government will 'set up special fund to support renewable energy projects, giving assistance to their research and development as well as favorable tax policies to relevant enterprises'); see also 'China Sets Up Special Fund for Renewable Energy,' *People's Daily Online* (14 June 2006), online, http://english.people.com.cn/200606/14/eng20060614_273831.html.

35 See D.M. Driesen, *The Economic Dynamics of Environmental Law* (Cambridge, MA: MIT Press, 2003) at 151–61.

36 See J.C. Dernbach et al., 'Stabilizing and then Reducing U.S. Energy Consumption: Legal and Policy Tools for Efficiency and Conservation' (2007) 37 Envt'l L. Rep. (Envt'l L. Inst.) 10003 (providing a review of policy tools useful for energy conservation); J.N. Swisher and M.C. McAlpin, 'Environmental Impact of Electricity Deregulation' (2006) 31 Energy 1067 at 1071, 1073, and 1077–8 (linking demand-side management to increased energy efficiency); R.C. Cavanagh, 'Least Cost Planning Alternatives for Electric Utilities and Their Regulators' (1986) 10 Harv. Envtl. L. Rev. 229 (explaining least cost planning and the rationale for it); R.C. Cavanagh, 'Responsible Power Marketing in an Increasingly Competitive Era' (1988) 5 Yale J. Reg. 331 at 337 (defining least cost planning as a model demanding choice of energy conservation when demand reduction proves cheaper than increasing supply to correct imbalances of supply and demand).

37 Cf. Driesen, 'Free Lunch,' *supra* note 1 at 79–81 (defending a focus on credits reflecting 'advanced technology').

38 See D.M. Driesen, 'The Changing Climate for United States Law' (2007) 1 Carbon & Climate L. Rev. 33 at 40 (describing the limits); see also C. Streck and T.B. Chagas, 'The Future of CDM in a Post-Kyoto World' (2007) 1 Carbon & Climate L. Rev. 53 at 58–9 (discussing environmental integrity problems in the offsets offered for credit through the Clean Development Mechanism); CDM Watch, 'The World Bank and the Carbon Market: Rhetoric and Reality' (2007) 16, online, http://www.cdmwatch.org/files/World%20Bank20paper20final.pdf; S. Greiner and A. Michaelowa, 'Defining Investment Additionality for CDM Projects-Practical Approaches' (2003) 31 Energy Pol'y 1007 at 1007 (linking the lack of targets for reductions in developing countries to potential problems with CDM's integrity).

39 See M.A. Mehling, 'Bridging the Transatlantic Divide: Legal Aspects of a Link Between Carbon Markets in Europe and the United States' (2007) 7 Sustainable Dev't L. & Pol'y 46 at 46 (associating linking of emissions trading markets with 'increased liquidity' and therefore 'reduced compliance costs'); J.B. Wiener, 'Global Environmental Regulation: Instrument Choice

in Legal Context' (1999) 108 Yale L.J. 677 at 748 (explaining that widening participation in emissions trading to include developing countries reduces abatement costs).
40 See D.M. Driesen and A. Sinden, 'The Missing Instrument: Dirty Input Limits' (forthcoming 2008); R.R. Nordhaus and K.W. Danish, 'Assessing the Options for Designing a Mandatory U.S. Greenhouse Gas Reduction Program' (2005) 32 Bost. Coll Envt'l Aff. L. Rev. 97 at 129–33 (discussing the model of an 'upstream' cap-and-trade program).
41 See Driesen and Sinden, *supra* note 40; see, e.g., Climate Stewardship Act of 2007, S. 280, 110th Cong., §§3(5), 121(a)(3).(b), 124(a) (1st Sess. 2007) (requiring refiners and importers to hold allowances representing the fossil fuel content of the fuel they supply).

11 A Comparative Evaluation of Different Policies to Promote the Generation of Electricity from Renewable Sources

DAVID G. DUFF AND ANDREW J. GREEN

1. Introduction

In developed and developing countries alike, electricity generation constitutes the single largest contributor to greenhouse gas (GHG) emissions, accounting for a quarter of GHG emissions in the European Union,[1] a third in the United States,[2] and over 40 per cent in China.[3] Although the figure is much lower in Canada, owing to the predominance of hydroelectric power in Canada's electricity mix, electricity generation is also the single largest contributor to Canadian GHG emissions, accounting for 17 per cent of emissions in 2004.[4]

Given these figures, it is not surprising that emphasis is increasingly placed on so-called 'new' renewable sources of electricity (wind, solar, geothermal, tidal, and wave power) as a key strategy to reduce GHG emissions.[5] Indeed, over the last two decades, many governments have already sought to encourage the growth of renewable energy source electricity (RES-E), relying for the most part on subsidies, price guarantees, and quantity-based regulations such as renewable portfolio standards (RPSs), and to a lesser extent on environmental taxes imposed on electricity from fossil fuels. Although the production of energy from these sources (particularly wind and solar) has grown much faster than other sources of energy during this period,[6] the generation of electricity from these renewable sources still represents a tiny percentage of global electricity generation, accounting for only 0.8 per cent of global electricity generation in 2004, compared to 39.8 per cent for coal, 19.6 per cent for natural gas, 16.1 per cent for hydro, 15.7 per cent for nuclear, and 6.7 per cent for oil.[7] Since the generation of RES-E varies considerably among different countries, however, public policies may have a significant impact on the development and diffusion of renewable energy technologies.[8]

Despite the importance of different policy instruments to meet environmental and other public policy goals, there has been little comparative analysis of the effectiveness of different policy instruments to promote RES-E. It is not clear which policies work to increase production from renewable energy sources, let alone aid in achieving the optimal level of such production. Nor is it clear whether environmental and other goals might be met more effectively and efficiently by other means. Not only has little work been done comparing across instruments, there has not been much analysis of the interaction among different policies when employed simultaneously.

This chapter examines different policy instruments to promote the generation of renewable source electricity, evaluating these instruments in terms of their environmental effectiveness, economic efficiency, distributional fairness, and political feasibility.[9] Part 2 reviews the main policy instruments that governments have employed to encourage the development of RES-E, considering subsidies, price guarantees, quantity-based regulations, and environmental taxes. Part 3 evaluates these policies according to the four criteria mentioned above. Part 4 concludes.

2. Policies to Promote RES-E

Although most countries that have attempted to encourage the development of RES-E have used of mix of different policy instruments, it is useful to describe and examine these instruments individually in order to evaluate their environmental effectiveness, economic efficiency, distributional effects, and political feasibility. The following sections review the main instruments that governments have used to promote the generation of electricity from renewable sources, considering subsidies, price guarantees, quantity-based regulations, and environmental taxes.

2.1. *Subsidies*

Among different policy instruments that governments have employed in order to encourage the development of RES-E, the most common is invariably government subsidies. In Denmark, Germany, and the United Kingdom, for example, the early development of wind energy technology was encouraged by government support for research and development.[10] More recently, in Germany, the installation of photovoltaic (PV) roof installations has been encouraged by loans at below-market rates under the 100,000 Roofs Program (*Hundert Tausand Dächer Programm* – HTDP).[11] In Spain, renewable energy projects have been encouraged through capital grants and loans for PV and wind installations under the country's Energy Saving Efficiency Plan (*Plan de Ahorro y Efiecienca Energética* – PAEE).[12] In the United States, investments in various kinds of renewable energy technologies are encouraged through accelerated depreciation for tax purposes as well as federal and state level tax credits, while production of electricity from renewable energy is encouraged through a Renewable Electricity Production Tax Credit.[13] In Canada, investments in renewable energy are encouraged by direct grants under a Renewable Energy Deployment Initiative (REDI) (refunding 25 per cent of the purchase, installation or addition costs of certain renewable energy systems),[14] accelerated depreciation for qualifying investments,[15] and the immediate deductibility of qualifying start-up costs.[16] Generation of electricity from renewable energy is encouraged by production subsidies available under a Wind Power Production Incentive and a Renewable Power Production Incentive (designed to provide production incentives per kWh for wind power and renewable energy, respectively).[17] Similar fiscal incentives have been employed in several other countries.[18]

As these examples suggest, fiscal subsidies may be delivered directly in the form of grants or loans, or indirectly as tax expenditures in the form of tax credits or deductions. Regardless of whether they are delivered directly or indirectly, moreover, these subsidies may be aimed at consumers (as is the case with Germany's 100,000 Roofs Program), at producers of renewable energy technologies (as is the case with research and development subsidies), or at producers of renewable electricity (as is the case with subsidies to encourage the installation of renewable generating equipment or subsidies based on production of electricity from renewable source energy).

2.2. *Price Guarantees*

After subsidies, guaranteed prices or feed-in tariffs (FITs) are the most widely used policy instrument to encourage the development of renewable energy. In the United States, for example, the initial development of wind power in California was encouraged by the *Public Utilities*

Policies to Promote the Generation of Electricity from Renewable Sources 225

Regulatory Act of 1978 (PURPA), which required utilities to purchase electricity from 'qualifying facilities' such as small renewable generators at the 'avoided cost' of other electricity – a concept that resulted in favourable rates in the early 1980s based on high oil prices at the time and expectations of future increases in these prices.[19] In Germany, the Electricity Feed-In Law (*Stromeinspeisungsgesetz* or StreG) of 1990 required electric utilities to connect RES-E generation to the grid and to purchase electricity at amounts ranging from 65 to 90 per cent of the average utility rate to final consumers, depending on the energy source.[20] Likewise, Denmark introduced a FIT in 1992, requiring utilities to purchase electricity from privately owned wind turbines at 85 per cent of the local average retail price for a household with a high annual consumption of 20,000 kWh.[21] In Spain, on the other hand, the FIT introduced in 1998 requires distributors to purchase electricity from renewable sources at a premium above the average market price of electricity,[22] with different premiums for different sources of renewable energy.[23] More recently, in Germany, the Renewable Energy Sources Act (*Eneuerbare Energien Gesetz* or EEG) of 2000 replaced these variable rates with a fixed tariff based on the type of renewable source energy and the size of the facility.[24] FITs have also been employed in Belgium, France, Italy, and Portugal,[25] and were recently adopted in the Netherlands.[26]

As these examples illustrate, FITs generally take one of three forms. Under cost-based FITs, like the U.S. PURPA, the rate at which utilities are required to acquire RES-E depends on the 'avoided cost' of conventional electricity, which may or may not result in competitive rates depending on how this amount is computed. Under variable rate FITs, such as Germany's original feed-in law (StreG) as well as the FITs adopted in Denmark and Spain, the rates at which utilities are required to purchase RES-E are based on market prices for electricity, with RES-E generators obtaining an implicit subsidy in the form premium pricing (as in Spain) or favourable rates based on a percentage of the market price (as in Denmark and Germany's StreG). Fixed-rate FITs, on the other hand, like Germany's more recent EEG, guarantee a fixed rate for RES-E production irrespective of variations in the market price of electricity.

In addition to these variations, FITs can differ in other important respects. While avoided cost schemes do not distinguish between amounts paid for RES-E from different technologies or different sites, fixed- and variable-rate FITs may vary prices according to the specific technology, site, or time (i.e., base or peak load, year, or season).[27] In Germany and Spain, for example, differentiated FITs provide greater levels of support for solar (photovoltaic) electricity than wind-generated electricity.[28] FITs may also vary according to the time period over which fixed or premium rates are guaranteed. Although Spain's original FIT guaranteed premium rates for a maximum period of four years, for example, subsequent amendments guarantee support levels during the useful life of the installations.[29] In Germany, rates are guaranteed for a period of 20 years from the date of the installation, but are gradually reduced over this period in order to account for expected increases in operational efficiencies from technological learning.[30] Finally, to the extent that a fixed- or premium-rate FIT provides an implicit subsidy to RES-E, this subsidy can be financed by consumers of electricity through additional charges (as in Spain and Germany),[31] or by taxpayers through a general subsidy from public revenues (as in Denmark).

2.3. *Regulated Quantities*

As an alternative to regulated prices for RES-E, several countries have introduced quantity-based regulatory regimes, requiring distributors or consumers of electricity to acquire a stipulated quantity of electricity from renewable energy sources. In the United Kingdom, for example, the Non-Fossil Fuel Obligation (NFFO) required regional electricity companies to purchase specified quantities of electricity from renewable sources, to be supplied by private generators under long-term contracts granted to the lowest-cost provider under a competitive bidding process.[32] A similar tendering system was adopted under Ireland's Alternative Energy Requirement.[33] Likewise, in Canada, provincial governments have issued Requests for Proposals (RFPs) inviting private generators to submit competitive bids to supply specified quantities of RES-E under long-term contracts.[34]

In addition to these tendering systems, several jurisdictions have introduced more sophisticated quantity-based regimes in the form of renewable obligations or portfolio standards, which require suppliers or consumers to obtain legislated quantities of electricity from renewable sources, but permit this obligation to be satisfied through the acquisition of renewable certificates that may be bought and sold.[35] In 1999, for example, the state of Texas established a renewable portfolio standard (RPS) based on stipulated increases in new renewable capacity, requiring electricity retailers to satisfy their respective shares of these increases by obtaining renewable energy certificates either from the direct acquisition of RES-E or from the purchase of certificates issued for RES-E generated within or delivered to the Texas grid.[36] In 2002, the United Kingdom replaced the NFFO with a Renewables Obligation (RO), requiring electricity suppliers to meet increasing percentages of their power from renewable sources, which is satisfied by the acquisition of renewables obligation certificates (ROCs) that may be bought and sold separately from the power itself.[37] Italy and Sweden introduced similar percentage-based quota obligations in 2002 and 2003,[38] and percentage-based RPSs have also been adopted in 20 U.S. states and the District of Columbia – with legislated percentages ranging from four per cent in Massachusetts by 2009 to 24 per cent in New York State by 2013.[39] In Canada, RPSs have been proposed or adopted in the provinces of New Brunswick, Nova Scotia, Ontario, and Prince Edward Island.[40]

As with FITs, a key issue in the design of a quota-based regime concerns the manner in which the implicit subsidy to RES-E is financed. Under the United Kingdom's NFFO, the additional cost of RES-E to regional electricity companies was reimbursed from a tax on electricity consumption called the Fossil Fuel Levy.[41] Under the more recent RO, renewable electricity generators are eligible for reward payments in addition to tradable ROCs, which are financed from a 'buy-out' penalty imposed on suppliers who fail to meet their targets under the RO.[42] Under other renewable portfolio standards, the costs of RES-E are generally passed on to consumers of electricity through higher prices.[43]

2.4. Taxes

In addition to subsidies, FITs, and quota requirements, some jurisdictions have also encouraged the generation of electricity from renewable sources through energy taxes that affect the relative prices of electricity from different sources by excluding RES-E from taxation. In 1996, for example, the Netherlands introduced a regulatory energy tax (REB) on the consumption of electricity that was specifically designed to encourage the generation of electricity from renewable sources by excluding RES-E from the tax.[44] In 2001, the United Kingdom introduced a similar energy tax on non-household consumers – called the Climate Change Levy (CCL) – which exempts RES-E.[45]

In other European countries which have introduced taxes on the consumption of electricity, however, the taxes generally do not distinguish between RES-E and electricity from fossil fuels.[46] In addition, most of these taxes include reduced rates for large industries, thereby reducing the impact that they might otherwise have on the development of RES-E. In Germany, for example, energy-intensive firms pay only 3 per cent of normal rates of energy taxes introduced as part of the ecological tax reform, while manufacturing firms have been granted a 60 per cent rate reduction.[47] Similarly, in the United Kingdom, energy-intensive industries are eligible for an 80 per cent reduction in the CCL if they satisfy negotiated energy efficiency targets.[48]

In the Scandinavian countries, where taxes based on the carbon content of fuels were introduced in the early 1990s, one might also have expected indirect encouragement to the development of RES-E through increases in the cost of electricity generated from fossil fuels. In practice, however, most of these taxes have exempted fossil fuels used in the production of electricity.[49]

3. Evaluation of Different Policies to Promote RES-E

In order to evaluate the merits of different policies to promote the generation of electricity from renewable sources, it is necessary to begin by specifying the criteria by which these policies should be evaluated. The first of these criteria (and the ultimate purpose of policies to promote RES-E) is their environmental effectiveness. To the extent that these policies are designed to encourage the generation of electricity from renewable sources, the most direct test of their environmental effectiveness is to measure their impact on the generation of RES-E. As the ultimate environmental purpose of these policies is to reduce GHG emissions, however, the ultimate impact of these policies on these emissions is presumably the best measure of environmental effectiveness.

While environmental effectiveness may be the ultimate purpose of policies to promote RES-E, this is not the only criterion by which they should be evaluated. In addition to environmental effectiveness, for example, the cost effectiveness of a policy is a crucial criterion for a comparative evaluation of different policies, since it is preferable to achieve a given environmental objective at the lowest possible cost.[50] For this purpose, it is important to consider not only direct costs to generate RES-E but also indirect costs such as administrative and compliance costs attributable to the policy.[51] In addition, since the costs to generate RES-E are not fixed but can decrease over time on account of technological innovation and learning-by-doing, it is also important to consider the likely effects of different policies on technological innovation.[52] More generally, different policies might also be evaluated by weighing their expected costs against their expected benefits, though it is widely acknowledged that cost-benefit analysis for mitigating environmental harms is often speculative and imprecise.[53] In this chapter, we consider these cost and cost-benefit criteria under the general category of economic efficiency.

In addition to environmental effectiveness and economic efficiency, distributional fairness is another common criterion for the evaluation of different policies. To the extent that different policies to promote RES-E impose different costs on particular groups or regions, for example, these distributive consequences should presumably be taken into account in a comparative evaluation of these policies.[54] Where different policies produce differential environmental or other benefits, moreover, a proper assessment of the distributive consequences of these policies should consider the distribution of these benefits as well as costs.[55] Although adverse distributional effects may weigh against a particular policy to promote the generation of electricity from renewable sources, it is also important to recognize that distributional concerns can often be addressed through other measures that either mitigate these effects or compensate groups or sectors that are unduly burdened by the particular policy.[56] Indeed, it is often argued that environmental policies should be designed to promote environmental and efficiency objectives alone, with adverse distributional effects offset through the tax and transfer system.[57]

A final criterion for evaluating different policies to promote the generation of electricity from renewable sources concerns the political feasibility of the policy in a given context. Although the political feasibility of a particular policy is at least to some extent a matter of political will, it is also influenced by various factors – social, economic, institutional, and ideological – that constrain politicians and policymakers in the range of policy options that are politically practicable. In a small open economy like Canada, for example, options for independent policymaking are apt to be more limited than they are for a much larger economy like the United States. Where domestic environmental policies can be integrated with international arrangements, on the other hand, the scope for domestic action may be greater. In assessing different policies to promote RES-E, it is important to recognize these constraints.

With these criteria in mind, we now turn to an evaluation of the various policies that we reviewed in part 2 of this chapter.

3.1. Environmental Effectiveness

In evaluating the impact of different policy instruments, several studies have considered the extent to which these instruments have encouraged the development of new renewable sources of electricity. In Denmark and Germany, for example, the early growth of wind power has been attributed in part to government subsidies for the research and development of wind turbines.[58] In the United States and Canada, more recent increases in installed wind capacity may be partly attributable to investments incentives in the form of accelerated depreciation and tax credits and production incentives in the form of the Renewable Energy Production Tax Credit in the United States and the Wind and Renewable Power Production Incentives in Canada.[59] In these and other countries, however, the success of these subsidies in promoting the development of new renewable electricity has also depended on other policy instruments, like FITs in Denmark and Germany and RPS regimes adopted by U.S. states and Canadian provinces. In Spain, for example, subsidies under the country's PAEE had a relatively modest impact on wind power installations until the FIT was introduced in 1998.[60] Likewise, in the United States tax incentives for RES-E appear to have had little impact on installed wind power capacities until the introduction of state-level renewable portfolio standards.[61] In each case, as Janet Sawin emphasizes, "supply-push" policies are much less effective than "demand–pull" policies to encourage the diffusion of renewable energy technologies.[62]

Of the two major demand-pull policies to encourage RES-E, experience suggests that price-based systems like the FITs in Germany, Denmark and Spain have been far more effective than quantity-based regimes like tendering systems or renewable portfolio standards.[63] In Germany, for example, installed wind capacity increased from 48 MW in 1990 to 4500 MW in 2000 when the EEG replaced the StreG, and further increased to 14,609 MW by the end of 2003.[64] In the United Kingdom, on the other hand, where wind resources are much better than Germany, installed wind capacity increased from 10MW in 1990 to 649 MW at the end of 2003.[65] More recently, a high FIT for solar (photovoltaic) power has caused Germany to surpass Japan as the world leader in PV-generated electricity.

To the extent that pricing regimes minimize the price risk of long-term investments in renewable electricity generating capacities, these results are perhaps not surprising,[66] though the inducement to specific technologies depends crucially on the level at which the FIT is set.[67] On the other hand, price risks may also be minimized by long-term contracts entered into under quantity-based regimes, suggesting that the practical differences between these policy instruments may be less than they appear.[68] Indeed, Texas experienced significant increases in installed wind power capacity as long-term contracts were entered into in order to satisfy the RPS adopted in 1999.[69] In addition to the RPS, however, this growth in Texas is also attributable to cost reductions resulting from the earlier diffusion of wind power technology in Europe and to financial support through the federal Renewable Energy Production Tax Credit.[70]

In contrast to price and quantity regimes, taxes appear to have had much less success when it comes to encouraging the development of renewable source electricity. In the Netherlands, for example, the exemption of electricity from renewable sources from the regulatory energy tax encouraged demand for RES-E but little in the way of increased domestic capacity, as domestic demand was satisfied by importing electricity from other countries.[71] Even if electricity or carbon taxes might encourage the development of RES-E, however, the design of these taxes has generally foreclosed this possibility by failing to differentiate among sources of electricity or exempting electricity generators from taxes on carbon inputs. As a result, environmental taxes have thus far played little role in the promotion of renewable sources of electricity.

Notwithstanding their impact on the development of RES-E, it is important to remember that the ultimate environmental objective of these policies is to reduce greenhouse gas emissions, not merely to increase the quantity of RES-E. Although the Dutch regulatory energy tax may not have encouraged the development of RES-E within the Netherlands, it is credited for a 15 per cent reduction in electricity consumption.[72] In the United Kingdom, the CCL is estimated to have reduced CO_2 emissions by approximately two per cent – notwithstanding that households are exempt from the tax and energy-intensive industries are eligible for an 80 per cent rate reduction if they satisfy negotiated energy efficiency targets.[73]

In contrast, even if subsidies increase the supply of RES-E, one might reasonably question their ultimate environmental effectiveness to the extent that they encourage increased production and consumption of electricity. Likewise, where price regimes like the Danish FIT subsidize the added costs of RES-E from public revenues, the environmental effectiveness of this policy instrument is diminished.[74] In contrast, where the additional costs of RES-E are explicitly financed by taxes on electricity consumption (as in Spain) or implicitly financed through increases in electricity prices (as in Germany), the environmental benefits resulting from the generation of renewable electricity are enhanced by the inducement to conservation from increased electricity costs. In these circumstances, FITs function as combined taxes on electricity consumption and subsidies for RES-E. A similar result occurs under quantity-based regulatory regimes like renewable portfolio standards, where additional costs to acquire RES-E are generally recovered through increased prices on all electricity.

Unlike a tax-based or cap-and-trade regime for pricing carbon emissions, however, that would not only encourage the generation of electricity from renewable sources but also discourage high-emission sources of electricity like coal, price-based and quantity-based regulatory regimes that encourage the development of RES-E do not distinguish among different non-renewable sources of electricity. To the extent that they cause renewable sources to displace non-renewable sources of electricity, therefore, they may 'crowd out' higher cost but lower emission generation from natural gas rather than lower cost but higher emission generation from coal. Nor do FITs or RPSs do anything to encourage alternative mitigation strategies like carbon capture and sequestration. For this reason, the environmental effectiveness of price-based or quantity-based regimes to encourage RES-E would be greatly enhanced either by a tax on carbon emission or through a cap-and-trade regime that would create a price for these emissions.

3.2. Economic Efficiency

With respect to the efficiency of different policy instruments to promote electricity from renewable sources, economic theory suggests that quota-based systems should encourage the generation of RES-E at least cost, either through competitive bidding under tendering systems like the United Kingdom's NFFO,[75] or through the tradability of certificates under quantity-based regimes like the renewable portfolio standard in Texas and the RO in the United Kingdom.[76] In contrast, price-based regimes are often regarded as more costly instruments to encourage the development of RES-E, since they are based on fixed prices rather than competitive prices and must be financed either through public revenues (Denmark) electricity taxes (Spain) or increased electricity rates

(Germany).[77] To the extent that price-based regimes reduce the risks of investment in renewable energy generation, however, they create an offsetting cost advantage compared to quantity-based regimes.[78] As well, where FITs are, like the German EEG, adjusted downward to account for technological developments and learning-by-doing, any cost differences between price-based and quantity-based regimes are significantly reduced.[79] FITs are also much less costly to administer than quantity-based systems, particularly those with tradable renewable certificates.[80] In practice, studies suggest that price-based regimes are no more costly than quantity-based regimes when differences in resource capacities are taken into account.[81]

Compared to quantity-based and price-based instruments, subsidies are a costly way to encourage the generation RES-E, since they invariably confer unnecessary benefits on so-called 'infra-marginal' economic actors who would have undertaken the desired activity without the subsidy. Where subsidies are delivered in the form of tax incentives, moreover, costs are difficult to predict and control as they depend on take-up rates by those eligible for the incentives. In contrast, while taxes on carbon or the consumption of electricity entail administrative and collection costs and impose costs on producers and consumers, they result in the collection of public revenues that can be used to reduce other taxes or finance incentives to further encourage RES-E. Although the ultimate cost of an environmental tax depends on the elasticity of producer and consumer responses to the imposition of the tax, available evidence on existing electricity and carbon taxes suggests that even imperfect taxes can significantly reduce energy demand and CO_2 emissions over the long run.[82] While environmental taxes are more costly to administer than FITs, these costs are not prohibitive. Germany's experience with ecological tax reform suggests that administrative and collection costs are a small percentage of taxes raised.[83] Administrative costs increase, of course, to the extent that environmental taxes include exemptions and rate reductions such as those under the U.K.'s CCL for industries that enter into climate change agreements.[84]

With respect to incentive effects, price-based regimes appear to be more efficient than quantity-based systems for two reasons. First, since FITs have stimulated much greater deployment of RES-E capacity than quantity-based regimes, they have facilitated dynamic cost reductions through economies of scale and learning-by-doing.[85] Second, since the surplus from technological innovation generally accrues to producers under a FIT and to consumers under a competitive quantity-based regime, producers have an incentive to invest in research and development and new technology.[86] For these reasons, it is not surprising that the leading producers of wind turbines are located in Denmark, Germany, and Spain,[87] and worth noting that reductions in RES-E costs in countries with quantity-based regimes have often resulted from the importation of technology (such as wind turbines) produced in countries which have stimulated the development of this technology through FITs.[88] To the extent that FITs are gradually reduced over time and regulated quantities are regularly increased over time, however, differences between the dynamic incentive effects of these instruments are reduced.

While subsidies like the U.S. Renewable Electricity Production Tax Credit and the Canadian Wind and Renewable Power Production Incentives may also have resulted in dynamic efficiencies through increased installation and learning by doing, it is not clear that these subsidies have encouraged the kind of sustained cost reductions attributable to the FITs in Denmark, Germany and Spain. As regulated prices tend to encourage the deployment of competitive or nearly competitive technologies (like wind power) rather than the early development of much higher cost technologies (like photovoltaics), however, subsidies may have an important role to play in a mix of policy instruments to encourage this early development. Environmental taxes and cap-and-trade regimes can also encourage dynamic efficiency by creating a consistent economic incentive to reduce greenhouse gas emissions in order to minimize the cost of the tax or emissions permits.

Regarding the social costs and benefits of different policies, analysis is necessarily more speculative.[89] While subsidies for RES-E may produce positive societal benefits in terms of cost reductions and reduced emissions, the revenues that must be raised (or foregone) in order to finance these subsidies necessarily impose social costs. To the extent that these subsidies decrease the cost of electricity, moreover, increased consumption can create additional environmental costs. In contrast, an optimal environmental tax is theoretically most efficient, since it requires market participants to account for the negative environmental externalities associated with the consumption of electricity from non-renewable sources.[90] The same may also be said for cap-and-trade regimes that establish a price for carbon emissions. Regulated prices and quantities, finally, can be regarded as mixed subsidies and taxes, which subsidize RES-E through FITs or renewables obligations, and allocate the costs of these subsidies either to taxpayers or electricity consumers. To the extent that these regimes increase the price of electricity to consumers, therefore, they may replicate some of the efficiency advantages of a tax or cap-and-trade regime. In Germany, for example, the declared purpose of the original feed-in law was to 'level the playing field' between renewable and non-renewable energy sources by taking into account the external costs of conventional power generation.[91]

3.3. *Distributional Fairness*

In addition to their environmental effectiveness and economic efficiency, policies to promote RES-E may also be assessed according to their distributive fairness. To the extent that a particular policy increases the price of electricity, for example, the relative effect of this price increase on low-income individuals or households may be regarded as unfair. Likewise, policies to promote RES-E may be considered inequitable where they have differential effects on different firms or regions.

In the case of individuals and households, studies confirm that those with lower incomes generally spend proportionately more on the consumption of goods and services, including the consumption of electricity – suggesting that policies that increase the price of electricity are typically regressive in their distributive impact, imposing a relatively higher burden on individuals and households with lower incomes than those with higher incomes.[92] Since taxes and quantity-based regulations tend to increase the price of electricity, therefore, their distributive impact is generally regressive.[93] The same may also be said of regulated prices, provided that they are not subsidized from public revenues as in Denmark. In contrast to these policies, the immediate impact of subsidies or subsidized FITs is either less regressive (or progressive) to the extent that they do not increase (or actually decrease) the price of electricity. Since these subsidies must be financed either from tax revenues or foregone tax revenues in the case of tax incentives, however, a proper assessment of the distributive impact of these subsidies should also consider the way in which they are financed.[94] Similarly, since energy or carbon taxes generate revenues that may be used to compensate for adverse distributive effects,[95] a full analysis of the distributive impact of this policy approach should also account for the use of these revenues.[96] In Denmark, Germany, and the Netherlands, for example, tax reforms that introduced or increased energy taxes were accompanied by reductions in personal income or payroll taxes, and (in Denmark) by increased transfer payments to pensioners and low-income households.[97] Adverse distributive effects may also be addressed through general exemptions or specific exemptions aimed at low-income households,[98] though these measures to mitigate the distributive impact of environmental taxes necessarily undermines their environmental effectiveness,[99]

As with individuals and households, different policies to promote RES-E can also have differential effects on firms or regions. Since tax deductions and credits have value only for firms with taxable income or tax liabilities to offset, for example, they tend to encourage investments by large and established firms rather than small and new firms with little or no income and tax otherwise payable. Not surprisingly, therefore, most of the benefits provided under the U.S. Renewable Electricity Production Tax Credit appear to have accrued to large firms with high tax liabilities.[100] Quantity-based regulations also encourage generation by larger enterprises, which are better able to diversify the price risk that is generally associated with this policy approach.[101] They may also favour investment in regions with better resources for renewable energy generation, though this benefit may be offset to the extent that this investment produces environmental or aesthetic burdens.[102]

In contrast to quantity-based regulations, the certainty associated with price guarantees is more likely to encourage investment by small and new enterprises. In Denmark, for example, a combination of subsidies and FITs appears to have encouraged widely dispersed ownership of wind generating capacity,[103] though it is important to acknowledge that the Spanish FIT has not had the same effect.[104] Where price guarantees do not differ according to the quality of available resources such as wind speeds, however, they can generate significant scarcity rents that favour regions and producers with access better renewable energy resources.

Although there is no obvious reason why energy or carbon taxes should differentiate among firms, except according to their use of fossil fuels, there does not appear to be any empirical work on the subject, presumably because the taxes that have actually been introduced impose little or no burden on energy-intensive enterprises.[105] If applied to the generation of electricity, however, these taxes would clearly impose a greater burden on regions that depend more heavily on fossil fuels than on regions with good renewable energy resources. As with taxes on individuals and households, however, revenues from environmental taxes may be used to offset these distributive affects through tax reductions or expenditures aimed at adversely affected regions.

3.4. Political Feasibility

In addition to other criteria, optimal policies to promote RES-E must be politically feasible. As a general rule, subsidies have the most political support of the policy instruments that we have reviewed, with many stakeholders viewing them as promoting both the environment and economic growth.[106] At the other extreme, political opposition tends to be greatest for taxes, which are generally viewed with hostility and distrust, make the costs of climate change policies acutely transparent, and generate concerns about the loss of domestic and global competitiveness.[107] Not surprisingly, therefore, energy intensive and declining industries generally press for exemptions, reduced rates, or the recycling of revenues back to industry, which reduce the environmental effectiveness of these taxes.[108] While addressing such political opposition may raise the administrative costs of environmental taxes, there are a number of ways to reduce opposition including providing exemptions, increasing spending, providing information and phasing in tax increases.[109]

Regulated price and regulated quantity policies have mixed political support. Both create scarcity rents and therefore tend to lead to rent-seeking by firms that may appropriate these rents. The public and environmental groups tend to support guaranteed prices because of their perceived effectiveness at promoting RES-E. As well, where FITs encourage widely dispersed ownership of renewable electricity generation, as occurred in Denmark, local opposition to the installation of generating capacity may be reduced.[110]

Despite these advantages, however, regulated price policies appear to be less popular in market-oriented economies than quantity-based regulations, since guaranteed prices contradict the market mechanism for determining prices, while the latter encourage competition to reduce costs.[111] For this reason, it is not surprising that the greatest public support for renewable energy instruments in the United States has been for renewable portfolio policies.[112] Notwithstanding this widespread support, however, there has been some opposition to regulated quantity policies because of competitiveness concerns resulting from increases in electricity prices, or the fear that some regions might not be able to meet the standard.[113]

4. Conclusion

Climate change policy should be focused on reducing emissions of greenhouse gases, not increasing the generation of RES-E for its own sake. Theory and experience suggest that the pursuit of this objective will demand a mix of instruments.[114] In Canada, however, policies have thus far tended to emphasize subsidies and voluntary initiatives.[115] In addition to these approaches, several jurisdictions have attempted to encourage RES-E through price guarantees and quantity-based regulations, while a few have also introduced energy or carbon taxes.

Subsidies and regulated price and quantity policies, while politically more palatable than taxes, are not particularly effective at reducing greenhouse gas emissions if they decrease electricity prices or crowd out relatively low emission sources such as natural gas. Nor are these instruments as economically efficient as a well-designed tax or cap-and-trade regime that would establish a price for greenhouse gas emissions and thereby create a continuing incentive for fuel switching as well as energy conservation. While energy and carbon taxes may be politically more difficult to implement than other policy instruments, and must be carefully designed or accompanied by other measures to offset distributional concerns, they are much more likely to promote the ultimate goal of reducing greenhouse gas emissions than subsidies, prices guarantees, and quantity-based regulations for the generation of RES-E. As evidence of ongoing climate change becomes increasingly apparent and public concern about the potential consequences of dangerous climate change builds, Canadian policymakers should shift their attention from policies that are designed simply to encourage RES-E to policies like taxes and cap-and-trade regimes that will target the ultimate cause of human-induced climate change – the emission of greenhouse gases themselves.

Notes

1 European Environmental Agency, *Greenhouse Gas Emissions Trends and Projections in Europe 2006* (Copenhagen: EEA, 2006) at 40, online, http://reports.eea.europa.eu/eea_report_2006_9/en/eea_report_9_2006.pdf.
2 Environmental Protection Agency, *Inventory of U.S. Greenhouse Gas Emissions and Sinks 1990–2005* (Washington, DC: EPA, 2007) at ES-14, online, http://www.epa.gov/climatechange/emissions/downloads06/07CR.pdf.

Policies to Promote the Generation of Electricity from Renewable Sources 239

3 F. Kahrl and D. Roland-Holst, 'China's Carbon Challenge: Insights from the Electric Power Sector,' Research Paper No. 110106, University of California Center for Energy, Resources, and Economic Sustainability (Berkeley, CA: November 2006) at 9, online, http://are.berkeley.edu/~dwrh/Docs/CCC_110106.pdf.
4 Government of Canada, *Canada's Fourth National Report on Climate Change*, (Ottawa: Environment Canada, 2006) at 32, online, http://unfccc.int/resource/docs/natc/cannc4.pdf.
5 See, e.g., D.M. Driesen, 'Renewable Energy under the Kyoto Protocol: The Case for Mixing Instruments,' in this volume.
6 International Energy Agency, *Renewables in Global Energy Supply* (Paris: Organisation for Economic Co-operation and Development, January 2007) at 4, online, http://www.iea.org/textbase/papers/2006/renewable_factsheet.pdf, reporting average annual growth rates from 1971 to 2004 of 48.1 per cent for wind energy, 28.1 per cent for solar energy, and 7.5 per cent for geothermal energy, compared to average annual growth rates during this period of only 2.2 per cent for total primary energy supply.
7 Ibid. at 5.
8 T.B. Johansson and W. Turkenburg, 'Policies for Renewable Energy in the European Union and its Member States' (2004) 8 Energy for Sustainable Development 5 at 18.
9 This list is based in large part on R. Revesz and R. Stavins, 'Environmental Law and Policy,' in A.M. Polinsky and S. Shavell, eds., *The Handbook of Law and Economics* (Amsterdam: North-Holland/Elsevier Science, Forthcoming) identifying three broad criteria for assessing policy instruments (cost-effectiveness, distributional equity, and political feasibility) and suggesting that questions of cost effectiveness should consider not only whether a policy instrument will achieve a stated environmental goal at least cost but also whether it will be flexible in the face of changing tastes and technology and whether it will create dynamic incentives for research, development, and adoption of new technologies.
10 See, e.g., G. Klassen, A. Miketa, K. Larsen, and T. Sundqvist, 'The Impact of R&D on Innovation for Wind Energy in Denmark, Germany, and the United Kingdom' (2005) 54 Ecological Economics 227 at 228–31.
11 V. Lauber and L. Mez, 'Renewable Electricity Policy in Germany, 1974 to 2005' (April 2006) 26 Bulletin of Science, Technology & Society 105.
12 European Renewable Energy Council, 'Renewable Energy Policy Review Spain,' *Review of Policy Initiatives within the EU* (Brussels: May 2006) at 11–12, online, http://www.erec-renewables.org/fileadmin/erec_docs/Projcet_Documents/RES_in_EU_and_CC/Spain.pdf.

13 See Database of State Incentives for Renewables and Efficiency, online, http://www.dsireusa.org. For a useful discussion of U.S. tax incentives for solar power, see L. Kreiser, B. Butcher, J. Sirisom and P.J. Lee, 'The Use of Environmental Taxation Incentives to Encourage Investments in Solar Power,' in A. Cavaliere et al., eds., *Critical Issues in Environmental Taxation: International and Comparative Perspectives III*, (Richmond, UK: Richmond Law & Tax Ltd., 2006) 461.
14 Natural Resources Canada, 'Background on REDI' (29 June 2005), online, http://www2.nrcan.gc.ca/es/erb/erb/english/View.asp?x=672. See also Natural Resources Canada, 'Market Incentives Program (MIP) for Distributors of Electricity from Emerging Renewable Energy Sources' (22 October 2004), online, ttp://www2.nrcan.gc.ca/es/erb/erb/english/View.asp?x=457.
15 See D.G. Duff, 'Tax Policy and Global Warming' (2003) 51 Can. Tax J. 2063 at 2103–4.
16 See the discussion of 'Canadian Renewable and Conservation Expenses' in D.G. Duff and A. Green, 'Wind Power in Canada,' in K. Deketelaere et al., eds., *Critical Issues in Environmental Taxation: International and Comparative Perspectives IV* (Oxford: Oxford University Press, 2007) 3 at 29–31.
17 Natural Resources Canada, 'Renewable Power Production Incentive' (16 September 2005) online, http://www2.nrcan.gc.ca/es/erb/erb/english/View.asp?x=681. For a discussion of the Wind Power Production Incentive, see Duff and Green, *supra* note 16 at 31–3.
18 See, e.g., P. Gillies and Z. Lipman, 'Producing Electricity from Renewable Energy Sources – Fiscal Measures in Australia,' in Cavaliere et. al., *supra* note 13 at 460.
19 International Energy Agency, *Renewable Energy: Market and Policy Trends in IEA Countries* (Paris: Organisation for Economic Co-operation and Development, 2004) at 87.
20 Lauber and Mez, *supra* note 11 at 106.
21 S. Auken, 'Answers in the Wind: How Denmark Became a World Pioneer in Wind Power' (Winter/Spring 2002) 26 *Fletcher Forum of World Affairs* 149 at 154. See also Niels Meyer and Anne Louise Koefoed, 'Danish Energy Reform: Policy Implications for Renewables' (2003) 31 *Energy Policy* 597.
22 See, e.g., P. del Rio and M.A. Gual, 'An Integrated Assessment of the Feed-in Tariff System in Spain' (2007) 35 Energy Policy 994 at 998.
23 European Renewable Energy Council, *supra* note 12 at 7 (reporting premiums of 2.17 to 3.01 eurocents per kWh in most cases, but 27.1 eurocents per kWh for electricity from photovoltaics).
24 J.P.M. Sijm, 'The Performance of Feed-in Tariffs to Promote Renewable

Electricity in European Countries' (Energy Research Centre of the Netherlands, November 2002) at 10, online, http://www.ecn.nl/docs/library/report/2002/c02083.pdf.
25. R. Haas et. al., 'How to Promote Renewable Energy Systems Successfully and Effectively' (2004) 32 *Energy Policy* 833 at 834–7.
26. S.N.M. van Rooijen and M.T. van Wees, 'Green Electricity Policies in the Netherlands: An Analysis of Policy Decisions' (2006) 34 *Energy Policy* 60 at 63.
27. See, e.g., del Rio and Gual, *supra* note 22 at 995.
28. See *supra* notes 23 and 24 and accompanying text.
29. del Rio and Gual, *supra* note 22 at 998–9.
30. N.I. Meyer, 'European Schemes for Promoting Renewables in Liberalised Markets' (2003) 31 *Energy Policy* 665 at 670–1.
31. In Spain, premium-rates are financed through an explicit tax on electricity consumption; see del Rio and Gual, *supra* note 22 at 998–9. Germany's EEG, on the other hand, requires utilities to distribute the costs of the fixed-rate FIT among electricity consumers in the form of higher prices; see Meyer, *supra* note 30 at 668.
32. See, e.g., C. Mitchell and P. Connor, 'Renewable Energy Policy in the UK, 1990–2003' (2004) 32 *Energy Policy* 1935 at 1936–38.
33. International Energy Agency, *supra* note 19 at 87.
34. See, e.g., Ontario Ministry of Energy, 'Renewable Energy' (2006) online, http://www.energy.gov.on.ca/index.cfm?fuseaction=english.renewable.
35. See, e.g., Meyer, *supra* note 30 at 669. For a useful survey of this policy instrument, see N.H. van der Linden, et. al., *Review of International Experience with Renewable Energy Obligation Support Mechanisms* (Amsterdam: Netherlands Energy Research Foundation, 2005) online, http://eetd.lbl.gov/ea/EMS/reports/57666.pdf.
36. See O. Langniss and R. Wiser, 'The Renewables Portfolio Standard in Texas: An Early Assessment' (2003) 31 *Energy Policy* 527 at 528. Under the RPS, new renewable capacity was to increase by 400 MW by 2003, 850 MW by 2005, 1400 MW by 2007 and 2000 MW by 2009.
37. van der Linden et. al., *supra* note 35 at 24. See also Mitchell and Connor, *supra* note 32 at 1939; and D. Toke, 'Are Green Electricity Certificates the Way Forward for Renewable Energy? An Evaluation of the United Kingdom's Renewables Obligation in the Context of International Comparisons' (2005) 23 Environment and Planning, C: Government and Policy 361 at 363–4. Under the RO, electricity suppliers were required to obtain ROCs for 3 per cent of their power in 2002, increasing to 10.4 per cent in 2010 and 15.4 per cent in 2015.

38 The Italian RPS requires fossil thermal producers to obtain Tradable Green Certificates (TGCs) corresponding to 2 per cent of their output. A. Lorenzoni, 'The Italian Green Certificates Market between Uncertainty and Opportunities' (2003) 31 Energy Policy 33. In Sweden, on the other hand, the obligation is imposed on all electricity consumers (other than energy-intensive industry), who are required to purchase a stipulated percentage (8.1 per cent in 2004) of their electricity from renewable sources. In practice, this obligation is handled by electricity suppliers. See van der Linden et. al., *supra* note 32 at 34–44.
39 Union of Concerned Scientists, 'Renewable Energy Standards,' online, http://www.ucsusa.org/clean_energy/clean_energy_policies/RES-climate-strategy.html.
40 Duff and Green, *supra* note 16 at 35–7.
41 L. Butler and K. Neuhoff, 'Comparison of Feed in Tariff, Quota and Auction Mechanisms to Support Wind Power Development,' *Cambridge Working Papers in Economics*, CWPE 0503 (December 2004) at 3, online, http://www.electricitypolicy.org.uk/pubs/wp/ep70.pdf.
42 Toke, *supra* note 37 at 364.
43 See, e.g., van der Linden et. al., *supra* note 35 at 36 (discussing the Swedish quota obligation system).
44 Meyer, *supra* note 30 at 672. See also van Rooijen and van Wees, *supra* note 26 at 62; and M. Boots, 'Green Certificates and Carbon Trading in the Netherlands' (2003) 31 Energy Policy 43 at 46.
45 S. Dresner, T. Jackson, and N. Gilbert, 'History and Social Responses to Environmental Tax Reform in the United Kingdom' (2006) 43 Energy Policy 930 at 932.
46 This is the case in Germany, for example, where the introduction of such a distinction is a topic of active discussion. C. Bauermann and T. Santarius, 'Ecological Tax Reform in Germany: Handling Two Hot Potatoes at the Same Time' (2006) 34 Energy Policy 917 at 926.
47 Organisation for Economic Co-operation and Development, *The Political Economy of Environmentally Related Taxes* (Paris: OECD, 2006) at 43.
48 Ibid. at 111.
49 J. Vehmas, 'Energy-Related Taxation as and Environmental Policy Tool – The Finnish Experience 1990–2003' (2005) 33 Energy Policy 2175 at 2176 (mentioning Denmark, Sweden, and Finland).
50 Revesz and Stavins, *supra* note 9.
51 See, e.g., D. Cole and P. Grossman, 'When Is Command and Control Efficient? Institutions, Technology and the Comparative Efficiency of Alternative Regulatory Regimes for Environmental Protection' [1999] Wisconsin

L.R. 887; and C. Rose, 'Rethinking Environmental Controls: Management Strategies for Common Resources' [1991] *Duke L.J.* 1 (both discussing administrative costs).
52 Revesz and Stavins, *supra* note 9.
53 See, e.g., F. Ackerman and L. Heinzerling, *Priceless* (New York: New Press, 2004) at 18 (arguing that environmental amenities are not commensurable with money and therefore cannot be priced). For alternative views, see Richard Posner, *Catastrophe: Risk and Response* (Oxford: Oxford University Press, 2004); C. Sunstein, *Risk and Reason*, (Cambridge: Cambridge University Press, 2002); and Revesz and Stavins, *supra* note 9 (both arguing that policy analysis should attempt to quantify costs and benefits to the extent possible in order to provide a framework and information for decision-making).
54 See, e.g., C. Sunstein, *supra* note 53.
55 OECD, *supra* note 47 at 135.
56 See, e.g., the discussion of mitigation and compensation measures in ibid. at 136–7.
57 See, e.g., Revesz and Stavins, *supra* note 9 at 48.
58 Klassen et al., *supra* note 10 at 237.
59 See, e.g., Duff and Green, *supra* note 16.
60 Installed capacity increased from 7 MW in 1990 to 377 MW by November 1997 before the FIT was introduced, and from 377 MW to 6202 MW by 2003 after its introduction. European Renewable Energy Council, *supra* note 12 at 5.
61 See, e.g., L. Bird et al., 'Policies and Market Factors Driving Wind Power Development in the United States' (2005) 33 Energy Policy 1397.
62 J.L. Sawin, 'National Policy Instruments: Policy Lessons for the Advancement & Diffusion of Renewable Energy Technologies Around the World,' International Conference for Renewable Energies (Bonn, January 2004) at 7, online, http://www.renewables2004.de/pdf/tbp/TBP03-policies.pdf at 2.
63 See, e.g., C. Mitchell, D. Bauknecht, and P.M. Connor, 'Effectiveness through Risk Reduction: A Comparison of the Renewable Obligation in England and Wales and the Feed-In System in Germany' (2006) 34 Energy Policy 397; Butler and Neuhoff, *supra* note 41; Sawin, *supra* note 62; and Meyer and Koefoed, *supra* note 21. See also Commission of the European Communities, *Communication from the Commission: The Support of Electricity from Renewable Energy Sources* (Brussels, 12 July 2005) at 6, online, http://ec.europa.eu/comm/energy/res/legislation/support_electricity_en.htm.
64 Butler and Neuhoff, *supra* note 41 at 5.
65 Ibid.

66 Mitchell, Bauknecht, and Connor, *supra* note 63.
67 del Rio and Gual, *supra* note 22 at 1000.
68 V. Lauber, 'REFIT and RPS: Options for a Harmonized Community Framework' (2004) 32 Energy Policy 1405.
69 By 2006, Texas had a cumulative total of 2370 MW of wind power capacity, making it the leader among U.S. states. American Wind Energy Association, 'Quarterly Market Report: Texas Overtakes California as Top Wind Energy State' (25 July 2006) online, http://www.awea.org/newsroom/releases/AWEA_Quarterly_Market_Report_072506.html.
70 Langniss and Wiser, *supra* note 36 at 534.
71 See, e.g., van Rooijen and van Wees, *supra* note 26.
72 J.P. Clinch, L. Dunne, and S. Dresner, 'Environmental and Wider Implications of Political Impediments to Environmental Tax Reform' (2006) 34 Energy Policy 960 at 962.
73 OECD, *supra* note 47 at 112.
74 Despite this objection, it is important to note that CO_2 emissions in Denmark fell by 11 per cent during the period 1990 to 2002, despite a 28 per cent increase in gross national product, Sawin, *supra* note 62 at 8.
75 Sawin, *supra* note 61 at 13. See also P. Mentanteau, D. Finon, and M.-L. Lamy, 'Prices versus Quantities: Choosing Politics for Promoting the Development of Renewable Energy' (2003) 31 Energy Policy 799 at 807–8.
76 See, e.g., Hass et. al., *supra* note 25 at 838. See also Johansson and Turkenburg, *supra* note 8 at 19.
77 Sawin, *supra* note 62 at 12.
78 Mitchell, Bauknecht, and Connor, *supra* note 63.
79 Mentanteau et al., *supra* note 75 at 807.
80 del Rio and Gual, *supra* note 22 at 1007.
81 See, e.g., Lauber, *supra* note 68; and Butler and Neuhoff, *supra* note 41.
82 OECD, *supra* note 47 at 50–1.
83 Ibid. at 146, reporting that collection costs were only 0.13 per cent of taxes raised, which compares favourably to much higher collection costs for income taxes.
84 Ibid. at 147.
85 See, e.g., Mentanteau et al., *supra* note 75 at 805 and 808; del Rio and Gual, *supra* note 22 at 1009; and S. Jacobsson and V. Lauber, 'The Politics and Policy of Energy System Transformation – Explaining the German Diffusion of Renewable Energy Technology' (2006) 34 Energy Policy 256.
86 See, e.g., Mentanteau et al., *supra* note 75 at 805 and 808.
87 del Rio and Gual, *supra* note 22 at 1007.
88 Sawin, *supra* note 62 at 13.

89 See, e.g., del Rio and Gual, *supra* note 22 at 1004–5, suggesting that support to renewable energy technologies in Spain from subsidies and regulated prices is efficient only for wind and small hydro, but emphasizing that this analysis does not account for possible long-term cost reductions in other technologies.
90 Mentanteau et al., *supra* note 75 at 800.
91 Lauber and Mez, *supra* note 11 at 106.
92 L.H. Goulder, I.W.H. Parry, and D. Burtraw, 'Revenue-Raising versus Other Approaches to Environmental Protection: The Critical Significance of Preexisting Tax Distortions' (1997) 28(4) RAND Journal of Economics 708. In theory, the impact of this price increase depends on the relative importance of electricity in the consumption bundles of low and high income households, the relative elasticities of demand between low and high income households in a particular country and whether there are any off-setting indirect impacts (such as where a broad based energy tax also raises the cost of other consumption goods that high income households consume in greater quantities). N. Johnstone and Y. Serret, 'Introduction,' in Y. Serret and N, Johnstone, eds., *The Distributional Effects of Environmental Policy* (Cheltenham, UK: Edward Elgar, 2006); and B. Kristrom, 'Framework for Assessing the Distribution of Financial Effects of Environmental Policy,' in Serret and Johnstone.
93 See, e.g., studies on the distributional impact of energy taxes discussed in OECD, *supra* note 47 at 134. Evidence also indicates that the degree of regressivity decreases once indirect effects from energy price increases are taken into account, since energy is an input into virtually all goods and services.
94 See, e.g., Johnstone and Serret, *supra* note 92; and Goulder, Parry, and Burtraw, *supra* note 92.
95 Goulder, Parry, and Burtraw, *supra* note 92. See also P.J. Clinch, L. Dunne, and S. Dresner, 'Environmental and Wider Implications of Political Impediments to Environmental Tax Reform' (2006) 34 Energy Policy 960.
96 See, e.g., Johnstone and Serret, *supra* note 92; and Kristrom, *supra* note 92 (reviewing the empirical literature on various tax/transfer swaps or offsets).
97 OECD, *supra* note 47 at 139–42.
98 Johnstone and Serret, *supra* note 92. In the United Kingdom, for example, the decision to exempt households from the Climate Change Levy was based in part on distributional concerns. OECD, *supra* note 47 at 114.
99 Ibid. at 143.
100 D. Toke, 'Are Green Electricity Certificates the Way Forward for Renew-

able Energy? An Evaluation of the United Kingdom Renewables Obligation in the Context of International Comparisons' (2005) 23 *Environment and Planning, C: Government and Policy* 361.
101 See, e.g., Sawin, *supra* note 62; and Mitchell, Bauknecht, and Connor, *supra* note 63.
102 Ibid.
103 Auken, *supra* note 21 at 150.
104 Toke, *supra* note 100.
105 See the discussion at *supra*, notes 44–9 and accompanying text.
106 U.S. Brandt and G.T. Svendsen, 'Fighting Windmills: The Coalition of Industrialists and Environmentalists in the Climate Change Issue' (2004) 4 International Environmental Agreements: Politics, Law, and Economics 327.
107 S. Dresner, T. Jackson, and N. Gilbert, 'History and Social Responses to Environmental Tax Reform in the United Kingdom' (2006) 34 Energy Policy 930; OECD, *supra* note 47; and F. Menz, 'Green Electricity Policies in the United States: Case Study' (2005) 33 Energy Policy 2398.
108 See, e.g., Clinch, Dunne, and Dresner, *supra* note 95; OECD, *supra* note 95; and T. Kåberger, Thomas Sterner, et al. 'Economic Efficiency of Compulsory Green Electricity Quotas in Sweden' (2004) 15(4) Energy & Environment 675.
109 OECD, *supra* note 47 at 152–4.
110 Johansson and Turkenberg, *supra* note 8 at 16; and D. Toke, 'Wind Power in UK and Denmark: Can Rational Choice Help Explain Different Outcomes?' (2000) 11(4) Environmental Politics 83.
111 N.H. van der Linden, M.A. Uyterlinde, et al., *Review of International Experience with Renewable Energy Obligation Support Mechanisms* (Petten, Netherlands: Energy Research Centre of the Netherlands, May 2005) online, http://eetd.lbl.gov/ea/EMS/reports/57666.pdf. In Denmark, for example, a more market-oriented government decided to switch from FITs to green certificates in part because of the fear of the cost of the FITs. Niels Meyer, 'Renewable Energy Policy in Denmark' (2004) 8(1) Energy for Sustainable Development 25.
112 van der Linden et al., *supra* note 117 at 50.
113 F. Sissine, 'Renewable Energy: Tax Credit, Budget, and Electricity Production Issues' (Congressional Research Service Issue Brief for Congress, 17 February 2006) online, http://fpc.state.gov/c4763.htm.
114 See, e.g., Driesen, *supra* note 5; and OECD, *supra* note 47.
115 See, e.g., J. Simpson, M. Jaccard and N. Rivers, *Hot Air: Meeting Canada's Climate Change Challenge* (Toronto: McClelland and Stewart, 2007).

12 Bringing Institutions and Individuals into a Climate Policy *for* Canada

ANDREW J. GREEN

1. Introduction

The optimal choice of instruments to address climate change is difficult given the costs that are entailed by reductions of the magnitude some say are necessary to stabilize temperatures.[1] Individuals in developed countries will have to make significant changes in their current production and consumption patterns if such stabilization is to occur. It is unfortunate then that much of the analysis of instrument choice in this area does not consider the role of institutions and takes a very thin view of individuals and how they make decisions. Institutions and individuals are central to designing an effective climate policy for each county. An optimal climate policy for Canada cannot be based on generalized notions of markets and governments, but must be based on how institutions and individuals actually operate in Canada.

Much of the analysis of climate change focuses on imposing a price on emissions of greenhouse gases. Canada has failed to stem the growth of emissions in part because it has used policies based on providing information and subsidies to individuals or industry.[2] Effective policies would include taxes, trading schemes, and/or regulations.[3] Such instruments could create strong incentives to reduce emissions and invest in the discovery of new technologies. The optimal set of policies would likely involve a mix of all these instruments. For example, Driesen argues that reliance on trading alone is unlikely to provide sufficient incentive for the increase in availability of renewable energies he views as necessary to address climate change.[4]

While price is important, there are other factors that must be considered in designing an optimal policy for Canada. An optimal policy must take into account the economic, political, and social context of Canada, including its institutional structure and how individuals make choices. Section 2 discusses how institutions, and in particular the constitutional division of powers, constrains the choice of instruments in Canada for addressing climate change. Section 3 turns to why a thicker view of individuals is important. It discusses both limitations on individuals' ability to choose 'rationally' and the importance of individuals' values.

2. Institutions as Constraints

Institutions create the framework within which individuals make choices and instruments operate. Each country has different institutions which determine what instruments may be chosen and how they are chosen. Not that these institutions are necessarily immutable. Some, such as constitutions, are hard to change. Others, such as administrative law arrangements for determining how decisions are made, can be changed much more readily. Failure to account for institutions, including whether or not they can be readily changed, will obstruct the development of effective policies.

For example, the Canadian constitution limits the types of policies that the federal and provincial governments may chose. The federal government does not obtain the power to implement an international treaty merely by signing onto or ratifying it. In implementing the obligations it accepts under the treaty, the federal government may use only the powers given to it under the constitution. The difficulty is that the constitution does not assign the power over the 'environment' as a category to either level of government. Any power to use a particular instrument to address climate change must be found in another, more general power.

Take a cap-and-trade system, for example. A cap-and-trade system can, in theory, have the same effect as a tax on emissions. It sets a price on emissions, with the price determined by the number of permits and the trading rules. The broader the trading system, the lower the cost of reducing emissions as there is a greater potential for parties that can reduce emissions cheaply to trade with those who face higher costs of reducing emissions. In theory, then, a national cap-and-trade system would allow the reduction of emissions at lower costs than provincial systems as there could be trading across a wider range of parties. It would also reduce the fear of industry moving from a province with stringent controls to one with less stringent controls.

However, it is not at all clear that the federal government has the constitutional jurisdiction to put in place a national system. The courts have been fairly generous in finding federal powers over the environment. For example, in *Hydro Quebec*, the Supreme Court of Canada examined whether the federal government had the power to implement toxic substances regulations under its criminal law powers.[5] For a valid use of the criminal law power, the federal government must put in place a prohibition backed by a penalty. The Court found that the toxic substances provisions were of the form of a prohibition backed by a penalty. However, the decision was close (five to four) and the majority did not particularly engage with the issue of 'regulation' versus a 'prohibition.'[6] It seems hard to argue that a trading system is a prohibition rather than regulation and, given the turnover on the Court since *Hydro-Quebec*, it is not certain that the federal criminal law powers would be found to provide a basis for a federal cap-and-trade system.

Similarly, in *Crown Zellerbach*, the Supreme Court upheld the federal government's power to regulate ocean dumping under its residual powers to make laws for the peace, order, and good government (POGG) of Canada.[7] It found that the ocean dumping requirements fit within the national dimensions aspect of POGG – that is, a matter that goes beyond local or provincial concerns and is of concern to the nation as a whole. However, the POGG power has limits. For example, the federal government can only act where the provinces are unable to deal effectively with extraterritorial effects and the matter must have 'a scale of impact on provincial jurisdiction that is reconcilable with the fundamental distribution of legislative powers under the Constitution.'[8] The minority took the majority to task, arguing that the majority's interpretation potentially expanded federal powers too far into provincial powers. An effective cap-and-trade system has the potential to infringe significantly on provincial powers, such as those over local matters or non-renewable resources. It is not difficult to imagine a challenge by industry or a province to a stringent federal cap-and-trade system. Whether or not it would survive a challenge as a valid use of the POGG power depends on the structure of the system as the system of allocating permits (determining the cost structure of the recipient industries) and trading regime could be viewed as regulating otherwise provincially-regulated industries.[9] Moreover, the requirement of provincial inability is unlikely to be met where some provinces have already begun taking action.

The provinces could set up their own cap-and-trade system under their broad jurisdiction over such matters as property and civil rights and non-renewable resources. However, given competitiveness concerns, each province faces incentives for weak rules if they alone institute a stringent cap-and-trade system. Further, many of the provinces have strong industries with strong attendant views, such as the energy industry which is opposed to stringent caps. Even if cap-and-trade systems spring up in different provinces, there is a concern that each province will design its system to favour its own industry. They may then be unwilling to give up concessions made to its industry to join a nation-wide or regional trading system.[10] Preferred design features in one system may lead it to be incompatible with other systems. For example, an intensity based system would be difficult to link with one that is premised on a hard cap.

On the other hand, the federal government does have very extensive taxation powers. Under section 91(3) of the constitution, it has the power of 'raising money by any mode or system of taxation' – that is, any direct or indirect tax. It has, therefore, broad powers to impose taxes on greenhouse gas emissions. The provinces face some limits on their taxation powers, in particular that their general taxing power is of 'direct taxation within the province [in order to the raising] of a revenue for provincial purposes' (section 92(2)). Provinces, therefore, can in general not impose direct taxes. The one exception is in the area of non-renewable natural resources, forestry resources, and electricity generation, for which the provinces have the power to raise both direct and indirect taxes (section 92A(4)).

The problem again is competitiveness. It is difficult to be the only province imposing a stringent carbon tax, given the fear of losing industry to other provinces or states. For example, Quebec recently imposed a carbon tax. However, it is not overly strong, likely at least in part because of industry concerns about competitiveness. Given these fears about competitiveness, it may be preferable for the federal government to impose a tax relating to greenhouse gas emissions. A federal tax would limit the leakage of industries within Canada from high tax to low tax provinces and would set one national price for emissions, which would, in theory, be more efficient.

However, the federal government would face stiff provincial opposition in imposing a carbon tax, particularly from provinces such as Alberta, which stand to lose from a tax. Simpson, Jaccard, and Rivers argue that to overcome provincial opposition to a federal tax, the federal government could return all funds raised in a particular province to that province.[11] Such a suggestion would at least overcome the fear that the federal government was using the tax as a revenue grab at the expense of the provinces. However, there would still be an impact on the industries in the provinces from the price effect of the tax. Provinces such as Alberta would still protest. The provinces have the power now to join in an agreement to tax their residents on greenhouse gas emissions and keep the revenue. They have not done so, in part because many provinces have concerns about the impact of any such tax on economic growth. A federal tax, even with a return of revenue to each province, would not overcome this concern.

Further, increased taxes are not popular with the Canadian population. The federal government would need strong political support to overcome the opposition of the provinces to the imposition of a tax. The need for such political support raises the issue of the thin view of individuals that lies at the heart of much instrument choice analysis. This issue is addressed in the next section.

3. Individuals, Values, and Institutions

Instrument choice analysis tends to view individuals as responding primarily to price signals. Sometimes, such analysis considers the importance of getting information to individuals so they can make appropriate choices. Further, in response to concerns about the need for political support for strong action on climate change, it has become popular to point to recent polls of Canadians that apparently show significant concern about the environment and climate change. Yet by and large, the instrument choice literature tends to work with a thin view of individuals and their motivations. Understanding how individuals actually choose is central to the choice of appropriate policies for addressing climate change for two reasons. First, individuals' direct actions, such as what type of car to drive, how much to drive, and what appliances to use, constitute approximately 30 per cent of emissions of greenhouse gases in Canada and the United States.[12] Impacting these choices should, therefore, be an important focus of climate policy. Second, individuals do not choose only as consumers but also as citizens – that is, they vote, may participate in policy decisions, and discuss shared goals and values with each other.[13] Any policy choices must have some support from citizens. Governments have to lead but how they lead and their success will depend in part on the impact of their policy and institutional choices on individuals' values.

There is a need, therefore, to think more carefully about how individuals actually do choose – both as consumers and as citizens – in order to provide a basis for pragmatic, sustainable climate policies. In this context, sustainable means capable of being maintained over a long period of time. There are two aspects of how individuals choose that are important in this respect: the limitations on how individuals choose, such as on their ability to rationally understand choices or to make choices; and the role of values in their decisions.

3.1. *How Individuals Choose*

Individuals do respond to prices. Such price responses will be extremely important in designing a climate change policy for Canada. However, individuals do not choose solely on the basis of price. In fact, at times there can be significant differences between what would be predicted to be a choice of an individual based on price and her actual choice. These differences have important implications for the choice across taxes, trading schemes, and regulations.

For example, President George W. Bush, in his 2007 State of the Union address, stated that Americans are 'addicted to oil.' If this is true, such an addiction has important implications for the use of taxes or prices to affect consumer choice, particularly in the near term. We may not be able to wean ourselves off of the use of cars, or of certain types of cars, quickly enough at a realistic tax rate to adequately reduce greenhouse gas emissions. Such limitations on choice or willpower may make regulations, or at least careful structuring of price mechanisms, more appropriate in the case of some decisions.[14]

Effective policies will require greater attention to how individuals actually make choices – such as the limits on their willpower or on their ability to understand choices.[15] Such attention is important, not least because, as noted above, individuals' direct choices make up about 30 per cent of emissions in Canada and the United States. Progress on climate change will require individuals to make substantial changes in these choices.

3.2. *Building Values*

Not only do individuals face constraints on their will or on their ability to understand, they also make decisions in part based on their values or norms. As noted above, polls show that individuals are now concerned more about the environment than other issues, including the economy and health care. Such an increased concern could lead to changes in individuals' choices both as consumers and as citizens.[16] However, this spike in interest is not sufficient in and of itself to anchor sustainable long-term actions to reduce climate change.[17]

Moreover, such a spike in public concern is not entirely new. There have been waves of interest in the past, including in the 1960s and in 1980s. Both times the spike in interest in the environment declined when the population was faced with other concerns, such as a failing economy.[18]

Is this current spike in interest part of a wave, a new higher baseline of public concern about the environment, or merely cheap talk such that individuals are willing to say they are concerned but in fact are unwilling to support policies to address climate change that impose significant personal costs? It would be nice to think it represents a new, higher baseline of concern. However, even if it does not, Canadian governments do have a brief window to put in place effective greenhouse gas policies.

Yet governments must take care to act in a manner that fosters, or at least does not diminish, current public opinion in favour of the environment. The choice of instruments may have an impact on the duration of the wave of interest and on what happens when the wave is over. If individuals come to care only about the price, it may diminish the possibility of future stringent climate policy. For example, if there are economic difficulties in the future, individuals may be unwilling to support an increase in any tax. Alternatively, if the government attempts to pre-commit to a strategy such as schedule of tax increases set out in legislation, there may be backsliding by governments through either the creation of loopholes or a failure to enforce the laws. As well, of course, legislation can be amended or repealed.

It is important, therefore, to consider how both instruments and institutions build and/or maintain individuals' values and concern about the environment. Education and information are an obvious (though not obviously effective) manner in which to attempt to influence individuals' values and perceptions. The more interesting connection, however, relates to price-based instruments such as taxes. Such instruments have several benefits, including potentially low costs of achieving targets, spurring of technological advances, and autonomy (the individual or industry may choose whether and how to reduce its greenhouse gas emissions). However, at the same time, there can be a trade-off with individuals' values. Placing a price on environmentally friendly behaviour can crowd out feelings of responsibility for the environment, leaving individuals to view the decisions about such behaviour as a choice whether to pay the price, as opposed to an obligation or issue of values.[19] Further, such a view can tie into long-term public support for climate change policies. For example, citizens may view fairness in policy as more about equality of options (such as no one can drive an SUV) rather than equality of price (in which case only some can afford to pay the tax). There may, therefore, be a trade-off between the efficiency and autonomy values of price-based instruments and the values (including fairness) that underlie support for climate policy.[20]

Institutions are also important to building or maintaining values about climate change. Discussion of climate policy tends to view individuals as in a sense independent of the issue of policy – or, as Sen has noted in the context of development, 'primarily as passive recipients of the benefits of cunning' programs.[21] Sen argues that individuals should be seen as agents (that is, capable of action and choice) rather than as 'patients' to be acted upon. Such agency points to the need for considering how decisions are made about policies. How decisions are made will aid in determining both how effective policies will be and what type of policies will be supported. The current policy process tends towards either very general public consultation or notice-and-comment processes, neither of which provides significant opportunities for the development of shared values.[22] It is difficult to determine precisely the contours of an appropriate institutional arrangement for development of climate policies when the need for shared values is taken into account. Some argue that the issue is essentially too difficult for most individuals to make rational decisions and that there is a need for the insulation of public policy makers from the irrationalities of public opinion.[23] Moreover, public consultation can be expensive and time consuming. However, building public values must be a central part in climate policies if they are to be 'sustainable' in the sense of policies that can survive over time.

4. It's Not Just the Price

Instruments that set a price for greenhouse gas emissions will be central to making progress on climate change. However, to develop more effective, and sustainable, policies for Canada, there is a need for a fuller account both of institutional constraints and of how individuals actually make consumer and political choices. In particular, political will and individuals' values must be taken into account. It has been argued that it will take a crisis to make individuals actually care about climate change and to provide the basis for strong policies.[24] However, to state the obvious – by then it may be too late. Canadian governments must base policy decisions on a fuller view of the Canadian context to make progress on climate change.

Notes

1 See, for example, N. Stern, *The Economics of Climate Change: The Stern Review* (Cambridge: Cambridge University Press, 2006) (arguing that developed countries should reduce greenhouse gas emission by 60–80 per cent by 2050 in order to stabilize greenhouse gas concentrations in the environment).
2 See, for example, J. Simpson, M. Jaccard, and N. Rivers, *Hot Air: Meeting Canada's Climate Change Challenge* (Toronto: McClelland and Stewart, 2007) and M. Jaccard and N. Rivers, 'Canadian Policies for Deep Greenhouse Gas Reductions' (Draft Chapter for IRPP Canadian Priorities Agenda Project, Institute for Research in Public Policy, May 2007).
3 Simpson, Jaccard, and Rivers *supra* note 2. See, more generally, Stern, *supra* note 1.
4 David Driesen, 'Renewable Energy under the Kyoto Protocol: The Case for Mixing Instruments,' chapter 10 of this volume.
5 *R. v. Hydro Quebec*, [1997] 3 S.C.R. 213.
6 The minority in *Hydro-Quebec* strongly dissented from the view that the toxic substances provisions constituted a prohibition given their broad nature.
7 *R. v. Crown Zellerbach Canada Ltd.*, [1988] 1 S.C.R. 401.
8 Ibid. at para. 18.
9 See, for example, P. Barton, 'Economic Instruments and the Kyoto Protocol: Can Parliament Implement Emissions Trading without Provincial Cooperation?' (2002) 40(2) Alberta Law Review 417 (arguing that each of the federal law power, the POGG power, and the trade and commerce power have difficulties as the basis for a federal cap-and-trade program, although POGG seems the strongest); and N.D. Bankes and A.R. Lucas, 'Kyoto, Constitutional Law and Alberta's Proposals' (2004–5) 42 Alberta Law Review 355.
10 See J. Weiner, 'Think Globally, Act Globally: The Limits of Local Climate Policies' (2007) 155 U. Penn. Law Review 1961.
11 Simpson, Jaccard, and Rivers *supra* note 3.
12 See M. Vandenburgh and A. Steinemann, 'The Carbon Neutral Individual'

(2007) 82 New York University Law Review (forthcoming) (arguing that emissions by individuals constitute approximately 30 per cent of U.S. greenhouse gas emissions). Similarly the Canadian government estimates such individual choices account for approximately 28 per cent of greenhouse gas emissions in Canada. Canada, *Project Green: Moving Forward on Climate Change* (April 2005) online, www.climatechange.ca. It is difficult to separate out individuals' 'direct' and 'indirect' choices and their impacts on the environment (see, for example, M. Vandenbergh, 'From Smokestack to SUV: The Individual as Regulated Entity in the New Era of Environmental Law' (2004) 57 Vanderbilt L.R. 515 at 537–40 (defining individual emissions)).

13 See, for example, A. Sen, *Rationality and Freedom* (Cambridge, MA: Harvard University Press, 2002) at 289 (arguing that the values of individuals are important for addressing environmental issues both for their influence on individual choices and for creating change politically); A. Green, 'Creating Environmentalists: Environmental Law, Identity and Commitment' (2006) 17 Journal of Environmental Law and Practice 1 at 5; and Vandenburgh, *supra* note 12 at 1107.

14 See A. Green, 'Self-Control, Individual Choice and Climate Change' (forthcoming) *Virginia Environmental Law Review* (discussing how limits on individuals' will (such as in the case of addiction or procrastination) impact the choice of instruments for addressing climate change).

15 For a discussion on limits on rationality, see, for example, C. Sunstein, *Laws of Fear: Beyond the Precautionary Principle* (Cambridge: Cambridge University Press, 2005).

16 For a discussion of the impact of values on the effectiveness of instruments, see A. Green, 'You Can't Pay Them Enough: Subsidies, Environmental Law and Social Norms' (2006) 30(2) Harvard Environmental Law Review 407 and Green, *supra* note 13.

17 For example, Simpson, Jaccard, and Rivers *supra* note 2 at 204 ('Polluting behavior must have a price, not in moral opprobrium but in financial terms'). See also Green, *supra* note 16 for a discussion of the relationship between values or norms about the environment (both internally and externally enforced) and consumer choices.

18 K. Harrison, *Passing the Buck: Federalism and Canadian Environmental Policy* (Vancouver: University of British Columbia Press, 1996).

19 Green, *supra* note 16.

20 Interestingly, the UK government climate plan contains a reference to removing barriers to behaviour change including that individuals fail to take action because of lack of information or because they put off action.

See Department of Environment, Food and Rural Affairs, 'UK Climate Change Programme: Annual Report to Parliament' (July 2007) online, www.defra.gov.uk/ENVIRONMENT/climatechange/uk/ukccp/index.htm.
21 A. Sen, *Development as Freedom* (New York: Anchor Books, 1999) at 11.
22 Green, *supra* note 13.
23 Sunstein, *supra* note 15.
24 See, for example, J.G. Speth, *Red Sky at Morning* (New Haven: Yale University Press, 2004).

PART FIVE

Canada's Energy Policy

13 Climate Change and Canadian Energy Policy

MARK S. WINFIELD WITH CLARE DEMERSE AND
JOHANNE WHITMORE

Introduction – Canada's De Facto Energy Policy

Energy and climate change policy are intimately connected. While Canada has gone through numerous articulations of its policies related to climate change over the past two decades, Canada has no formally articulated national or federal energy policy. However, a considerable de facto federal energy policy framework has emerged since the end of the National Energy Policy in the early 1980s. This de facto energy policy framework is strongly oriented towards the development and export of conventional, non-renewable energy resources, such as coal, oil, natural gas, and uranium. The framework consists of a number of specific elements.

Non-renewable energy exploration and development activities receive extensive fiscal support from the federal government. Although direct subsidies or equity investments in energy projects by Canadian governments have become less common, extensive support is provided through the federal tax system. The most recent available estimates of this support to the oil and gas sector, for example, conservatively estimate its value in the range of $1.4 billion per year.[1] Additional supports are provided to other non-renewable energy sources, including an operating subsidy of approximately $100 million per year to Atomic Energy of Canada Ltd (AECL). Substantial export development assistance for foreign sales of nuclear reactors has been provided to AECL as well.[2]

In addition, with a few exceptions where aboriginal interests are involved, federal environmental assessment or other environmental approval processes have been applied as weakly as possible to major energy projects.[3] A major-projects office was established within Natural

Resources Canada in October 2007, with a specific mandate to facilitate federal project approvals.[4] The policy role of the National Energy Board, for its part, has been significantly reduced. The board's focus is now on the facilitation of energy exports and non-interference with energy markets.[5] Further, the overall conventional resource export orientation of Canadian energy policy was strongly embedded into the 1994 North American Free Trade Agreement.[6]

In comparison to the extensive policy infrastructure in place to support the development and export of conventional, non-renewable energy resources, the policy frameworks related to energy efficiency and low-impact renewable energy sources, such as wind and solar energy, remain weak and largely symbolic. Natural Resources Canada's Office of Energy Efficiency,[7] itself a survivor of the original 1980 National Energy Policy, provides some public information programs, such as the EnerGuide labelling program, and funding for the EcoENERGY retrofit program (a revised version of the EnerGuide for Homes program). Although the office has authority over the establishment of energy efficiency standards under the federal *Energy Efficiency Act*, it is largely focused on the provision of leadership in intergovernmental forums that deal with energy efficiency issues, particularly the Canadian Council of Energy Ministers. In this context, the office relies strongly on provincial and territorial governments to actually implement energy efficiency standards and other initiatives.[8]

The Wind Power Production Incentive (WPPI) was introduced by the federal government in 2002, providing an incentive payment of one cent per kilowatt-hour (kWh) for the first 10 years of operation of eligible wind-power projects. Total expenditures under the program were to be $250 million over five years.[9] In April 2007 the program was renamed ecoENERGY for Renewable Power, and expanded to cover other renewable energy sources, with a commitment of up to $1.48 billion over the period 2005–2011.[10] The program is the federal government's principal initiative to promote renewable energy.

With respect to climate change policy per se, an explicit withdrawal from the *Kyoto Protocol* has been ruled out by the federal minister of the environment, John Baird.[11] However, the government will make no attempt to reach the *Kyoto Protocol*'s 2008–2012 target, and will not pursue the purchase of Clean Development Mechanism credits. Rather the federal government's climate change policy is clearly to back away from Canada's commitment under the *Kyoto Protocol* or any short-term targets for reductions in greenhouse gas (GHG) emissions.[12] The federal government's April 2007 Regulatory Framework for Air Emissions[13] has been widely criticized as being unlikely to result in significant near-term reductions in GHG emissions.[14]

Sub-National Governments and Energy Policy

In this context of the continuation of the orientation of federal energy policy towards conventional non-renewable source development and exploitation, perhaps the most interesting activities on energy and climate change policy in Canada are occurring at provincial level. The emerging situation in Canada parallels developments in the United States where, in the absence of significant federal action on climate change issues, states and local governments have become the key sources of energy policy innovation.[15] California, in particular, has emerged as a major leader on state-level climate change initiatives. These developments at the state and provincial levels are strong reminders of what students of federalism tell us about the role of sub-national governments as alternative forums for policy development and innovation.[16]

In the Canadian context, the potential scope for provincial action on energy policy and climate change is very broad. Canadian provinces have the potential to exercise substantial influence in a number of areas that have been identified as major sources of GHG emissions. The provinces, for example, exercise effective control over energy policy, particularly electricity and non-renewable resource development. Electricity production and oil and gas production and distribution accounted for more than 36 per cent of Canada's GHG emissions in 2005.[17] In addition, passenger cars and trucks accounted for 10 per cent of Canada's 2005 GHG emissions.[18] Provincial jurisdiction over land use and responsibility for transportation policy at the local and regional levels provides extensive opportunities to influence urban form and, by implication, transportation patterns. Increased spending on public transit, and the reorientation of land-use planning policies towards the redevelopment of existing urban areas and the mixing of land uses may result in reductions in automobile use, with the implication of reductions in transportation-related GHG emissions.

Agricultural activities are the source of 7.6 per cent of Canada's emissions GHG emission.[19] The provinces again have considerable potential to influence behaviour in the sector through fiscal tools, outreach and education initiatives, and land-use policies.

Finally, the provinces have primary responsibility for waste management policy and the regulation of the operation of waste disposal facilities. Emissions from landfills, principally methane produced by decaying wastes, contributed 3.7 per cent of Canada's 2005 emissions.[20] Strategies to reduce the landfilling of organic wastes, and to capture and use the methane gas produced by landfills as fuel, could reduce GHG emissions from the sector significantly.

Provincial Climate Change Initiatives

The Pembina Institute completed a survey of key provincial initiatives on climate change in August 2007.[21] The survey found that most of the provinces studied have articulated some sort of GHG reduction targets. However, with the exception of Quebec, there is no clear indication of how those targets will be achieved. A range of policy initiatives have been announced, but very few detailed explanations of how these initiatives will result in the targeted reductions in GHG emissions have been provided. Indeed, given the relatively minor nature of many of the announced initiatives, it seems unlikely that they will result in major reductions in GHG emissions.

Perhaps the most significant of the provincial initiatives to date has been the introduction of a modest carbon tax in Quebec.[22] A similar proposal is under consideration in British Columbia. In addition, some provinces, particularly Ontario and British Columbia, have initiated discussions with U.S. states that have launched regional GHG emission cap-and-trade systems, with British Columbia joining the Western Climate Initiative in April 2007.[23] In Ontario, the province has committed to a 2014 regulatory deadline for the phase-out of coal-fired electricity generation. Although largely driven by air quality concerns, a coal phase-out would account for nearly half of the province's GHG reduction target of a 6 per cent reduction in emissions relative to 1990 levels by 2014.[24]

On the whole, the actual measures adopted to date at the provincial level have largely been fiscal incentives and investments in research, supplemented by some public awareness and education initiatives. The fiscal initiatives are typically relatively small scale, and often consist of short-term or one-time-only expenditures.

Some provinces have put in place increasingly substantial financial incentives around the development of renewable energy, although in many cases this has been more a function of overall energy policy than specific GHG reduction strategies. Ontario's standard offer contract programs for renewable energy[25] and cogeneration[26] are particularly noteworthy in this regard, although it is unclear if these initiatives are of a sufficiently ambitious scale to fundamentally change the province's existing dependence on large scale, centralized non-renewable energy sources.

A number of provinces, including British Columbia, Quebec, Ontario, and New Brunswick, have announced regulatory initiatives to strengthen the energy efficiency provisions of their building codes and energy efficiency performance standards for goods and appliances. However, these programs have been presented as one-time initiatives. Provincial governments have not committed to the regular and ongoing upgrading of energy efficiency standards and codes. This is despite the fact that many U.S. states, including California, identify the regular upgrading of codes and standards as being the crucial factor in their success to date in reducing energy consumption.[27]

Over the past four years, Ontario has undertaken an extensive redrafting of land-use planning legislation and significantly increased its expenditures on public transit. The intention behind these changes is to reduce urban sprawl and automobile dependency, particularly within the Greater Golden Horseshoe Region.[28] Although these initiatives have been largely undertaken for growth management purposes, rather than GHG emission reductions, if successful they may result in significant reductions in transportation-related GHG emissions. A number of provinces, including British Columbia, Quebec, New Brunswick, and Nova Scotia, have indicated support for California's proposed vehicle emission standards for GHGs. Ontario, the centre of the automobile industry in Canada, has pointedly declined to support the California initiative.

Finally, a number of provinces have adopted regulatory requirements regarding the capture and combustion of the methane gas produced by decaying waste in landfills.

Conclusion

Although the emergence of provincial GHG reduction strategies is an important development, it is far from clear that provincial initiatives can replace an effective federal strategy for GHG reductions. Most, but not all provinces have articulated GHG reduction target. In fact, in some cases, like Quebec and Manitoba, these provincial targets come close to being consistent with Canada's *Kyoto Protocol* target of a six per cent reduction in GHG emissions relative to 1990 by 2008–2012. Unfortunately, none of the provinces have articulated, to date, any real plans for meeting the targets they have established, although Quebec's efforts come closest to a comprehensive plan in this regard.

Perhaps the most significant gap at the provincial level is the issue of the reduction of emissions from the large final emitters (LFEs). These major industrial sources, ranging from coal fired power plants to pulp and paper mills, account for just under 50 per cent of Canada's GHG emissions.[29] The Regional Greenhouse Gas Initiative (RGGI), a Northeastern U.S. state-level GHG emission cap-and-trade initiative being considered by Ontario, is focused exclusively on the electricity sector, although the Western Climate Initiative – in which British Columbia and Manitoba participate – may be broader in scope. Alberta has adopted legislation for the regulation of GHG emissions intensity from LFEs. The Alberta initiative has been widely criticized, however, as being likely to result in continuing growth in absolute GHG emissions.[30]

In addition to the major gap around the reduction of emissions from LFEs, consideration also has to be given to the inertial effects of the existing de facto energy policy framework at the federal level. As long as the federal government maintains an extensive policy infrastructure that supports and promotes the development of conventional non-renewable energy sources, it will be difficult, if not impossible, to re-orient Canada's energy production and consumption patterns in a more sustainable direction.

Rather, the achievement of significant reductions in Canada's GHG emissions will require significant changes in the direction of long-standing policies by the federal government. In particular, the federal government needs to establish an effective regulatory framework for reducing GHG emissions from LFEs. Perhaps even more importantly, the federal government needs to develop an overall strategy to re-orient Canada's energy path away from conventional non-renewable energy development and export and towards greater energy efficiency and reliance on low-impact renewable energy sources.

Highlights of Provincial Greenhouse Gas Reduction Plans: August 2007

	British Columbia	Alberta	Saskatchewan
Climate Change Plan	– 2004: *Weather, Climate and the Future – BC's Plan* (BC 2004) online, http://www.env.gov.bc.ca/air/climate/cc_plan/pdfs/bc_climatechange_plan.pdf; – BC's 2004 CC plan has been superseded by commitments made in the February 2007 Throne Speech (SFT); http://www.leg.bc.ca/38th3rd/Throne_Speech_2007.pdf – BC is currently developing a new climate plan, to be released 'soon,' which will 'aspire to meet or beat the best practice in North America' for reducing GHGs (SFT at 14–15); – BC also released an energy plan on 27 February 2007, entitled *The BC Energy Plan: A Vision for Clean Energy Leadership* (Energy Plan), online, http://www.energyplan.gov.bc.ca/PDF/BC_Energy_Plan.pdf.	– October 2002: *Albertans and Climate Change: Taking Action* (AB 2002) online, http://environment.gov.ab.ca/info/library/6123.pdf – Alberta is currently developing a new five-year plan. The draft plan is scheduled for release in late summer 2007, with the final version expected by late fall 2007.[31]	June 2007: Saskatchewan *Energy and Climate Change Plan* (Sask 2007) online, http://www.saskatchewan.ca/green.
Targets for Provincial GHG Emissions	– 10% below 1990 by 2020 (SFT at 14) – Commitment to develop 2012, 2016, and 2050 targets (SFT at 14, 16)	– 50% reduction in emissions intensity below 1990 by 2020, equivalent to a 20–35% increase in absolute emissions relative to 1990 levels (AB 2002 at 10).	– 'Stabilize the absolute level of greenhouse gas emissions by 2010' (Sask 2007 at 4; the plan does not specify at what level emissions will be stabilized).

Highlights of Provincial Greenhouse Gas Reduction Plans: August 2007 *(Continued)*

	British Columbia	Alberta	Saskatchewan
		– 'Alberta recognizes that more significant emission reductions will be required over the longer term (2050)' (AB 2002 at 12).	– 32% below 2006 by 2020 (Sask 2007 at 4). – 80% below 2006 by 2050 (Sask 2007 at 4).
Do Measures Add up to Target?	No. Although BC has announced a 2020 target and several associated measures, it does not yet have a climate change plan. The Government's Speech From the Throne refers to a 'soon-to-be released' climate change plan; in the interim, BC does not have a full list of measures that correspond to its targets.	No. Measures in the plan have no emission reduction targets.	The plan provides a graph representing emission reduction wedges that add up to the government's 2050 target (Sask 2007 at 8), but the wedges represent technologies, not measures. Most measures in the plan have no emission reduction targets or estimates.
Extent of Funding for Measures?	As noted above, BC has made some notable initial commitments, but does not yet have a full climate change plan nor the allocation of funds to support such a plan. The government has made only modest initial funding commitments, notably: – $4 million for 2007–08 to support actions to reduce GHGs and improve the assessment of the	Alberta has funded various initiatives under its 2002 climate change plan (see partial list, below) and is operating an intensity-based regulatory system for heavy industry. As noted above, Alberta is in the process of developing a new climate plan; the level of funding that plan receives will be an important test of its credibility. – $3.6 million for climate change for	The majority of initiatives in Saskatchewan's plan are not tied to specific funding; the plan's action items consist mainly of targets, announcements of processes, and initiatives or commitments without specific dollar figures attached. The plan does not lay out funding requirements. When the plan was released, the news release noted that 'The 2007–08 Budget contains $48 million to support

Highlights of Provincial Greenhouse Gas Reduction Plans: August 2007 *(Continued)*

British Columbia	Alberta	Saskatchewan
impacts of climate change on BC (Budget 2007[32] at 35) – $10 million over 2006–2010 for the purchase of hybrid vehicles as part of the government's fleet (Budget 2007 at 35) – Commitment to establish a $25 million Innovation Clean Energy Fund 'to encourage the commercialization of alternative energy solutions and new solutions for clean remote energy' (SFT at 18). The government has subsequently announced a plan to raise money for the Fund through a '0.4 per cent levy on sales of electricity, natural gas, grid propane and fuel oil that are non-transportation related, with a cap of $500,000 per year for high-use energy customers. (For an average home, the government estimates that the levy will cost about $8 per year.) The levy is expected to be implemented in the summer of 2007.[33]	2006–07: $3.688 million estimate for 2007–08; $3.744 million budgeted for 2008–09; $3.804 million budgeted for 2009–10[34] – Budget 2007:[35] $18 million in Energy Innovation Fund Initiatives to be invested in R&D focused on energy supply and protection of the environment. – Budget 2007: $41 million for the Bio-Fuel Initiative in 2007–08 (see 'Transportation' section below). – $25 million monitoring and evaluation project on the long-term reliability of storing CO_2 in geological formations (AB 2002 at 27).[36] – Up to $200 million in royalty relief for enhanced oil recovery-type initiatives from 2007–2011 (AB 2002 at 26).[37] – $30 million in interest-free loans to 60 municipalities for energy-efficiency projects between 2003 and 2006.[38]	various climate change initiatives. Crown Corporations will spend an additional $49 million. The premier announced today an additional $44.4 million over three years to fund emission reduction initiatives similar to projects that will receive support under the federal trust fund for clean air and energy efficiency projects.'[39] Some of the climate change funding contained in Budget 2007–08[40] includes: – '$7.5 million will be spent on green and climate change initiatives' (Budget 2007 at .78) – 'In 2007–08, the budget for the ethanol fuel tax rebate will increase by $3.3 million to $21 million as industry production expands.' (Budget 2007 at. 79)

Highlights of Provincial Greenhouse Gas Reduction Plans: August 2007 (*Continued*)

	British Columbia	Alberta	Saskatchewan
Measures			
Electricity	– BC to be 'electricity self-sufficient by 2016' (SFT at 16). – Zero-GHG electricity production: 'All new and existing electricity produced in BC will be required to have net zero greenhouse gas emissions by 2016' (SFT at 17). – 'Ensure that clean or renewable energy generation continues to account for at least 90 per cent of total generation' (Energy Plan at 13). – Requirement for 100% carbon sequestration for any coal-fired power project (SFT at 17). – Commitment to develop a cap-and-trade system with U.S. states under the Western Regional Climate Action Initiative, to be in force by 2012 (SFT at 18–19).[41]	– As of 1 July 2007, Alberta has regulations in effect that set a 12% intensity reduction target for all existing large industrial facilities emitting over 100,000 tonnes CO_2e. New facilities (those in operation after 2000) have a three-year grace period before being subject to regulated targets. The government expects about 12Mt of annual industrial emissions to be subject to the new regulations,[42] but companies can meet their targets by making payments of $15/tonne into a Climate Change and Emissions Management Fund as well as through on-site reductions or purchasing offset credits from within the province.	– All SaskPower's new and replacement electricity generation facilities to be carbon neutral (Sask 2007 at 4). – Develop demand-side management (DSM) practices to reduce consumer demand for SaskPower's electricity by 300MW by 2017, relative to business-as-usual levels. The plan states that Saskatchewan will use DSM 'as an alternative to the construction of new facilities' (Sask 2007 at 10). – Expand the eligibility for Saskatchewan's Investment Tax Credit for Manufacturing and Processing to certain types of renewable energy and energy conservation equipment used to generate electricity (Sask 2007 at 10).
Other Industry	– Commitment to develop a cap-and-trade system with U.S. states under the Western Regional Climate Action Initiative (see 'Electricity' section above). – Commitment to reduce emissions	– Same as 'Electricity' section above	– Before the end of 2008, work with the oil and gas industry to prepare recommendations on reducing emissions from venting and flaring along with fugitive emissions (Sask 2007 at 14).

Highlights of Provincial Greenhouse Gas Reduction Plans: August 2007 (*Continued*)

	British Columbia	Alberta	Saskatchewan
	from the oil and gas sector to 2000 levels by 2016 (SFT at 17). – Proposed requirement for zero flaring at producing wells and production facilities (SFT at 17). – Beehive burners to be eliminated (SFT at 18; no timeline).		– Invest up to $20 million in 'pipeline expansions,' and up to $12 million with industry 'to participate in the development of flare gas processing opportunities' (Sask 2007 at 14). – Establish a Saskatchewan Technology Fund to 'receive voluntary payments from Saskatchewan industry as a method of complying' with the federal GHG regulations (Sask 2007 at 18).
Transportation	– Commitment to phase in California's vehicle emission standards between 2009 and 2016; this is expected to reduce CO_2 emissions from automobiles 'by some 30%' (SFT at 21; the government does not provide the year by which 30% reduction will occur, nor the baseline level it is using for the 30% comparison). – Commitment to establish a low-carbon fuel standard which is expected to 'reduce the carbon intensity of all passenger vehicles by at least 10% by 2020' (SFT at 21; the government	– In Budget 2007, the Government of Alberta increased its funding for biofuel initiatives to $41 million in 2007–08, up from $5 million in the 2006–07 forecast. This funding will be invested in 'bio-energy development projects and initiatives, including biofuel commercialization and marketing, infrastructure development and producer credits.'[43] Alberta does not have its own biofuel requirement, choosing to rely on a national regulatory approach.[44]	– Saskatchewan already mandates ethanol blending in gasoline; their climate plan calls for an increase by 2010 (the level of the increase is not specified) from the existing average blend of 7.5% ethanol (Sask 2007 at 12–13). – Develop a 1.4 billion litre biofuel industry (Sask 2007 at 13; the plan does not give a timeline for reaching the 1.4 billion litre level, this presumably represents an annual production target). – Work with other governments on a

Highlights of Provincial Greenhouse Gas Reduction Plans: August 2007 (*Continued*)

	British Columbia	Alberta	Saskatchewan
	does not provide the baseline level it is using for the 10% comparison). – $2000 sales tax exemption on new hybrid vehicles to be extended (SFT at 21). – $1.9 billion public-private partnership in building the rapid-transit Canada Line (linking downtown Vancouver, Richmond, and the Vancouver airport) by 2010; the government estimates that the project will reduce 'net GHG emissions by up to 14,000 tonnes by 2021' (SFT at 20). – Federal-provincial partnership to invest $89 million in fuelling stations and 'the world's first fleet of 20 fuel cell buses' (SFT at 19). – $40 million LocalMotion Fund to help local governments build walkways, cycling paths, and disability access.		nation-wide E85 corridor and with industry on the development of E85 corridors in Saskatchewan (Sask 2007 at 13).
Buildings	– New BC Green Building Code to be developed 'over the next year' (SFT at 22). – Commitment to update minimum energy efficiency standards for	– Commitment to incorporate an energy efficiency requirement into Alberta's Building Code (AB 2002 at 32).	– Over the next two years, implement new province-wide energy efficiency building standards in consultation with industry (Sask 2007 at 11).

Highlights of Provincial Greenhouse Gas Reduction Plans: August 2007 (Continued)

	British Columbia	Alberta	Saskatchewan
	equipment (BC 2004 at 19–20). – Commitment to review energy performance standards for houses in the BC Building Code (BC 2004 at 20).		– Invest $28 million to extend Sask EnerGuide for Houses program to 2011 (Sask 2007 at 20). – Extend Home Energy Improvement Program, whose mandate is to reduce energy use in low and moderate income households, to 2011 (Sask 2007 at 20). – Train specialized trades people for energy efficiency retrofits, solar and wind installation and biomass applications (SK 2007 at 19).
Other sectors (agriculture, forest management, landfills, government operations, etc.)	– Provide support to the BC Agriculture Council to implement best management practices on farms and ranches (BC 2004 at 27). – BC has made a bilateral agricultural agreement with the federal government to reduce GHG emissions from agricultural operations 'to a target of 2.4 Mt by 2008,' using a range of management activities (BC 2004 at 27). – Commitment to develop a policy framework to support the creation of	– Commitment to reduce GHGs from government operations by 26% below 1990 levels by 2005 (AB 2002 at 21). – As of 2005, more than 90% of the electricity used in government-owned facilities in Alberta comes from green power sources. Alberta Infrastructure signed purchase agreements in 2003 for approximately 210,000 MWh annually, split equally between ENMAX and Canadian Hydro. The power comes	– Encourage farmers to establish agricultural soil sinks to remove 25Mt CO_2/year by 2012, 37 Mt CO_2/year by 2050 (Sask 2007 at 13). – Develop the management practices and technology to make a 20% reduction in agricultural emissions intensity, per animal of livestock production and per acre of crop production, by 2030 (Sask 2007 at 14). – Reforest 20,000 hectares of 'not sufficiently regenerated land' by 2017. The government estimates that this

Highlights of Provincial Greenhouse Gas Reduction Plans: August 2007 (*Continued*)

British Columbia	Alberta	Saskatchewan
incremental forestry sinks (BC 2004 at 25). – Commitment to develop legislation 'over the next year' to 'phase in new requirements for methane capture' from landfills; the speech suggests using the landfill gas for clean energy (SFT at 18). – Commitment to making the Government of BC carbon neutral by 2010 (SFT at 15). – Beginning in February 2007, all new cars leased or purchased by the BC government to be hybrids (SFT at 21).	from a wind farm at McBride Lake and from biomass combustion at a new facility in Grande Prairie.[45]	will sequester about 4.9 Mt of CO_2 'over the life of the plantation' (Sask 2007 at 13). – Use green power to meet 36% of total electricity requirements for government buildings by 2008; 50% by 2009; and 90% by 2010 (Sask 2007 at 20). – Reduce energy consumption of the government's core building by 20% through the replacement of building components (Sask 2007 at 10). – Implement a 'Government and Crown vehicle purchase policy' that requires all vehicles to be either hybrid-electric, alternative/flex-fuel or within top 20% efficiency in their class (Sask 2007 at 10). – Establish a voluntary, provincially certified Emission Offset Fund to allow organisations or the public to offset their emissions by supporting emission reduction initiatives in Saskatchewan (Sask 2007 at 18).

Highlights of Provincial Greenhouse Gas Reduction Plans: August 2007 *(Continued)*

	British Columbia	Alberta	Saskatchewan
Further Information			
Legislation	*Energy Efficiency Act*, online, http://www.em.gov.bc.ca/AlternativeEnergy/EnergyEfficiency/Energy_Efficiency_Act.htm.	– *Climate Change and Emissions Management Act*, online, http://www.qp.gov.ab.ca/documents/Acts/C16P7.cfm?frm_isbn=9780779723386. – *Specified Gas Emitters Regulation*, online, http://www3.gov.ab.ca/env/air/pubs/Specified_Gas_Emitters_Regulation.pdf.	
Other		Accomplishments to date according to Alberta Environment, online, http://www3.gov.ab.ca/env/climate/accomplishments.html	
Contact	Karen Campbell, Pembina Institute, 604-874-8558, ext. 225 Ian Bruce, David Suzuki Foundation, 604-306-5095	Jaisel Vadgama, Pembina Institute, 403-807-6566 Nashina Shariff, Toxics Watch Society of Alberta, 780-915-8946	Ann Coxworth, Saskatchewan Environmental Society, 306-665-1915

Highlights of Provincial Greenhouse Gas Reduction Plans: August 2007

	Ontario	Québec	New Brunswick
Climate Change Plan	– Yet to be released. However, the government has made a series of climate change announcements (all of which are available on the premier of Ontario's website, http://www.premier.gov.on.ca/home/default.asp?lang=EN).	– June 2006: *Quebec and Climate Change - A Challenge for the Future, 2006–2012 Action Plan* (QC 2006) online http://www.mddep.gouv.qc.ca/changements/plan_action/2006-2012_en.pdf – June 2007: *Quebec and Climate Change - A Challenge for the Future, 2006–2012 Action Plan – First Year Results*, online, http://www.mddep.gouv.qc.ca/changements/plan_action/bilan1-en.pdf	June 2007: *New Brunswick Climate Change Action Plan – 2007–2012* (NB 2007) online, http://www.gnb.ca/0009/0369/0015/0001–e.asp.
Target for Provincial GHG Emissions	– 6% below 1990 by 2014[46] – 15% below 1990 by 2020[47] – 80% below 1990 by 2050[48]	– 6% below 1990 by 2012 (QC 2006 at 14). – There is a growing scientific and government consensus that global warming of more than 2°C constitutes 'dangerous' climate change which must be prevented. Québec 'welcomes the idea of limiting warming to under the 2°C threshold,' noting that 'the threshold is already likely too high' for northern latitudes (QC 2006 at 9).	– 1990 levels by 2012 (5.5 Mt reduction in annual emissions below business-as-usual in 2012) (NB 2007 at 1). – 10% below 1990 by 2020 (NB 2007 at 1)

Highlights of Provincial Greenhouse Gas Reduction Plans: August 2007 *(Continued)*

	Ontario	Québec	New Brunswick
		The measures in Québec's plan are estimated to produce a reduction in annual emissions of 10 Mt CO_2e below business as usual in 2012, which would get the province to 1.5% below the 1990 level in that year. To reach the 'Kyoto level' of 6% below 1990[49] (13.8 Mt below business as usual), Québec says it will need funding from the Government of Canada for the final 3.8 Mt (QC 2006 at 3).	
Do Measures Add up to Target?	The government has broken down the emission reductions that each relevant sector is expected to contribute to the achievement of its 2014 and 2020 targets.[50] However, the policies that will lead to the targeted reductions in each sector have yet to be identified in most cases. Although the Government of Ontario has made a series of announcements about climate change policy measures, it has yet to tie these together into a cohesive plan that demonstrates how the province can achieve its targets.	In its 2006 plan, Québec provides a table that shows the cost of each measure and the anticipated GHG reductions from each (QC 2006, Appendix 1). Québec estimates that the 24 measures in its 2006 climate plan will lead to reductions to 1.5% below 1990 emission levels by 2012. To reach its target of 6% below 1990 levels by 2012, Québec asked the federal government for additional funding. In February 2007, the federal government committed $349.9 million to	– New Brunswick's climate plan breaks down the emission reductions that each sector is expected to contribute to the achievement of its 2012 target. In addition, the plan proposes to create a Climate Change Secretariat in the Department of the Environment to track and report on the implementation of the plan (NB 2007 at 32). Sectoral targets are as follows (all for reductions in annual emissions below business-as-usual in 2012) (NB 2007 at 11):

Highlights of Provincial Greenhouse Gas Reduction Plans: August 2007 *(Continued)*

	Ontario	Québec	New Brunswick
		climate change action in Québec through a federal-provincial trust fund. According to the federal press release, as a result of this funding, 'the Government of Quebec has indicated that it will be able to reduce greenhouse gas emissions by 13.8 million tonnes of carbon dioxide or equivalent below its anticipated 2012 level.'[51] Some of the provincial projects which may receive funding from the federal initiative include (among others): – investments to improve access to new technologies for the trucking sector; – a program to develop renewable energy sources in rural regions; – a pilot plant for production of cellulosic ethanol; and – the promotion of geothermal heat pumps in the residential sector.[52]	– 2.2 Mt from energy efficiency and renewable energy measures; – 1.2 Mt from transportation measures; – 1.2 Mt from waste management measures; – 0.7 Mt from Industrial sources (this reduction depends on the implementation of federal policy, as New Brunswick does not plan an independent regulatory policy for GHG emissions from industry); – 0.2 Mt from 'other' actions, including 'Government leading by example' and 'Partnerships and Communications.'
Extent of Funding for Measures	Ontario has dedicated funding to renewable energy, energy efficiency,	– Québec plans to allocate $200 million per year to a Green Fund to	New Brunswick's Climate Change Action Plan does not state how much

Highlights of Provincial Greenhouse Gas Reduction Plans: August 2007 (*Continued*)

Ontario	Québec	New Brunswick
green R&D (tied to the auto sector), ethanol production and public transit capital spending. However, as the province has yet to produce a comprehensive climate change plan (see above), the extent of funding to support that plan remains to be seen. Examples of Ontario climate-related funding to date: – Ontario received $338 million for climate change through the 2005 Canada-Ontario Agreement (funds are allocated for the years 2006–07 to 2008–09).[53] – $650 million Next Generation Jobs Fund over five years, starting from 2007.[54] – $150 million Home Retrofit Program And Solar Initiatives over five years, starting from 2007.[55] – $220 million Municipal Eco Challenge Fund for investments in GHG-reduction projects and infrastructure ($20 million devoted to grants, and $200 million to loans).[56] – $17.5 billion rapid transit action plan	implement its climate change plan, for a total of $1.2 billion (QC 2006 at 3). The money for the Green Fund will come from a modest levy on hydrocarbons, which will be calculated on the basis of CO_2 equivalents for each form of energy' (QC 2006 at 28). – Additional funding of $350 million (one-time allocation) from the federal trust fund (see above).	funding will be needed to implement it. The Plan itself contains very few spending commitments (most of the actions described are targets, strategies, programs and commitments). In the 2007–08 Budget, the main new spending related to climate change was a commitment to provide 'additional funding of $15.3 million ... to the New Brunswick Energy Efficiency and Conservation Agency for comprehensive home energy conservation initiatives, along with new programs for the commercial and industrial sectors.'[59] New Brunswick's plan was released after the 2007–08 budget. A key test of the plan's credibility will be whether sufficient funding is allocated to fulfilling its commitments by Budget 2008.

Highlights of Provincial Greenhouse Gas Reduction Plans: August 2007 *(Continued)*

	Ontario	Québec	New Brunswick
	for the Greater Toronto Area and Hamilton for a 12-year new transit capital building program starting in 2008; 52 transit projects have been identified as recipients for the funding. The project depends on about $6 billion in federal funding (35% of project costs).[57] – $520 million, 12-year Ethanol Growth Fund to help Ontario producers meet the province's Renewable Fuels Standard, which requires 'an average of five per cent ethanol in all gasoline sold in Ontario by January 1, 2007.'[58]		
Electricity	– Of the reductions in annual emissions below business-as-usual needed to reach the provincial 2014 target, 44% are projected to come from the electricity sector, notably through the phase-out of coal-fired electricity, the use of renewable energy and other electricity policies.[60] – Intention to ban sale of inefficient	– Commitment to negotiate voluntary agreements for GHG reductions in industrial sectors; expected reduction in annual emissions of 0.94 Mt below business-as-usual in 2012 (QC 2006 at 24, 40). – 2004 regulation banning CFCs and halons; expected reduction in annual emissions of 0.7 Mt in 2012 (QC 2006 at 24, 40)	– Use of demand-side management programs and the province's Renewable Portfolio Standard (see below) are expected to achieve a reduction of 2 Mt below 2003 emission levels by 2020 (NB 2007 at 20). – Energy efficiency grants and loans available for residential, commercial and industrial sectors through Efficiency NB, an agency whose man-

Highlights of Provincial Greenhouse Gas Reduction Plans: August 2007 (Continued)

Ontario	Québec	New Brunswick
light bulbs beginning in 2012.[61] – Suite of initiatives to encourage homeowners to increase energy efficiency via retrofits and to invest in renewable energy. The policy package includes incentives for home retrofits, retail sales tax exemptions for Energy Star products and renewable energy equipment, and several consumer information programs.[62] – Commitment to develop new renewable energy projects to meet 5% of Ontario's electricity capacity by 2007 (1350 MW) and 10% (2,700 MW) by 2010. Approved projects have included 'waterpower,' landfill gas, biogas, and wind.[63] – Standard offer contracts in place for renewable energy (including wind energy, small hydro, and small solar photovoltaic installation); standard offer under development for cogeneration. This program provides a favourable pricing regime and a streamlined qualifying process to help get small renewable energy electricity projects onto the grid in Ontario.[64]		date is to 'promote energy efficiency measures in the residential, community and business sectors of New Brunswick.'[66] – Targets for the thermal electricity sector: an absolute reduction of 25% from 2003 levels by 2020, and an anticipated 65% reduction in emissions (baseline not given) by 2050 (NB 2007 at 20). – Under the *Renewable Resources Regulation* of the *Electricity Act*, '10% of electricity sales must come from new renewable sources by 2016.' Under this standard, NB Power announced an expression of interest to provide 400 megawatts (MW) of renewable electricity generation,' and 96 MW for 2008 are already contracted for (NB 2007 at 14). – The province plans to refurbish the Point Lepreau nuclear power station (NB 2007 at 20). – Commitment to implement a forest biomass policy, and to study the feasibility of new small hydro, tidal, and other renewable opportunities (NB 2007 at 15).

Highlights of Provincial Greenhouse Gas Reduction Plans: August 2007 (Continued)

	Ontario	Québec	New Brunswick
	– Commitment to reduce peak electricity demand in Ontario by 5% by 2007. The government estimates that this measure will reduce annual emissions by 1Mt.[65]		
Other industry	– Amendments to Ontario's Refrigerants Regulations 'will phase out the use of chlorofluorocarbons (CFCs) in large refrigeration equipment and chillers, and ensure surplus stocks are properly handled.' CFCs are greenhouse gases that deplete the ozone layer.[67] – $14.4 million over four years to encourage facilities in the industrial/commercial sector to convert solar thermal heat. Under the program, facilities would receive 25% of the cost of the installation of a solar thermal heating system from the province, to a maximum of $80,000. The government estimates that this program will generate 500 installations over four years.[68] – $650 million, five-year Next Genera-	As noted above, Québec is in the process of imposing a modest levy on hydrocarbons, 'which will be calculated on the basis of CO_2 equivalents for each form of energy.' (QC 2006 at 28). Although the levies will help Québec to raise money for its GHG-reduction activities, the charge is likely too small to encourage consumers to cut their fuel use. (Notably, Québec does not attribute any direct GHG reductions to its carbon tax measure.)	'The federal government has indicated its intent to take a leadership role in regulating greenhouse gas emissions from industrial facilities' (NB 2007 at 5). Thus, New Brunswick plans to leave the field of industrial GHG emission regulation to the federal government, citing the advantages of 'Canada-wide' regulations that 'result in fair treatment of industry sectors' (NB 2007 at 19). New Brunswick has opted to 'work with the federal government to address industrial facilities and focus on emission reductions from operations through energy efficiency and fuel-switching actions' (NB 2007 at 5) and also to 'work with the federal government to ensure that forest management carbon offset credit opportunities' are 'fully recognized' under the federal system (NB 2007 at 19).

Highlights of Provincial Greenhouse Gas Reduction Plans: August 2007 *(Continued)*

	Ontario	Québec	New Brunswick
	tion Jobs Fund provides support to companies that build 'green cars and auto parts,' produce clean fuels, or work on the development of clean technologies and products.[69]		
Transportation	– Ontario will 'push for the development of a harmonized, continental approach to vehicle fuel-efficiency standards',[70] but has ruled out adopting California's tailpipe emissions standard.[71] – Signed a Memorandum of Understanding with California that will require producers to 'reduce carbon emissions from transportation fuels by 10 per cent by 2020' through a low-carbon fuel standard; the government estimates that this policy will spur emission reductions equivalent to 'removing 700,000 cars from the roads.'[72] – Funding for capital investments in public transit (see above). – Next Generation Jobs Fund (see above) will invest in green cars and clean fuels.	– Commitment to adopt California vehicle emission standards by 2010; expected reduction in annual emissions of 1.7 Mt in 2012 (QC 2006 at 21, 40). – Québec 'aims to have gas distributors include a minimum 5% of ethanol in their total fuel sales by 2012.' Québec hopes to 'encourage local production of ethanol from biomass, agriculture and municipal waste' through a $30 million incentive. Expected reduction in annual emissions of 0.78 Mt in 2012 (QC 2006 at 22, 40) – Support program for the marketing of technological innovations in energy efficiency in the transportation sector. Expected reduction in annual emissions of 0.9 Mt in 2012 (QC 2006 at 40).	– In partnership with other provinces – QC, NS, and BC – and U.S. states, set vehicle standards in New Brunswick that are 'stringent with respect to energy consumption and consistent with California's low emission vehicle standards' (NB 2007 at 16–17). – Commitment to develop a public transit strategy for the province (NB 2007 at 15). – Commitment to provide plug-in electrical rest stations for truckers, so that they can run essential equipment without operating their engines (NB 2007 at 16). – In partnership with Québec, 'implement a strategy of limiting truck speeds to 105 km/hour' (NB 2007 at 17).

Highlights of Provincial Greenhouse Gas Reduction Plans: August 2007 *(Continued)*

	Ontario	Québec	New Brunswick
	– Ethanol Growth Fund to support the production of ethanol fuel in Ontario.[73] This fund supports producers in reaching Ontario's Renewable Fuels Standard, which set a target of 5% ethanol blending in sales of gasoline by Jan 1, 2007.[74]	– $120 million of the $200 million/year Green Fund will be allocated to public transit initiatives. This amount will be additional to Québec's average annual investment of $350 million/year in public transit; expected reduction in annual emissions of 0.1 Mt in 2012 (QC 2006 at 21, 40). – Québec plans to adopt legislation requiring the mandatory use of a speed-timing device on all registered heavy-duty trucks that sets the maximum speed for these vehicles at 105 km/h. Expected reduction in annual emissions of 0.33 Mt in 2012 (QC 2006 at 23, 40).	– Supporting the federal commitment for ethanol-blended gasoline and conduct R&D with a view to a future 5% biodiesel requirement (NB 2007 at 17).
Buildings	– Ontario's 2006 Building Code contains new energy efficiency standards that will be phased in between 2006 and 2012. The Code will require builders to meet the Ener-Guide for Homes 80 standard for new homes by December 2011.[75]	– Amend Québec's Building Code with new efficiency standards by 2008; expected reduction in annual emissions of 0.05 Mt in 2012 (QC 2006 at 40, 20).	– Expand the retrofit incentives offered through Efficiency NB for all types of buildings (no detail provided; NB 2007 at 14). – 'Adopt an energy performance standard that goes beyond the federal Canadian model building code, for both new and renovated building in

Highlights of Provincial Greenhouse Gas Reduction Plans: August 2007 (*Continued*)

	Ontario	Québec	New Brunswick
			the residential and commercial markets, to be implemented in increments beginning in 2009' (NB 2007 at 14). – Phase-in of energy efficiency standards at Energy Star levels for appliances and equipment (NB 2007 at 14; no dates provided for the phase-in period).
Other sectors (agriculture, forest management, landfills, government operations, etc.)	– $9 million Biogas Systems Financial Assistance program will help farmers/rural businesses conduct feasibility studies for biogas systems, and will also cover up to 40% of construction costs for biogas digesters (up to a maximum of $400,000 in total funding). Biogas systems use renewable materials like manure and crops to produce electricity or heat.[76] – Reduce the Ontario government's own electricity use by at least 10% by 2007 (the government does not provide a baseline for this target).[77]	– 2005 regulatory requirement for the capture and incineration of biogas from landfills; expected reduction in annual emissions of 0.5 Mt in 2012 (QC 2006 at 24, 40). – Implement $18 million/year financial support for the capture of biogas from landfills not covered by the regulation; expected reduction in annual emissions of 2.5Mt in 2012 (QC 2006 at 24). – Implement incentive for waste treatment and energy recovery of biomass; expected reduction in annual emissions of 0.3 Mt in 2012 (QC 2006 at 25, 40). – Improve energy efficiency of public	– 'Encourage projects that capture methane gases from landfills and produce energy, where it is feasible to do so' (NB 2007 at 18). – The target for reducing emissions from government operations is a 25% reduction below 2001 emission levels by 2012. Expected measures include Energy Star procurement policies, use of low-emission vehicles and biofuels, and the use of LEED or other green building standards in construction (NB 2007 at 21–2).

Highlights of Provincial Greenhouse Gas Reduction Plans: August 2007 *(Concluded)*

	Ontario	Québec	New Brunswick
		buildings by 10–14% over 2003 level and fuel consumption of government departments and public organizations by 20% relative to 2003 levels by 2010; expected reduction in annual emissions of 0.15 Mt in 2012 (QC 2006 at 25, 40).	
Further Information			
Legislation	– Bill 104 of the *Greater Toronto Transportation Act* creating the greater Toronto Transportation Authority specifically references reduction of GHGs as a goal. – *Renewable Fuels Standard* (Ontario Regulation 535/05).	– *Règlement relatif à la redevance annuelle au Fonds Vert, Loi sur la Régie de l'énergie* (L.R.Q., c. R-6.01, a. 85.36 et 114, 1er al., par. 9°; 2006, c. 46, a. 48 et 51). – *Règlement sur l'enfouissement et l'incinération de matières résiduelles.* – *Loi sur la qualité de l'environnement* (L.R.Q., c. Q-2, a 32–33).	– *New Brunswick Conservation and Energy Efficiency Agency Act.* – *Renewable Energy Regulation – Electricity Act.* – *Energy Efficiency Act.* – *Coastal Area Protection Regulation* – *New Brunswick Community Planning Act.*
Other	Ministry of the Environment backgrounder (2005) online, http://www.ene.gov.on.ca/en/news/2005/120601mb.pdf.		Efficiency NB website, http://www.efficiencynb.ca/about-e.asp.
Contact	Cherise Burda, Pembina Institute, 416-644-1016	Jean-François Nolet, Équiterre, 418-522-0006, ext. 2261	David Coon, Conservation Council of New Brunswick, 506-458-8747

Notes

The author wishes to thank Clare Demerse and Johanne Whitmore, climate change policy analysts with the Pembina Institute, Gatineau, for their help with this chapter.

1 A. Taylor, M. Winfield, and M. Bramley, *Government Spending on Canada's Oil and Gas Industry: Undermining Canada's Kyoto Commitment* (Drayton Valley: The Pembina Institute, 2005).
2 On the subsidization of ACEL, see D. Martin, *Canadian Nuclear Subsidies: Fifty Years of Futility* (Sierra Club of Canada, 2002) online, http://www.cnp.ca/resources/nuclear-subsidies-at-50.pdf.
3 See, for example, The Pembina Institute, 'Federal Government a No-Show at Crucial Oil Sands Expansion Hearing,' press release, 13 July 2006, online, http://www.oilsandswatch.org/media-release/1257; and The Pembina Institute, 'Alberta Tar Sands Project May Face Supreme Court Challenge,' press release, 27 March 2006, online: http://energy.pembina.org/media-release/1227.
4 Natural Resources Canada, 'Canada's New Government Launches Major Projects Management Office,' press release, 1 October 2007, online, http://www.nrcan.gc.ca/media/newsreleases/2007/200794_e.htm.
5 B. Doern and M. Gattinger, *Power Switch: Energy Regulatory Governance in the Twenty-First Century* (Toronto: University of Toronto Press, 2003) (see, in particular, chapter 4 'The National Energy Board').
6 See R. Roff, A. Krajnc, and S. Clarkson 'The Conflicting Economic and Environmental Logics of North American Governance: NAFTA, Energy Subsidies, and Climate Change,' presentation to the Second Symposium on Understanding the Linkages between Trade and the Environment, North American Commission for Environmental Cooperation, March 2003. Online, http://www.cec.org/files/PDF/ECONOMY/Session-1-3-Roff-Krajnc_en.pdf.
7 See http://oee.nrcan.gc.ca/english/index.cfm.
8 See, for example, Council of Energy Ministers, *Moving Forward on Energy Efficiency for Canada: A Foundation for Action* (September 2007), online, http://www.nrcan-rncan.gc.ca/com/resoress/publications/cemcme/cemcme-eng.pdf.
9 Natural Resources Canada, 'Wind Power Production Incentive,' backgrounder, 17 March 2003, online, http://www.nrcan.gc.ca/media/archives/newsreleases/2003/200312a_e.htm.
10 Government of Canada, 'ecoENERGY for Renewable Power,' online,

http://www.ecoaction.gc.ca/ecoenergy-ecoenergie/power-electricite/index-eng.cfm.
11 M. Brewster, 'Canada Won't Formally Withdraw from Kyoto, Baird Says,' *The Globe and Mail*, 20 October 2007.
12 Government of Canada, 'Speech from the Throne,' 16 October 2007, online, http://www.sft-ddt.gc.ca/eng/media.asp?id=1364.
13 Environment Canada, *Regulatory Framework for Air Emissions* (Ottawa: Minister of Supply and Services, 2007) online, http://www.ec.gc.ca/doc/media/m_124/toc_eng.htm.
14 See, for example, National Round Table on Environment and Economy, *Interim Report to the Minister of the Environment* (Ottawa: NRTEE, 2007) online, http://www.nrtee-trnee.ca/eng/media/media-releases/20070627-ECC-Interim-Report-eng.htm.
15 M. Bramley, *Comparison of Current Government Action on Climate Change in the U.S. and Canada* (Ottawa: Pembina Institute 2002).
16 See, for example, G. Skogstad and H. Bakvis, eds., *Canadian Federalism: Performance, Effectiveness and Legitimacy* (Toronto: Oxford University Press, 2007) chapter 1.
17 Environment Canada, *National Inventory Report: Greenhouse Gas Sources and Sinks in Canada, 1995–2005* (Ottawa: Environment Canada, 2007).
18 Natural Resources Canada, *Energy Use Data Handbook Tables (Canada)*, online, http://oee.nrcan.gc.ca/corporate/statistics/neud/dpa/handbook_tables.cfm.
19 Environment Canada, *supra* note 17.
20 Environment Canada, *supra* note 17. Methane is a GHG twenty-one times more potent that carbon dioxide.
21 J. Whitmore and C. Demerse, *Highlights of Provincial Greenhouse Gas Reduction Plans* (Ottawa: Pembina Institute, 2007).
22 Sidhartha Banerjee, 'Quebec's Carbon Tax Gets Green Light despite Fears about Who Will Have to Pay,' *Canadian Business* (20 September 2007) online, http://www.canadianbusiness.com/markets/headline_news/article.jsp?content=b093008A.
23 See www.westernclimateinitiative.org.
24 See Government of Ontario, *Ontario Greenhouse Gas Emission Targets: A Technical Brief* (Toronto, June 2007) online, http://www.gogreenontario.ca/docs/061807-TechnicalBrief.pdf.
25 See Ontario Power Authority, 'Standard Offer Program,' online, http://www.powerauthority.on.ca/SOP/.
26 See Ontario Power Authority, 'Clean Energy Standard Offer Program,'

online, http://www.powerauthority.on.ca/Page.asp?PageID=924&SiteNodeID=245.
27 See R. Peters, A. Ballie, and M. Horne, *Successful Strategies for Energy Efficiency* (Drayton Valley: Pembina Institute, 2006).
28 See, generally, M. Winfield, *Building Sustainable Urban Communities in Ontario* (Toronto: Pembina Institute, 2006).
29 Environment Canada, *supra* note 17.
30 See M. Bramley, *An Assessment of Alberta's Climate Change Action Plan* (Ottawa: Pembina Institute, 2002).
31 See http://www3.gov.ab.ca/env/climate/index.html.
32 BC Budget 2007; online, http://www.bcbudget.gov.bc.ca/2007/pdf/2007_Budget_Fiscal_Plan.pdf.
33 BC Ministry of Energy, Mines and Petroleum Resources. 'Amendment Supports Innovation, Clean Energy Technology,' press release, 23 April; online, http://www2.news.gov.bc.ca/news_releases_2005-2009/2007EMPR0023-000504.htm.
34 Government of Alberta, *Environment Business Plan, 2007–10*; online, http://www.finance.gov.ab.ca/publications/budget/budget2007/envir.pdf.
35 Alberta Budget 2007; http://www.gov.ab.ca/budget2007/index.cfm?page=1651.
36 See http://www3.gov.ab.ca/env/climate/accomplishments.html.
37 This is intended primarily to boost royalty revenues through increased production. See http://www.energy.gov.ab.ca/2858.asp.
38 See http://www3.gov.ab.ca/env/climate/accomplishments.html.
39 Government of Saskatchewan, 'New Plan Attacks Climate Change in Saskatchewan,' press release, 14 June 2007; online, http://www.gov.sk.ca/news?newsId=78e66c74-c0a2-4041-813f-54e666cdf591.
40 Saskatchewan Budget 2007–2008; online, http://www.gov.sk.ca/adx/aspx/adxGetMedia.aspx?DocID=799,1,Documents&MediaID=972&Filename=07-08-Finance-BudgetSummaryBook-En.pdf.
41 Government of BC, 'B.C. joins Western Regional Climate Action Initiative,' press release, 24 April 2007; online, http://www2.news.gov.bc.ca/news_releases_2005-2009/2007OTP0053-000509.htm. For information about the Western Regional Climate Action Initiative, see http://www.azclimatechange.gov/download/022607wrca.pdf.
42 According to the Government of Alberta's backgrounder 'Climate Change: Strategy for Reduced Emissions' (available online at http://www3.gov.ab.ca/env/Climate/docs/Strategy_for_Reduced_Emissions.pdf), the

290   Mark S. Winfield

estimated annual cost of compliance with Alberta's regulations is $177 million. At $15/tonne, that means the government expects about 12 Mt of annual emissions to be covered by the system.

43 Government of Alberta, 'Budget 2007 Addresses Alberta's Price of Prosperity,' press release, 19 April 2007; online, http://www.gov.ab.ca/acn/200704/213150AE0FA3F-E1E5-ECC6-AD21AE143266D665.html.
44 See http://www1.agric.gov.ab.ca/$department/newslett.nsf/all/gm10921
45 Government of Alberta, 'Alberta Leads Country in Purchase of Green Power,' press release, 12 March 2003; http://www.gov.ab.ca/acn/200303/14035.html.
46 Government of Ontario, *supra* note 24.
47 Ibid. at 7.
48 Ibid. at 7.
49 Strictly speaking, 6 per cent below 1990 by 2012 is not the Kyoto level. Canada's *Kyoto Protocol* target is more demanding: to reduce annual emissions to 6 per cent below 1990 levels *on average between 2008 and 2012*.
50 Government of Ontario, *supra* note 24 at 8–9.
51 Government of Canada, 'Prime Minister Unveils New Canada EcoTrust,' press release, 12 February 2007; online, http://www.ecoaction.gc.ca/news-nouvelles/20070212-eng.cfm.
52 Ibid.
53 Ontario Budget 2006 at 44; online, http://www.fin.gov.on.ca/english/budget/ontariobudgets/2006/pdf/papers_all.pdf.
54 Government of Ontario. 'McGuinty Government Is Creating Jobs by Going Green,' press release, 19 June 2007; online, http://www.premier.gov.on.ca/news/Product.asp?ProductID=1400.
55 Government of Ontario, 'McGuinty Government Helps Ontarians Go Green At Home,' press release, 20 June 2007; online, http://www.premier.gov.on.ca/news/Product.asp?ProductID=1409&Lang=EN.
56 Government of Ontario, 'McGuinty Government Investing in Green Communities,' press release, 13 June 2007; online, http://www.premier.gov.on.ca/news/Product.asp?ProductID=1371&Lang=EN.
57 Government of Ontario, 'McGuinty Government Action Plan for Rapid Transit Will Move the Economy Forward,' press release, 15 June 2007; online http://www.premier.gov.on.ca/news/Product.asp?ProductID=1383&Lang=EN&offset=5.
58 Government of Ontario, 'Ethanol Growth Fund is Good News for Farmers, Rural Communities and the Air We Breathe,' press release, 15 June 2005; online, http://www.premier.gov.on.ca/news/Product.asp?ProductID=94.
59 Government of New Brunswick, 'Financial Challenge Addressed Leading

to Balanced 2007–2008 Budget,' press release, 13 March 2007; online, http://www.gnb.ca/cnb/news/fin/2007e0324fn.htm.
60 Government of Ontario, *supra* note 24 at 8.
61 Ontario Ministry of Energy. 'McGuinty Government to Ban Inefficient Light Bulbs by 2012,' press release, 18 April 2007; online, http://www.energy.gov.on.ca/index.cfm?fuseaction=english.news&body=yes&news_id=148.
62 Government of Ontario, *supra* note 56.
63 Ontario Ministry of the Environment, 'Ontario is Clearing the Air on Climate Change,' backgrounder, 6 December 2005; http://www.ene.gov.on.ca/en/news/2005/120601mb.pdf.
64 See http://www.powerauthority.on.ca/sop/.
65 Ontario Ministry of the Environment, *supra* note 64.
66 For program information, see Efficiency NB's website, http://www.efficiencynb.ca/contacts-e.asp.
67 Ontario Ministry of the Environment, 'Ontario Phasing out Use of Potent Ozone Depleting Substances,' press release, 8 May 2007; online, http://www.ene.gov.on.ca/en/news/2007/050802mb.php.
68 Ontario Ministry of Energy, 'McGuinty Government Home Energy Retrofit and Solar Power Initiatives,' backgrounder (2007); online, http://www.energy.gov.on.ca/index.cfm?fuseaction=english.news&back=yes&news_id=156&backgrounder_id=122.
69 Government of Ontario, 'Green Technology Key to Future of Ontario's Economy,' press release, 19 June 2007; online, http://www.premier.gov.on.ca/news/Product.asp?ProductID=1401.
70 Government of Ontario, *supra* note 24 at 10.
71 Rob Gillies, 'Ontario Won't Limit Emissions as Much as California, Premier Says,' *Associated Press*, 29 May 2007.
72 Government of Ontario, 'McGuinty and Schwarzenegger Team Up to Curb Climate Change, Boost Stem Cell Research. Press release, 30 May 2007; online, http://www.premier.gov.on.ca/news/Product.asp?ProductID=1281.
73 See http://www.omafra.gov.on.ca/english/policy/oegf/index.html.
74 Government of Ontario. 'Cleaner Gasoline for Healthier Ontarians and Healthier Economy.' Press release, 26 November 2004; online, http://www.premier.gov.on.ca/news/Product.asp?ProductID=249.
75 Ontario Ministry of Municipal Affairs and Housing, 'Ontario Households and Businesses to Save Energy and Money,' press release, 29 June 2006; online, http://ogov.newswire.ca/ontario/GPOE/2006/06/29/c7090.html?lmatch=&lang=_e.html.

76 Ontario Ministry of Agriculture, Food and Rural Affairs, 'Ontario Biogas Systems Financial Assistance Program,' press release, 26 July 2007; http://www.omafra.gov.on.ca/english/engineer/biogas/index.html.
77 Ontario Ministry of the Environment, *supra* note 64.

14 Integrating Climate Policy and Energy Policy

IAN H. ROWLANDS

1. Introduction

The purpose of this chapter is to explore the relationship between energy and climate change in the context of Canada's participation in international regimes to advance sustainability. With climate-change negotiations set to resume in Bali, Indonesia, in December 2007, and with the Intergovernmental Panel on Climate Change's 2007 assessment underscoring the severity of the issue, it is critical that this country's role in international efforts to address climate change be debated comprehensively. To help contribute to this debate, this chapter examines the extent to which energy and climate change are linked, may continue to be linked, and, with policy changes, could be linked in the future.

More specifically, this chapter is divided into seven main sections. After this short introduction, the links between energy and climate change are briefly outlined in the second section – it is shown not only that energy production and use contribute to global climate change, but also that global climate change itself has implications for the continuing operation of the energy system. In the third section, the focus turns explicitly to energy, reviewing Canada's position on the energy issue, with particular emphases on energy production, use, and the associated greenhouse gas emissions. The fourth section considers the future, reviewing two major studies on projected energy production and use in Canada. In the fifth section, climate-change trends – considered in many of the other chapters in this book – are juxtaposed with these same energy trends, and the case is made for an integrated strategy in response. The sixth section presents some initial ideas as to how this integrated strategy might be developed in a Canadian context, while the seventh section concludes the chapter with a brief summary and a number of key recommendations.

This chapter is not meant to provide an exhaustive overview of the impact of Canada's energy system upon the global climate. Instead, it is intended to identify some of the most pivotal links, and to highlight some representative strategies to move the issue forward – that is, to encourage this country to think further about an energy strategy that would not only be 'climate-friendly,' but that would also ultimately serve to advance sustainability.

2. Energy and Climate

Energy production and consumption is the most critical contributor to global climate change. The United Nations Framework Convention on Climate Change (UNFCCC) reported that, for industrialized countries (that is, Annex I countries), 82.8 per cent of greenhouse gas emissions were energy related;[1] the corresponding figure for developing countries that reported (that is, those countries not part of Annex I) was 63.9 per cent.[2]

It is also important to recognize, however, that global climate change also has consequences for the global energy system – in at least two important ways. First, as the global climate changes, patterns of energy demand will probably also change. For example, with increased global climate change, there will be, in many locations, more intense and prolonged heat waves during the summer. Accordingly, the demand for air conditioning could well rise; hence, energy demand will rise in the wake of higher temperatures. (And, to take it one step further, note that, should the electricity for this air conditioning be provided by fossil fuels, then that will, in turn, serve to exacerbate global climate change – a positive feedback loop will inadvertently have been established.)

Second, the supply patterns of energy may also be affected by global change. For one, these same periods of higher temperatures during the summer could cause increased evaporation from dam reservoirs, which may well mean that there is less hydroelectricity production potential. For another, higher temperatures could affect the condition of the land in the northern reaches of the planet (melting permafrost, for example), which could affect workers' abilities to reach locations for fossil fuel production – while the melting of 'ice roads' may make it more difficult to get there; once there, the softening of the soil may make it easier to drill. In any case, the relationship between energy and climate is close and multidimensional.[3]

3. Canada's Energy Systems

Given that the provision of energy services is vital to human well-being, it is not surprising that energy plays an important role in Canadian society. Additionally, however, Canada is a significant energy producer – in both absolute and relative terms – so the importance of the sector is all the more significant in this country.

In absolute terms, Canada was the world's fifth largest producer of energy in 2004.[4] Particularly noteworthy within the broad portfolio of our energy production was hydroelectricity, where we are the world's second largest producer in absolute terms.[5] Significant, as well, is natural gas, where we rank third.[6] It should be noted, however, that we are also a major oil producer – ranking seventh in 2006.[7] In dollar terms, the 'energy industry accounted for almost six per cent of Canada's Gross Domestic Product (GDP)' in 2006.[8] Even more importantly, it accounted for '22 percent of the total value of Canadian exports' in the same year;[9] surprising for some, perhaps, is that the value of crude oil exports surpassed the value of natural gas exports in this year.[10] Finally, we are also a world leader in nuclear energy, ranking seventh.[11]

Canada is also a significant consumer of energy. In absolute terms, we ranked seventh in the world in 2006.[12] Per capita figures are even more dramatic. The World Resources Institute reported that this country comes ninth, of the 134 countries it ranked, in terms of 'total energy consumption per capita.' (The Canadian figure is 8,301 kg oil equivalent per person, while the global average is 1,674 kg oil equivalent per person.) While that is significant in itself, perhaps all the more noteworthy is that many of the countries ranked above us are small oil-producing entities (for example, Qatar and Bahrain); amongst the 30 OECD countries, we rank third, behind only Iceland and Luxembourg.[13]

Canada's energy consumption can be broken down in various ways. In figure 14.1, we see the kinds of fuels used to meet the demand for energy services. And in table 14.2, the differences in those energy services demanded by Canadians are noted. Of course, note that it is problematic to think of 'Canada' as one, single 'energy entity.' Our vast land mass means that different parts of the country are endowed with different kinds (extents) of energy resources. Add to this the fact that individual provinces have substantial (legal and political) standing on energy

Table 14.1. Domestic energy production by energy source (petajoules)

	2002	2003	2004	2005	2006
Petroleum	6,049	6,365	6,517	6,404	6,739
Natural gas	6,660	6,462	6,524	6,373	6,588
Hydroelectricity	1,245	1,198	1,207	1,289	1,271
Nuclear	824	817	986	1,009	1,090
Coal	1,430	1,326	1,476	1,494	1,554
Renewable and other	631	633	657	681	707
Total	16,839	16,801	17,367	17,250	17,949

Source: National Energy Board, *Canadian Energy Overview 2006: An Energy Market Assessment* (Calgary: NEB, May 2007) at 3.

Notes
1 Figures for 2006 are estimates.
2 'Petroleum' includes crude oil and gas plant natural gas liquids.
3 'Renewable and other' includes steam, solid wood waste, spent pulping liquor, and annual firewood.

issues, and the resulting mosaic that is the Canadian physical and regulatory energy profile should come as little surprise. Details about Canadian energy, at a provincial level, can be found elsewhere.[14]

All of this energy activity, not surprisingly, leads to significant greenhouse gas emissions. In terms of fossil-fuel carbon dioxide emissions rates (per capita), we are ranked tenth out of 207 countries listed; again, however, small countries make up the vast majority of those who are ranked ahead of us. Of the other OECD countries, only the United States (at ninth) has a higher ranking (5.61 tonnes of carbon for the United States, as compared to 5.46 tonnes for Canada).[15]

4. Canada's Energy Future

In 2006, Natural Resources Canada developed a report outlining 'a reference outlook for Canadian energy supply and demand up to 2020.'[16] The principal assumptions included a population growth rate of approximately 0.7 per cent annually, a real increase in gross domestic product of 2.4 per cent annually, and a gradually declining real price of oil (to US$45 per barrel, in 2003 dollars) by 2010, and remaining constant thereafter.[17]

Given these assumptions, it is anticipated that total energy demand will grow by 1.3 per cent annually. In terms of supply, the mix of energy

Figure 14.1. Total energy consumption in Canada by type (2004)

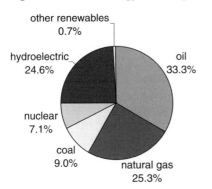

Source: Energy Information Administration, *International Energy Annual, 2004* (Washington, DC: EIA, 2006).

resources in Canada is anticipated to be approximately the same, though diminishing conventional oil production will be compensated for by increased output from the oil sands; also noteworthy is that our exports of natural gas will fall significantly. In terms of demand, the residential and commercial sectors will experience growth in energy demand below the overall figure of 1.3 per cent noted above, while the transportation and industrial sectors will experience higher growth rates. Finally: 'Growing energy demand and a changing energy production mix lead to growth in [greenhouse gas] emissions from 758 megatonnes (Mt) in 2004 to 828 Mt in 2010 and 897 Mt in 2020. The 2010 figure is 265 Mt above Canada's Kyoto target (6 per cent below 1990 levels).'[18] Noteworthy is that not only has economic growth increased the emissions, but so too has the changing profile of supply: 'Only the refining sector increases its emissions intensity, due to synthetic crude production and the processing of generally heavier crude oils.'[19]

The Energy Information Administration within the United States Department of Energy produces annual reports with long-term projections of energy supply, demand, and prices. The most recent of these reports – one focusing upon the United States, the other looking at the world as a whole – examined developments out to the year 2030. In addition to a 'reference scenario,' which assumes 'baseline economic growth of 2.9 percent per year from 2005 through 2030 and includes other assumptions about world oil prices and technology,'[20] the report

298   Ian H. Rowlands

Table 14.2. Domestic energy consumption (petajoules)

	2002	2003	2004	2005	2006
Space heating	1,970	2,065	2,032	2,074	2,105
Transportation	2,250	2,242	2,346	2,383	2,357
Other uses	3,164	3,298	3,312	3,399	3,499
Non-energy	894	903	1,018	1,020	1,015
Electricity generation	1,911	1,850	2,029	2,068	1,973
Total	10,189	10,358	10,737	10,944	10,950

Source: National Energy Board, *Canadian Energy Overview 2006: An Energy Market Assessment* (Calgary: NEB, May 2007) at 4.

Notes
1 All figures include consumption of imported energy.
2 Figures for 2006 are estimates.
3 'Other uses' includes energy used for space cooling and ventilation, appliances, water heating, as well as a variety of uses in the industrial sector.
4 'Non-energy' includes energy used for petrochemical feedstocks, anodes/cathodes, greases, lubricants, etc.
5 'Electricity generation' includes producer consumption and losses as well as nuclear energy conversion requirements.

presents a number of other scenarios, varying assumptions about economic growth, prices, and technological developments. While one of the key conclusions of these studies is that the energy consumption in developing countries is predicted to grow at a faster rate than in developed countries, noteworthy for the purposes of this chapter are the references to Canada therein.

Many of the findings mirror those of the Canadian study just identified (following many of the same assumptions). In particular, total energy consumption is expected to rise by 1.2 per cent per year from 2004 through 2030.[21] On the supply side, the profile is largely the same, although declining use of oil in power stations and increasing use of natural gas in the same, means that growth of the former is below this 1.2 per cent overall average while growth of the latter is above. Moreover, the residential and transportation sectors are below this overall average, commercial is at it, while industrial is above. Turning to carbon dioxide emissions, they rise, on average, 1.0 per cent a year from 2004 to 2030, starting at 584 Mt in 2004, and rising to 648 Mt in 2010, 694 Mt in 2020, and 750 Mt in 2030.[22] Canada's carbon dioxide intensity (measured in terms of emissions of carbon dioxide per unit gross domestic product) is expected to be the highest among OECD countries in 2030 at 410 tonnes per US$million (2000), which is, nevertheless, an improvement on the 2004 value of 581 tonnes per US$million (2000).[23] See figure 16.2 for further information about projected greenhouse gas and carbon dioxide emissions.

5. Energy and Climate Policies in Canada

Hence, two trends appear to be on some kind of tragic collision course: climate-change trends (as noted in other chapters of this book) and energy supply and demand trends (as briefly outlined above). 'Collision course' may not actually be the correct metaphor, for the two trends are in some ways mutually reinforcing, just not in a sustainable manner. What is unequivocal is that if Canadian energy trends continue as anticipated, Canada will continue to be contributing to a climaticly changed world. If the anticipated changes in climate do indeed occur, the consequential environmental, economic, and social impacts upon this country could well be devastating. What is critical, therefore, is the development of an integrated energy/climate policy strategy.

To date, however, we have not seen much evidence of an integrated energy/climate policy strategy in Canada.[24] Indeed, the extent to which we have seen even distinct energy policy or climate policy strategies developed in a comprehensive manner in this country has been limited. Let me briefly review each here.

In terms of energy policy, Canada's last major foray into large-scale energy strategizing was the National Energy Program (NEP). Initiated in 1980, the NEP was designed to increase Canadian control over, and ownership of the energy industry. It aimed to do this through a series of price controls on oil, in particular, as well as new taxation strategies on the oil and gas industry as a whole. The fall-out from that program – not least the resentment created among many in the West, who felt that their wealth was being unfairly plundered[25] – means that any subsequent efforts to implement large-scale energy policies in this country have been seen to have substantial political costs and/or risks associated with them.

Thus, while we have certainly continued to have, at times and in places, significant provincial-level leadership on energy issues, the national level has been much more timid.[26] While there may have been rhetoric about sustainable development in general and climate change in particular, especially during the Chrétien/Martin years, the reality

Figure 14.2. Canadian greenhouse gas and carbon dioxide emissions (Mt), actual (1990–2004) and projected (2010–2030)

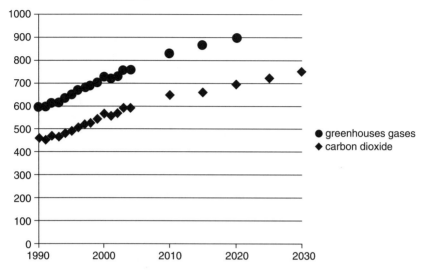

Source: Environment Canada, National Inventory Report 1990-2004, *Greenhouse Gas Sources and Sinks in Canada, The Canadian Government's Submission to the United Nations Framework Convention on Climate Change* (Ottawa: Environment Canada, April 2006); National Energy Board, *Canadian Energy Overview 2006: An Energy Market Assessment* (Calgary: NEB, May 2007); and Energy Information Administration, *International Energy Outlook, 2007* (Washington, DC: EIA, May 2007).

for many is that there has been a vacuum in Canadian energy policy. This vacuum, Brownsey maintains, 'has been filled by industry associations, the energy producing provinces and the United States.'[27] Gattinger also notes an important shift in the development of energy policy in this country from the primary commitment of 'government' to a new system of 'multi-level associative governance.' She uses this latter term to refer to instances 'where multiple public, private and civic actors interact across the sub-national, national and international levels.'[28] In turn, all of these actors exert different pressures: 'The United States wants a secure supply of energy, the companies want to maintain and increase profits, and the provinces want greater revenues and to protect their constitutional jurisdiction from what several of them view as an encroaching federal government.'[29]

The change of government in early 2006 meant some change in the direction of Canadian energy policy. The largely rhetorical link to sustainable development had what little substance there was attached to it mostly taken away – the quick 'shutting down' of the popular 'EnerGuide for Houses' program that same year is but one example.[30] A consolidation of the already evident interest in advancing 'North American energy integration' is another slight shift;[31] the desire for Canada to be an 'energy superpower,' or even a 'clean energy superpower,' has also entered the discourse with the new government.[32] Much, however, remains the same. The Commissioner of the Environment and Sustainable Development, for example, observed that there is an 'absence of an overall direction for energy development in Canada.'[33] And, this year, Pavlich and Fogarassy maintained that 'A unified, coherent Canadian energy policy simply does not exist.'[34]

Other chapters in this book highlight the fact that the 'climate policy' story in this country has had a similar recent history. Canadian policy strategy initially relied upon a 'voluntary approach' – agreed to in 1995, the *National Action Program on Climate Change*, for example, centred around 'principles' and 'strategic directions';[35] similarly, the *National Implementation Strategy on Climate Change*, released in 2000, relied upon 'encouraging action' in others.[36] With emissions continuing to grow, however, the debate around some kind of set of regulatory policies grew increasingly tense. *A Climate Change Plan for Canada*, launched by the Liberal government in 2002, made little headway; the legacy of a second plan – entitled *Moving Forward on Climate Change: A Plan for Honouring our Kyoto Commitment* and released in 2005 (also known as Project Green) – was cut short by the change in government in 2006.[37] The new Conservative government did not have any aspect of the environment as part of its five top priorities when it came to office in January of that year. Increasing public pressure, however, prompted the tabling of *Canada's Clean Air Act* in October 2006. The assessment of those plans, as they have subsequently evolved, has been covered elsewhere in this book. For our purposes, it is useful to highlight that they are somewhat skeletal at best, irrespective of their anticipated impact.[38]

Thus, comprehensive and effective policies for the distinct issues of energy and climate change have yet to emerge in the Canadian context. As a result, it is probably not surprising that a policy framework that successfully acknowledges the links between these two sets of issues does not exist at the federal level in this country. Of course, this absence has been noticed by many in recent years. VanNijnatten and Mac-

Donald argue that there has been a 'failure' on the part 'of the federal government to develop one coherent energy and climate change package.'[39] Doern and Gattinger flag the absence of any integration, instead maintaining that they have seen 'regulatory stacking' of policies to address different aspects of the energy and sustainability sets of issues.[40] Moreover, in his review of the Chrétien/Martin years, Brownsey concludes that: 'Ottawa has been left on the sidelines, unable or unwilling to formulate much beyond contradictory commitments to sustainability and the Kyoto accord while promising to sell Canada's energy to the United States and other potential customers in emerging Asian economies.'[41] More recently, the Commissioner of the Environment and Sustainable Development has urgently called for energy and climate-change policy to be integrated: 'The government cannot effectively address climate change without considering changes in the way Canadians produce, distribute, and consume energy. And we cannot secure our energy future without considering climate change – these issues are unavoidably linked. Any new approach must confront this reality.'[42]

Similar sentiments have been expressed by those based outside of the country. In 2004, a review team from the International Energy Agency (IEA) recognized that: 'Within the IEA member countries, Canada is in a unique and challenging situation with a strong commitment to climate change mitigation and large fossil fuel reserves currently being developed and large associated emissions growth.'[43] The review team went on to comment: 'However, living up to Canada's commitment to the Kyoto Protocol, moving to a less emission-intensive economy and at the same time ensuring continued growth is the biggest single economic and political challenge for Canadian energy policy in the coming years.'[44] To achieve this, they recommend the following: 'Increase cooperation with provinces and territories to implement the National Climate Change Plan, and in particular to develop the range of market incentives based on climate change policies. Promote the integration of energy and greenhouse policy objectives across federal and provincial governments.'[45]

These various observations would seem to be well warranted. Not only do we see that recent greenhouse gas emissions in Canada continue to rise, but also that ministerial statements sometimes seem to be at cross purposes. In February 2007, for example, the environment minister declared that 'the science is clear. The evidence is conclusive. It is important that we accept what the experts are telling us. We must now focus our attention on solutions to protect our planet's fragile ecosystem. And we must take real action on climate change.'[46] Just one month later, however, the minister for natural resources noted: 'Let me begin by stating the obvious, and that is just how strong the economy is here in Calgary and Alberta. You know a lot of people in this room would say it's bursting at the seams, and the big reason is without question the oil and gas industry. But I also think it's very important that we don't take it for granted. Yes, the economy is strong, but we have to keep it strong. Between 1996 and 2005, $45 billion was invested in Canada because of the oil sands. And from now until 2015, we're looking at another $125 billion in new projects. And I think it's also worth pointing out that the investments in the conventional oil and gas sector are three times that of the oil sands.'[47] Thus, it would appear that integrated thinking in this area is urgently required.

6. Promoting Integrative Thinking

There are a number of places to look for inspiration regarding how thinking about the integration of energy and climate policies could be developed – or, at least, could be promoted. In this section, I highlight a few policy approaches (looking at various parts of the world) and a few scenario perspectives (again, drawing insights from different parts of the world). By identifying and briefly highlighting these initiatives and proposals, I aim to stimulate thinking about how movement towards an integrated national strategy – one that is critical in light of growing greenhouse gas emissions, most of it energy-sector driven – could be encouraged.

Attempts to integrate energy and climate are evident in the development of Ontario's electricity system plans. In efforts led by the Ontario Power Authority (OPA), a body created in 2004, the province's Integrated Power System Plan has striven to integrate sustainability into its plans,[48] using for this purpose Gibson's criteria for sustainability.[49] While criticisms can – and, indeed, should – certainly continue to be levelled at the way in which the criteria have been applied and the data used to analyse them, these efforts have served to stimulate new thinking about how principles of 'sustainability assessment' could serve to develop and to encourage integrated planning – integrating, more specifically, aspects of energy (in this case, electricity) with aspects of sustainability (including climate change).

Moving beyond the assessment of long-term options, and towards the implementation of regulations and policies, we turn our attention south of the border – more specifically to California. While California has some natural advantages in its efforts to integrate climate and energy concerns – in particular, a mild climate and significant roles for high technology and services in its economy – its achievements are nevertheless still impressive. While average energy consumption in the United States is 341.5 million Btu per capita, in California it is 233.4 million Btu per capita – making the state the 48th ranked energy-consuming state per capita in the country.[50] Since the mid-1970s, the state has adopted a variety of policy approaches – regulations, pricing, and so on – in order to improve energy efficiency.[51] And its actions more recently – for example, AB32, the *California Global Warming Solutions Act* of 2006 – have placed California 'on the vanguard of state-level climate change action.'[52] Hence, a review of its efforts would help to inform discussions in this country.

Finally, another approach involves 'visioning' – that is, imagining a future that is desired, and then developing the means to get there. 'Backcasting' is another term that is used to refer to the same process; the term 'normative scenarios' has also been used to describe this.[53] The IEA, for one, has developed an explicitly normative scenario about the future, with a concern for climate-change mitigation being at the top of its list of driving factors – more specifically, 60 per cent of world energy supply would be of a 'zero-carbon' nature by 2050.[54] Perhaps not surprisingly, nuclear power, renewables, and carbon capture all play important roles in this scenario. When the calculations are completed, it is revealed that this scenario would lead to a stabilisation of carbon dioxide concentrations in the atmosphere at 550 ppm at the end of the 21st century.[55] Although this scenario does not depend on unprecedented technological innovation, it does depend on unprecedented levels of policy commitment (though it is also noted that limited experience with the requisite policy instruments has already been gained). To achieve, for example, the goals laid out for renewable energy, what would be needed is the following:

> Deployment and market development policies should be implemented as soon as possible to prepare for the acceleration that should take place, especially for non-biomass renewables, after 2020. Market deployment could be promoted through economic incentives, feed in tariffs, renewable portfolio standards and green certificate trading for power from renewables, as well as taxation of environmental externalities. In parallel, public

R&D policies would be needed to improve conversion and power or energy storage technologies, reduce the intermittence problems and lower the overall cost of these systems. Finally, adequate technology transfer and investment policies are required if the potential for renewables development is to be tapped where it is largest, i.e. in developing countries.[56]

Canadian exploration of similar themes also exists. In 1998, Natural Resources Canada initiated its own scenario-building program in order to assist with planning and policy development. Of the four scenarios, the one that achieves the greatest reduction in greenhouse gas emissions brings those levels, in 2050, to less than half of what they were in 1990. It is the scenario entitled 'Come Together,' and it is described in figure 14.3.

Again, renewables, nuclear energy, and carbon capture play important roles in achieving those reductions. High technology (for example, high-speed computing and nano-technologies) also plays an important role; so, too, do biotechnologies (for example, in strategies to reduce the carbon intensity of particular industrial operations, of oil sands operations, and of site remediation). There are also instances where these two sets of technologies are linked and, together, serve to reduce greenhouse gas emission intensities through 'DNA computing, sensors & controls and artificial intelligence and robotics.'[57]

More recently, the National Energy Board has developed two scenarios for the future of Canada's energy supply. It is recognised that these are not predictions, but instead means to generate debate and discussion by presenting 'descriptions of plausible alternative futures.'[58] These scenarios are differentiated by their anticipated level of progress on two axes: the pace of technological development and the degree of action on the environment. When both are low, a future is described in the 'supply push' scenario – one that largely mirrors those projections laid out in section four of this chapter. When both are high, a future called 'techno-vert' is created. For stimulating debate and discussion regarding integrated decision-making for Canada, it is the one that holds the most promise.

This scenario is laid out in figure 14.3. The resulting energy portfolio has increased shares of renewables and nuclear power as compared with the 'supply push' alternative. Overall energy use falls by almost nine per cent. Oil use falls significantly (by more than 30 per cent), as does coal (falling by just under 30 per cent), but natural gas use rises by 10 per cent. While solar rises dramatically – by more than fourfold – it is

Figure 14.3. Scenarios for the Future

> 'Come Together': This is a [greenhouse gas] responsive world, with open markets, rapid innovation, and high levels of environmental etiquette. There is a strong cohesion of views amongst government, industry, and the public over environmental issues, and this is shared internationally. Multinational companies exert political as well as economic power, and press governments on environmental, trade and monetary issues. The expanded interconnectedness of the world allows new technology to be openly developed, traded, and applied in innovative ways across all sectors. Canada was well placed in this global market, and Canadian business has been able to capitalise on its expertise and products, and to re-invest in improvements and establish new areas of competence readily demanded in these open and expanding global markets. This world, despite its strong environmental ethic, is mid-green in colour. Reduced GHG emissions realised through technology gains are somewhat offset by expanded industrial operations, extensive international distribution systems, and energy using leisure activities.

Source: Natural Resouces Canada, *Energy Technology Futures* (Ottawa: NRC, 1999) at 11.

> The Techno-Vert (TV) scenario is a world in which technology advances more rapidly. In addition, Canadians take broad action on the environment. The main theme of this scenario is the heightened concern for the environment and the accompanying preference for environmentally friendly products and cleaner-burning fuels. Consumers are also willing to pay more for these products and consider both financial and environmental costs when making purchasing decisions. While governments assist with research and development program funding, reliance is primarily placed on market solutions. Technological breakthroughs and the adaptation of improved technologies ('best practices') result in the development of diverse energy sources and energy efficiency improvements. The new technologies produce and deliver goods and services more efficiently and cost-effectively. Consumers and producers embrace the new products and equipment which result from the technological advances. Productivity is higher in all sectors due to higher pace of technological improvements and the widespread application of new capital and new products. As a result, this scenario generates higher economic growth than the [supply push scenario].

Source: National Energy Board, *Canada's Energy Future: Scenarios for Supply and Demand to 2025* (Calgary, AB: NEB, 2003) at 15.

important to recognize that its absolute total is still just 1.89 petajoules, representing only 0.02 per cent of total energy demand in Canada. Thus, while an improvement upon the alternative scenario, greenhouse gas emissions (though not evaluated) would probably still be significant.

7. Summary and Recommendations

This concluding section presents a summary of the main findings and recommendations for the future.

1. *Energy and climate are inextricably linked.* Not only does increased energy use (as conventionally supplied) contribute to global climate change, but global climate change affects patterns of both supply of and demand for energy.
2. *Canadian energy use is presently high and projected to remain high.* A resource-rich country, Canadians have become used to having relatively low-cost energy resources readily available. Indeed, we rank amongst the world's leaders in terms of both absolute and per capita energy production and consumption, across a variety of fuel types. Moreover, with investment being pumped into the oil sands in Alberta, and continuing demand for our natural gas resources by our neighbour to the south, we appear set to continue on a path of high energy production and consumption, with largely the same respective shares of different kinds of fuels.
3. *Canadian energy policy and climate-change policy – modest as they are individually – have been developed in isolation from each other.* Each for their own reasons (though these reasons are linked, to some extent), successive federal governments have shied away from developing comprehensive policies for either a national energy policy or a national climate-change policy. Given this pattern, it is not particularly surprising that what little elements of policy that have emerged in each realm appear not to speak to the other. As a result, we have pronouncements on energy that seem largely oblivious to similar pronouncements on climate change (and vice versa).
4. *Given future climate-change projections, energy policy and climate-change policy must speak to each other – that is, they must be integrated.* These two sets of policies, what little we have seen of them at any rate, are working at cross-purposes. Moving forward, we need an integrated approach to policy making in these areas.
5. *Policy development in this area could benefit from the analysis of activities*

taking place elsewhere, both within and outside of Canada. Ideas about sustainability assessment, policy implementation, and scenario development, for example, must be introduced into the mix so as to invigorate the national debate regarding the integration of energy policy and climate-change policy.

As Canada moves forward at both the international and domestic levels on global climate change, the country needs to integrate discussions regarding its present energy production and consumption, as well as the future of the same. The purpose of this chapter has been to stimulate debate and discussion regarding the ways in which this integration could serve to advance Canadian goals – one of the most important of which must be global sustainability.

Notes

1 UNFCCC, *National Greenhouse Gas Inventory Data for the Period 1990–2004 and Status of Reporting* (Bonn: UNFCCC, FCCC/SBI/2006/26, 19 October 2006) at 9; based on 2004 data.
2 UNFCCC, *Sixth Compilation and Synthesis of Initial National Communications from Parties Not Included in Annex I to the Convention* (Bonn: UNFCCC, FCCC/SBI/2005/18/Add2, 25 October 2005) at 18; based on 1994 (or closest year for reporting country) data.
3 See the discussion in Intergovernmental Panel on Climate Change (IPCC), *Climate Change 2001. Working Group II: Impacts, Adaptation and Vulnerability* (Cambridge: Cambridge University Press, 2001) at section 7.3.1.
4 Energy Information Administration, *Country Analysis Briefs: Canada* (Washington, DC: Department of Energy, May 2007); in terms of electrical energy, Canada ranked sixth, based on 2005 data from International Energy Agency (IEA), *Key World Energy Statistics, 2007* (Paris: IEA, 2007) at 27.
5 IEA, *supra* note 4 at 19 (based on 2005 data); British Petroleum (BP), *BP Statistical Review of World Energy* (London: BP, 2007) based on 2006 data.
6 IEA, *supra* note 4 at 13, based on 2006 data.
7 Ibid. at 7, based on 2006 data.
8 National Energy Board (NEB), *Canadian Energy Overview 2006: An Energy Market Assessment* (Calgary: NEB, May 2007) at 1.
9 Ibid.
10 Ibid.

11 IEA, *supra* note 4 at 17 (based on 2005 data); BP, *supra* note 5 (based on 2006 data).
12 BP, *supra* note 5.
13 See online: World Resources Institute EarthTrends: The Environmental Information Portal, http://earthtrends.wri.org; based on 2003 data.
14 See, for example, National Resources Canada (NRCan), *Canada's Energy Outlook: The Reference Case 2006* (Ottawa: Analysis and Modeling Division, 2006) at Appendix V.
15 G. Marland, T.A. Boden, and R.J. Andres, *Trends: A Compendium of Data on Global Change* (Oak Ridge: Carbon Dioxide Information Analysis Center, Oak Ridge National Laboratory, U.S. Department of Energy, 2007).
16 NRCan, *supra* note 14 at 3.
17 Ibid. at 7, 8, and 11.
18 Ibid. at iv.
19 Ibid. at 55.
20 Energy Information Administration (EIA), *Annual Energy Outlook, 2007, with Projections to 2030* (Washington, DC: EIA, February 2007) at 212.
21 EIA, *International Energy Outlook, 2007* (Washington, DC: EIA, May 2007) at 83.
22 Ibid. at 93.
23 Ibid. at 76 and 77.
24 To see this discussion placed within the broader context of Canadian energy policy and sustainable development, see the contributions in G.B. Doern, ed., *Canadian Energy Policy and the Struggle for Sustainable Development* (Toronto: University of Toronto Press, 2005).
25 See, for example, G.B. Doern and G. Toner, *The Politics of Energy: The Development and Implementation of the National Energy Program* (Toronto: Methuen, 1985).
26 For an historical review, which nicely sets the context, see G.B. Doern and M. Gattinger, *Power Switch: Energy Regulatory Governance in the Twenty-First Century* (Toronto: University of Toronto Press, 2003) particularly chapters 1 and 2.
27 Keith Brownsey, 'Canadian Energy Policy: Supply, Sustainability, and a Policy Vacuum' in G.B. Doern, ed., *How Ottawa Spends, 2006–2007: In From the Cold – The Tory Rise and the Liberal Demise* (Montreal: McGill-Queen's University Press, 2006) 73 at 88.
28 Monica Gattinger, 'Canada-United States Electricity Relations: Policy Coordination and Multi-level Associative Governance' in G.B. Doern, ed., *How Ottawa Spends, 2005–2006: Managing the Minority* (Montreal: McGill-Queen's University Press, 2005) 143 at 156.

29 Brownsey, *supra* note 27 at 89.
30 Paul Parker and Ian H. Rowlands, 'City Partners Maintain Climate Change Action Despite National Cuts: Residential Energy Efficiency Program Valued at Local Level,' *Local Environment* (forthcoming, 2007).
31 Keith Brownsey, 'Energy Shift: Canadian Energy Policy under the Harper Conservatives' in G.B. Doern, ed., *How Ottawa Spends, 2007–2008: The Harper Conservatives – Climate of Change* (Montreal: McGill-Queen's University Press, 2007) at 143–60.
32 Prime Minister Stephen Harper, quoted in CBC, 'Canada Must Be a Clean Energy Superpower: PM' (22 March 2007) online, http://www.cbc.ca/canada/story/2007/03/22/harper-green.html, accessed on 11 October 2007.
33 Office of the Auditor General of Canada, *Report of the Commissioner of the Environment and Sustainable Development to the House of Commons, The Commissioner's Perspective – 2006* (Ottawa: Minister of Public Works and Government Services Canada 2006) at 12.
34 Jonathan Pavlich and Tony Fogarassy, *Canadian Energy Policy Overview* (Toronto: Clark Wilson LLP, 14 August 2007) at 12.
35 Government of Canada, *Canada's Second National Report on Climate Change: Actions to Meet Commitments Under the United Nations Framework Convention on Climate Change, 1997* (Ottawa: Environment Canada, November 1997) at 33.
36 Government of Canada, *Canada's Third National Report on Climate Change: Actions to Meet Commitments Under the United Nations Framework Convention on Climate Change, 2001* (Ottawa: Environment Canada, 2001) at 58.
37 Government of Canada, *Canada's Fourth National Report on Climate Change: Actions to Meet Commitments Under the United Nations Framework Convention on Climate Change, 2006* (Ottawa: Environment Canada, 2006) at 55–6.
38 See, as well, G.B. Doern, 'The Harper Conservatives in Power: Emissions Impossible,' in Doern, *supra* note 31 at 3–22.
39 Debora L. VanNijnatten and Douglas MacDonald, 'Reconciling Energy and Climate Change Policies: How Ottawa Blends,' in G.B. Doern, ed., *How Ottawa Spends, 2003–2004: Regime Change and Policy Shift* (Don Mills: Oxford University Press) 72–88 at 85.
40 Doern and Gattinger, *supra* note 26 in chapter 8.
41 Brownsey, 'Canadian Energy Policy,' *supra* note 27 at 88.
42 Office of the Auditor General of Canada, *supra* note 33 at 12.
43 IEA, *Energy Policies of IEA Countries: Canada, 2004 Review* (Paris: IEA, 2004) at 62.
44 Ibid. at 7.

45 Ibid. at 63.
46 Environment Canada, 'Speech: Statement by the Honourable John Baird, Minister of the Environment on the release of the Intergovernmental Panel on Climate Change 4th Assessment Report, Paris, France (2 February 2007),' online, http://www.ec.gc.ca, accessed 11 October 2007.
47 National Resources Canada, 'Notes for a Speech by The Honourable Gary Lunn, P.C., M.P., Minister of Natural Resources at the Economic Club of Canada, Calgary, Alberta, March 12, 2007,' online, http://www.nrcan-rncan.gc.ca/media/speeches/2007/200721_e.htm, accessed 11 October 2007.
48 In the interests of full disclosure, note that the author served as a member of an Advisory Panel on Sustainability for the Integrated Power System Plan (IPSP). Note, as well, that the OPA acknowledges that the final version of the IPSP does not necessarily reflect the views of any contributing members of the advisory panel.
49 R.B. Gibson, *Sustainability Assessment: Criteria and Processes* (London: Earthscan, 2005).
50 EIA, *State Energy Data, 2004* (Washington, DC: EIA) at Table R2.
51 Howard Geller et al., 'Policies for Increasing Energy Efficiency: Thirty Years of Experiences in OECD Countries' (2006) 34 Energy Policy 556 at 568–70.
52 Erwin Chemerinsky et al., 'California, Climate Change, and the Constitution' (2007) 37 Environmental Law Reporter at 10653.
53 IEA, *Energy to 2050: Scenarios for a Sustainable Future* (Paris: IEA, 2003) at 16.
54 Ibid. at 112–14.
55 Ibid. at 130.
56 Ibid. at 142.
57 NRCan, *Energy Technology Futures* (Ottawa: NRCan, 1999) at 11. Online, http://www.bctia.org/files/PDF/energy_scenerios/Energy_Technology_Scenarios_-_NRCan_-_2004.pdf
58 NEB, *Canada's Energy Future: Scenarios for Supply and Demand to 2025* (Calgary: NEB, 2003) at 7.

PART SIX

Policy Obstacles and Opportunities

15 A Proposal for a New Climate Change Treaty System

SCOTT BARRETT

Addressing global climate change requires international agreement for a simple reason: it is a global problem, riddled through with transnational externalities. The *Kyoto Protocol* – the first major international agreement on climate change – is a poor first attempt. It suffers from three deficiencies: First, it addresses only one dimension of the problem (reducing greenhouse gas emissions) and fails to change the incentives that cause this externality. A more comprehensive approach is required, one that embraces alternative forms of mitigation and is designed to overcome freerider incentives.

Second, Kyoto offers only a short-run remedy to a centuries-long problem. Addressing climate change will eventually require the adoption of breakthrough technologies worldwide. Kyoto provides modest incentives for innovation, but it provides little or no incentives for countries to carry out fundamental research. Substantial investment in research and development (R&D) is needed; it should be undertaken now.

Finally, Kyoto mistakenly ignores developing countries. It should have aimed to put the fast growing developing economies onto a different kind of development path and assisted the most vulnerable poor countries to adapt to inevitable climate change. Rich countries should finance both of these efforts, in addition to reducing their own emissions.

The scale of the effort required is without precedent. How should we meet this challenge?[1]

Public Goods for Climate Change

To address global climate change, five different kinds of public good must be provided:

First, global emissions of greenhouse gases must be reduced (relative, at least, to a 'business as usual' benchmark). Reductions from any country are a public good since greenhouse gases disperse themselves evenly worldwide. Emissions reductions will require some combination of energy conservation, fuel substitution, a shift to renewable energy, and carbon capture from the exhaust gases of power plants that burn fossil fuels.

Second, the knowledge of how to do all of these things on a massive scale is essential. This knowledge is a public good. Fundamental new energy and related technologies are needed, the discovery of which will require investment in basic research (to correct the 'innovation market failure').

Third, we need to begin to consider the possibility of removing carbon dioxide (CO_2) directly from the atmosphere, by means of new industrial processes. CO_2 can be removed by planting trees, which eat CO_2 as they grow, but large scale tree planting will have environmental consequences, and trees may reduce albedo in the higher latitudes, causing temperatures there to rise. It is more important to prevent deforestation, especially in the tropics. CO_2 can also be removed by fertilizing the oceans with iron, to stimulate growth of CO_2-eating phytoplankton, but current research suggests that this cannot reduce concentrations by very much.

Fourth, we must also contemplate the possibility of reducing the amount of solar radiation that strikes the Earth, to counteract the effects of increasing atmospheric concentrations of greenhouse gases. The eruption of Mount Pinatubo in 1991 did this naturally, lowering the Earth's surface temperature by about 0.5°C the next year. 'Geoengineering' would essentially fabricate a similar effect; the most developed proposals would throw sulphate or engineered particles into the stratosphere, where they would linger for a few years before being 'rained out' over the poles. Such an intervention would introduce new risks, but it may help to reduce the risk of abrupt and catastrophic climate change.

Finally, societies will have to adapt to climate change at the local, national, and regional levels, and some adaptation will involve the supply of public goods (for example, augmenting the Thames Barrier to protect London from flooding is a local public good).

All these public goods are interrelated. Emission reductions and the knowledge of how to produce and distribute energy without releasing greenhouse gases are complements. By contrast, efforts to reduce atmospheric concentrations of greenhouse gases, geoengineering, and adaptation are all substitutes. Because of these connections, the provision of these different public goods needs to be coordinated. We don't need one international agreement; we need a system of interlocking agreements, all gathered together under a revised Framework Convention that recognizes the higher need to reduce climate change risk.

Crucially, the provision of these different public goods also involves very different incentives. I discuss these next.

Emission Reductions

Reductions in global emissions depend on the aggregate efforts of all countries. Provision of this kind of global public good is especially vulnerable to free riding. Its provision requires enforcement.

The difficulty of enforcing an agreement is expressed in three different ways. First, participation in a treaty is voluntary, and some countries may decline to participate. The failure of the *Kyoto Protocol* to get the United States on board illustrates this problem. Second, an agreement must make it in the interests of states to comply after having signed on to (ratified) the international agreement. Canada signed on to the *Kyoto Protocol*, yet its emissions are on course to exceed Kyoto's prescribed level by 45 per cent or more. The government has just announced a policy to cut emissions, but by the government's own admission that policy will not meet the Kyoto targets. Finally, the demands made by an agreement can be diluted so as to ensure that countries participate and comply. China and India are both parties to the *Kyoto Protocol*, and they are sure to comply with the agreement, but that is only because the agreement does not require that they limit their emissions.[2]

How to address the enforcement problem? Joseph Stiglitz has proposed using trade restrictions as a means of enforcing participation in the Protocol.[3] So did Dominique de Villepin when he was Prime Minister of France. It is hard to see how substantial mitigation can be achieved without such a mechanism. However, it is just as hard to understand how trade restrictions could be adopted. Doing so would require an amendment (or a new agreement). To pass, the amendment would have to be adopted by three-quarters of the parties to the *Kyoto Protocol*. Even then, the amendment would apply only to the countries that ratified it.

To be effective, enforcement would have to apply to compliance as well as to participation – otherwise countries would simply choose to participate, to avoid being subject to a trade restriction, and yet fail to change their emissions. The existing parties may be reluctant to adopt such a measure, especially as at least some of the existing parties are at risk of not complying.

Should trade restrictions be adopted, it cannot be assumed that the rest of the agreement will be left unaltered. The minimum participation level might be increased; the emission reduction obligations might be weakened. Changes like these are to be expected because, to be effective, trade restrictions must be severe, and yet as trade restrictions become more severe they also become less credible (they would hurt the countries imposing them as well as those on the receiving end). Further, trade restrictions would be difficult to apply as a practical matter, since the carbon emitted in the manufacture of a good cannot easily be determined.

Finally, the response of the countries targeted by the trade restrictions must also be considered. These countries might respond by acceding to the agreement – the desired response. But they might also respond by applying trade restrictions of their own. The use of trade restrictions to enforce the *Kyoto Protocol* could provoke a trade war.

Given these difficulties in bringing about trade restrictions, and the reliance of the *Kyoto Protocol* on such restrictions, it seems likely that the Kyoto approach can only succeed in reducing emissions by modest amounts, supported mainly by domestic enforcement. To do more will require a different approach.

More might be achieved by shifting attention away from emission limits and towards new technologies. Provided these technologies exhibit certain characteristics, such as network externalities or substantial domestic benefits independent of their effects on climate change, it is possible that their adoption can be promoted without the need for international enforcement. Currently, technologies of this kind do not exist. This is just one reason why more knowledge is needed.

Knowledge

The *Kyoto Protocol* provides little or no incentive for countries to discover fundamental knowledge (the kind that cannot be patented) concerning new technologies that can produce energy without releasing greenhouse gases.

Knowledge is a different kind of public good. Discrete knowledge can be supplied by a single best effort, making it less vulnerable to free riding. An example is the ITER project, which aims to sustain a nuclear fusion reaction at full scale – an essential ingredient into demonstrating the technology's scientific and technical feasibility. The ITER is now being built in France, financed by the European Union, China, India, Japan, South Korea, Russia, and the United States.

Though the example of nuclear fusion is relevant, it is also somewhat special; it offers benefits that are unrelated to climate change. The value of knowledge for addressing climate change will in most cases depend on the prospects of this knowledge helping ultimately to reduce emissions. Hence, the willingness of countries to finance R&D will be linked to the likelihood of any new technologies emerging from the R&D being diffused. If the technologies emerging from R&D have little chance of being adopted, the incentive to invest in R&D will be dulled. For this reason, R&D must not only develop technologies that reduce emissions; it must develop technologies that reduce emissions and that are likely to be adopted globally.

Air Capture

The idea of capturing CO_2 from the air by means of an industrial process (a related but different idea from removing CO_2 from the emissions of power plants) and then storing it in some way has yet to be tried, even as a pilot project. However, the concept has the potential to transform the problem of stabilizing atmospheric concentrations. Air capture can reduce atmospheric concentrations even if global emissions continue to rise. Air capture also has the advantage of being decoupled from our energy systems; the technology can be located anywhere. Finally, it can be undertaken unilaterally or by a small 'coalition of the willing.'

Preliminary estimates of the economics of air capture suggest that it will not play a role in the near future – it is too expensive. Should current obstacles be overcome, however, the challenge would be political. If individual countries or small groups of countries had the wherewithal and the incentive to reduce atmospheric concentrations independently, should they be allowed to do so? This is a question of governance.

Geoengineering

The economics of geoengineering – which I take here to mean 'solar radiation management' – are much more attractive, making its governance a more pressing matter. A number of countries may, in the next few decades, have an incentive to undertake it unilaterally, or as part of a small coalition. The problem is that the consequences of using this technology would have implications for other countries – and not all of them favorable. Geoengineering would entail a large-scale experiment, not unlike the one it is meant to address (climate change caused by rising atmospheric concentrations of greenhouse gases). Some countries may benefit from climate change, at least in the medium run, and geoengineering would harm them (alternatively, these countries may seek to engineer a warmer rather than a cooler climate). Geoengineering may also alter regional climates, even as it stabilizes the global average. Finally, geoengineering would not address the related environmental problem of ocean acidification.

So, which countries should decide whether geoengineering ought to be tried? Ironically, geoengineering has the opposite problem of emission reductions. The latter is limited by free rider incentives. The former is not.

Adaptation

The incentive for countries to adapt to climate change – another substitute for mitigation – is also strong. Indeed, much adaptation will occur 'automatically,' via the market mechanism. Some adaptation, however, will require the provision of local, national, and even regional public goods, like sea defences, dikes, and large-scale irrigation projects.

Poor countries are relatively the most vulnerable to climate change for three reasons. First, they tend to be located in the low latitudes. In a sense, these countries are already 'too warm,' and climate change would make them even warmer. Second, poor countries depend on the natural environment for a larger share of their livelihoods. Agriculture as a share of income is much higher in poor countries than in rich ones, and agriculture is especially vulnerable to climate change. Finally, poor countries typically have weak domestic institutions. They are the least able to supply the local, national, and regional public goods of adaptation.

The *Stern Review* on climate change concluded that today's rich countries should cut their emissions (and finance cuts in the emissions of today's poor countries) dramatically in the short run, to assist today's poor countries a century or more from now.[4] However, this reasoning fails to link mitigation to adaptation assistance. Rich countries should also invest today in the adaptation needed to help today's poor countries not only in the distant future but sooner. The most vulnerable countries need to be more robust to climate change.

Consider as an example the connection between climate change and malaria. Climate change is expected to increase malaria prevalence in the future, mainly by extending the range of the mosquito vector to higher elevations. Malaria might increase 5 per cent a century from now because of climate change. Mitigation could reduce this increase a little bit, but investment in the R&D needed to discover and develop a malaria vaccine could reduce malaria prevalence across-the-board – and sooner. Similarly, R&D into new agricultural technologies could lift agricultural productivity throughout the tropics and, in the bargain, make the countries in these regions less vulnerable to climate change.

From the perspective of self-interest, the incentives for rich countries to assist poor countries to adapt are weak. However, the moral imperative for them to do so is strong.

A System of Agreements

How to proceed? The existing international arrangement consists of a framework agreement establishing the objective of limiting atmospheric concentrations and a protocol intended to make a start in meeting that objective, to be succeeded by a sequence of follow-on agreements that achieve even more. This arrangement is too narrow in focus. Atmospheric concentrations do need to be limited, but more needs to be done. The overall objective should be to limit climate change risk, and achieving this will require not a linear sequence of agreements but a system of interlocking agreements.

Though the *Kyoto Protocol* cannot be enforced internationally, many of its parties will take steps to reduce their emissions. Eventually, all of these actions will probably be organized under an international framework. That framework, however, will do little more than coordinate the activities of different states. To achieve more will require a change in strategy. It will require, in particular, a focus on new technologies.

Poorer countries – especially the fast-growing ones like China and India – must play a part in reducing emissions. Kyoto's project-based clean development mechanism was intended to help with this, but it is burdened by significant transactions costs. It makes more sense for countries to agree to establish technology standards for new investment, and for the richer countries then to finance the adoption of these new technologies in poor countries. The main task for an international agreement on technology transfer will be to specify the technologies to be adopted. Agreement is also needed on cost sharing, but there are precedents for this; cost-sharing arrangements should not be hard to negotiate.

R&D should be a priority, since our ability to reduce emissions substantially in the long run depends on it succeeding. Agreements to produce discrete knowledge already exist – an example mentioned earlier being the ITER. Many more agreements like this will have to be added, with participation in individual agreements depending on the overall cost and the interests of states in particular technologies.

Air capture and geoengineering must also be contemplated. Compared with emission reductions, both of these interventions have certain advantages. Both also introduce new risks. The challenge in these cases is not so much free riding as governance.

Climate change is inevitable, and as mentioned before, the poorest countries are especially vulnerable. They must be helped. Technologies like a malaria vaccine, if provided by the rich countries, would make it easier for poor countries to adapt to climate change, and contribute to their development as well – a kind of compensation for expected climate change damages.

Notes

This chapter was previously published in *Economists' Voice*, www.bepress.com/ev, October 2007, and is reprinted by permission. I am grateful to Aaron Edlin and Larry Goulder for comments on an earlier draft.

1 See, for more detail, S. Barrett, *Why Cooperate? The Incentive to Supply Global Public Goods.* (Oxford: Oxford University Press, 2007).
2 See S. Barrett, *Environment and Statecraft: The Strategy of Environmental Treaty-Making* (Oxford: Oxford University Press, 2005) (paperback edition).
3 J.E. Stiglitz, 'A New Agenda for Global Warming' (2006) 3(7) The Economists' Voice Art. 3. Available at http://www.bepress.com/ev/vol3/iss7/art3.
4 N. Stern, *Stern Review: The Economics of Climate Change* (Cambridge: Cambridge University Press, 2007).

16 Climate Change and Global Governance: Which Way Ahead?

JOHN DREXHAGE

1. Introduction

Climate change poses serious challenges to traditional global environmental governance models. It is, therefore, a fascinating issue on a number of fronts. For one, it represents a strong challenge to traditional (neo-)realist paradigms of international order, which assume state or national hegemony in an anarchic world. The staying power of the neo-realist model in frustrating real progress on climate change should not be underestimated, however. This dimension will not be addressed in this paper.[1] Second, the issue represents a concrete manifestation of sustainable development. While at its core climate change remains an environmental issue, the responses required to effectively address it lie far beyond traditional environmental challenges (such as ozone depletion or acid rain). So far beyond, in fact, that legitimate questions arise as to whether the appropriate policy and/or negotiating forums for addressing climate change should be left in the hands of environment ministers. This article argues that to address the multi-faceted climate challenge we face, governance efforts must evolve beyond the current global regime-building model, and that environmental and development policies must become much better integrated.

2. Rethinking Climate Governance

In the 20-plus years that climate change has been a subject of serious international negotiations, we have seen a trend of broadening participation in those deliberations, but, for the most part negotiations continue to be led by environment departments and environmental constituencies. Initially, when the science of climate change was the dominant topic, the discussions were, not surprisingly, dominated by climate and meteorological specialists who sometimes were based in environment departments and sometimes not. In Canada, for example, the initial group responsible for negotiating the *United Nations Framework Convention on Climate Change* (UNFCCC) was Atmospheric Environmental Services Canada, the sector of Environment Canada responsible for weather forecasts and atmospheric sciences. This was fairly typical of most countries, with the notable exception of the United States, where all international negotiations – including those on environmental issues – have been led by the State Department.

Much of the reason the environment departments took such a predominant position in all matters relating to climate change – including mitigation and adaptation – is rooted in the establishment of the Intergovernmental Panel on Climate Change (IPCC). It was founded by the World Meteorological Organization (WMO) and the United Nations Environment Programme (UNEP) and was mandated to assess – on a comprehensive, objective, open, and transparent basis – the scientific, technical, and socio-economic information relevant to understanding the scientific basis of the risk of human-induced climate change, its potential impacts, and options for adaptation and mitigation.[2] What was interesting right from the start was that, despite the fact that the IPCC was very much the brainchild of UNEP and WMO, it ventured into areas far beyond their particular area of expertise. While Working Groups (WG) I and II, focusing on the science and impacts of climate change, clearly implicated the climate and environmental scientific community, WG III, focusing on adaptation – in the First Assessment Report – and mitigation activities, required expertise far different from climate or environment. Hence it was made up of development economists, energy specialists, agriculturalists, foresters, and so forth.

And yet, WG III continued to be led by individuals and staffed by secretariats that first had extensive experience in climate change matters. Even when it did begin to engage specialists from these other fields, it usually was dominated by those who had a strong background in the climate change field – that is, more often than not, they would be energy specialists from an environmental department rather than pure energy specialists.

It must be noted that this issue has been recognized by the IPCC in its development of the Fourth Assessment Report. WG III, under the leadership of Bert Metz of the Netherlands,[3] actively sought out sector experts and industry specialists in the development of the report, and also held a series of outreach sessions with industry around the world to ensure that they had ample opportunity to contribute to the development of the final report.

Nor, of course, does the panel work in a vacuum. It served as the credible, independent source of scientific information for the development of the UNFCCC and later the *Kyoto Protocol*.[4] In particular, the role of the first and second assessment reports, respectively, cannot be underestimated in laying the groundwork and support for the convention and the protocol.

It is hoped that the Fourth Assessment Report will work in the same manner, providing momentum for the successful launch of the post-2102 negotiations at Bali. In this interaction between the IPCC and the UNFCCC, it must be kept in mind that many of the government reviewers of the Summaries for Policy Makers and the Synthesis Reports were also, in fact, negotiators at Rio and Kyoto. What we constantly have to be on the watch for, then, is having policy developed by a tightly knit climate change community that does not sufficiently reach out to the mainstream of policy-making.

Of course, one of the other critical institutional features of the UNFCCC process was, and continues to be, an extremely competent and (despite formally answering to the UNEP) autonomous secretariat. One of the unintended impacts of this feature had been to further marginalize the UNEP as an effective international champion for sustainable development – the very issue that has helped raise the environment to the top of the global agenda was the one issue that the UNEP has had the least direct control in managing (at least at the international policy/management side). Having such a competent secretariat in place also had the unintended effect of limiting cross-fertilization with other UN multilateral institutions in championing sustainable development. This fact has not only helped to further marginalize the UNEP but also made more difficult coordination with other agencies and institutions that were not able to demonstrate the same degree of commitment and/or capacity.

All of these factors played a role in determining where we ended up with Kyoto. What at the end of the day was the outcome of Kyoto? And can we learn lessons from that experience so that the post-2012 regime captures a broader group of emitters with more realistic targets, particularly for countries whose economies are rapidly growing and rely on natural resources for a large part of that growth? Kyoto played a critical and necessary role in establishing a global value for carbon and in sending positive investment signals, directly and indirectly, for clean-energy investments worldwide.[5] This alone is a tremendous achievement and, in the view of the author, more than justifies the treaty coming into force. In addition, it set in place the critical architecture for responding to climate change: covering reporting, monitoring, verification, and compliance regimes (as weak as the latter are), and coordinating market mechanisms under the clean development mechanism (CDM), joint implementation (JI), and international emissions trading (IET). Kyoto's primary weakness, in my view, was the politically charged, top-down process by which targets were established, with all too little thought by country leaders whether those targets were achievable and how we could go about achieving them. I would submit that now is *the* time, starting at Bali, to get that dynamic right and not be so panicked about having ever more stringent reduction targets (accepted by fewer and fewer participants) in order to maintain fealty to environmentally correct thinking. The targets established at Kyoto were much more the result of an agreement amongst G8 leaders trying to 'outgreen' one another than any rigorous analysis.

There is a growing consensus that, at a minimum, global greenhouse gas (GHG) emissions will need to be reduced at least 50 per cent by the middle of this century. Clearly, achieving such a goal will require the engagement of all major economies and, just as clearly, those same countries need some time and opportunity to seriously figure out what they can do domestically; how regional or international coordination can help; and what the potential contributions of discrete sectors are in that formulation. Give the economies some time to address these questions *seriously* – publics won't let them do otherwise – and then revisit the possibility of a globally binding regime by the end of this decade, assuming countries are now much more informed and engaged on what they can actually accomplish.

What does this mean for the short- and long-term international governance of climate change? The UNFCCC should continue to play the critical environmental role as the home and protector of Article 2 – the ultimate objective of ensuring that anthropogenic interference does not permanently damage the global environment. It also needs to continue reviewing countries' actions and should expand that activity to review the effectiveness of regional and bilateral efforts to reduce greenhouse gas emissions. In other words, it should be the 'bellwether,' in cooperation with the IPCC, in notifying where emissions are going and what the likely concentration-impacts of that would be. It should also serve as the pressure point in clearly identifying what emissions reductions would be expected at these different levels and report on the extent to which those targets are being met.

In addition, a more effective regime clearly needs to be established on the mitigation side – and here we could look at the possibility of establishing expert roundtables or forums where industry, academics, and governments can work together and seriously commit to ways in which they can cooperate to alter the course of development in climate-friendly and clean-energy directions. This in no way should be considered as an out for addressing climate change as an urgent issue. We have an increasingly limited time to get this right – anywhere from 120 to 200 months to stabilize global GHG emissions if we want to avoid the risk of serious environmental and social damage. In that respect, while some may want to use the Asia-Pacific Partnership on Clean Development and Climate (AP7) model as an example of how to proceed on the mitigation side, I am proposing a model with some important differences. For one thing, these sector groups need to agree on what their contributions to a global reduction should be. Initially, if sectors can't come to agreement, one may simply use their current emissions profile in determining what would be an appropriate contribution. But it might also be possible to determine contributions in other formats areas – for example, in the case of carbon capture and storage (CCS), by setting target dates for a certain percentage of coal-based plants in a country or region to have CCS implemented on-site. Another example might be supporting renewable energy by setting targets for the penetration of renewables in a country or region's energy profile. This means, of course, that any such initiative must have some real and significant money behind it, with clear programs of action. And, of course, it must have a broader constituency comprising all major economies, including that of the European Union.

One requirement for such a regime would be a new maturity in industry whereby, and this is particularly the case for the energy industry, it will need to depart from its parochial ways and truly seek solutions in its sectors that work for the common good. Industry must do more than either justify current practices or set about focusing on why their particular technology represents *the* answer to all the world's woes. A tall order, I know, but an absolutely critical one for industry's more serious and active participation to have any credibility.

Does this activity need to take place strictly under UN auspices? Perhaps I might try to reformulate a legendary Canadian policy response to fit this particular debate. During World War II, Canada's Quebec-Anglo relations were severely tested on the issue of mandatory military service, with Quebec strongly opposing any such measure and English Canada supporting it as strongly. Prime Minister William Lyon McKenzie King's answer to the question (before the government had decided in favour of conscription): 'Conscription if necessary, but not necessarily conscription.'

I would submit we are in the same kind of sensitive quandary on the question of an international regime on GHG mitigation activities – outside the UNFCCC if necessary, but not necessarily outside the UNFCCC. This proposition is in no way intended as a slight to the already mentioned unparalleled competence of the secretariat. In fact, the secretariat needs to be strongly commended, particularly under the current leadership of Yvo De Boer, for seriously exploring innovative ways in which non-governmental actors, including industry, can play a more effective role in the multilateral process.

My argument has more to do with the current reluctance of major economies – including three of the top four global emitters – to submit their GHG emission activities to strict, internationally binding commitments. If, for example, a mitigation regime strictly under the UN means further delay in the United States on a post-2012 agreement, due to its Senate being unable to ratify such an agreement, then why not try and set up an alternative structure, even if only as an initial step? Or, given the challenges faced in ratifying any internationally binding agreement in the U.S. Senate, could we actually envision a situation where the UN regime would apply everywhere *but* the United States? And if so, what would motivate major developing-country economies to agree to submit to a system the United States had refused?

These are all extremely difficult questions of course. What can the UN system do to build more confidence on the part of major economies to submit to an internationally binding GHG-emission-cap regime? The UN system certainly needs to continue to find new ways in which to engage actors other than states in their particular areas of expertise. Indonesia's leadership is calling for separate meetings of finance and trade ministers, respectively – another initiative that needs strong support as a way of demonstrating that the solutions to climate change lie well beyond the brief of environmental policy. Is it perhaps time, as the Pew Centre, the World Business Council for Sustainable Development (WBCSD), and others are proposing, to relax the reins a bit in the

UNFCCC when it comes to absolute, legally binding targets? Advocates argue that this may provide an opportunity for a serious re-visiting of what can actually be accomplished, and, once that is clear, major economies may be less reluctant to commit to an internationally binding regime. Others are legitimately concerned that such an approach would play into the hands of the disingenuous who will have found yet another effective tactic to delay making any real progress.

However, there may be a new trump card on the global policy horizon which will force the issue – global public opinion. Unlike at Kyoto, the public won't let us get away with smoke and mirrors any longer, so now governments have little choice but to seriously address the issue. The goal over the longer term, certainly by the end of the next decade, is a mitigation regime that finds a home in a reformed UN – one that has managed to make itself less a state-centric institution (at least on the issue of climate change) while being able to effectively engage economic, natural resource, and energy decision-makers in both the public and private sectors.

Some thought also needs to be given to the international carbon market and whether the UN should continue to be the home for the international registry, recording and approving all individual transactions under IET, CDM, and JI. Again, to this point the UNFCCC has played an invaluable role in getting these mechanisms off the ground. But as international standards on the different modes of carbon market transactions become codified, thought should be given towards transferring these tasks to an external entity (or entities), whereby an independent body would oversee a range of national and private registries.

3. Development and Climate Change[6]

A sustainable-future regime also has to be closely tied to an aggressive development agenda for developing countries and least-developed countries (LDCs) consistent with the agreement or 'bargain' struck in drafting Agenda 21 at the Rio Summit in 1992: namely, that developing countries will shoulder important environmental responsibilities and, in return, developed countries will take on serious commitments to help fund and support them in that process. At the very least, it was expected that member governments of the Organisation for Economic Co-operation and Development (OECD) would meet their commitments to provide 0.7 per cent of their GDP to provide development assistance. Other modalities of this deal have been the subject of debate ever since, but an important area of unexploited potential lies in global trade and investment.

In the area of trade, the current Doha Development Agenda is in danger of coming to a standstill, in no small part because countries cannot agree on what constitutes appropriate trade measures to foster development in developing and least-developed countries. This question is an important one to get right; these negotiations have more implications for development than any number of bilateral initiatives by official development agencies, even if countries were to meet their 0.7 per cent commitment (which, it is clear, the vast majority of OECD countries will not). A good deal on the Doha track could ease negotiations around a post-2012 climate change regime, by creating the necessary good will; by fostering the requisite economic growth and restructuring that will make developing countries better able to contribute; and through targeted reforms and provisions specifically designed to help trade law and policy combat climate change and contribute to the deployment of clean energy systems and technologies.

In the area of investment, there is no obvious institutional home for international efforts to foster the critically needed flows of clean energy investment in developing countries, and to help ensure that they foster development. In part, this is because the investment regime is scattered among more than 2,500 bilateral investment treaties and a growing number of investment chapters in free trade agreements. One possibility for international efforts in this area might be the Energy Charter Treaty[7] – a pan-European and Asian treaty designed to foster increased energy investment and trade – which could house a new initiative explicitly based on the principles of clean energy investment.

Nor has the issue of climate change been effectively integrated into the mainstream activities of development agencies. Developing countries have, for the most part, not identified climate change as an issue of concern to development agencies. A number of analyses have indicated that, while there have been some successful initiatives, particularly those related to supporting the G77 and China in their national communications under the UNFCCC and, to a lesser extent, helping them develop national adaptation strategies, these successes have not spread into 'normal' technical assistance. In other words, the strong linkages that do exist between the threat of climate change and poverty eradication and development are still not appreciated at the field level. A challenge on the donor side is to engage finance and development planners effectively in the climate policy discussion, whereas recipients have to acknowledge that they need more effectively to identify climate change in their development planning activities.

At the end of the day, the most critical component in developing a global regime on climate change with the full engagement of developing countries is a much more effective basis and means of complementarity among official development assistance (ODA), foreign direct investment (FDI), and sustainable development. Surely ODA and private financial resources can play more effective complementary roles than is currently the case. The efforts of Bretton Woods institutions, including the World Bank (WB) and the International Monetary Fund (IMF), have been making progress, but more needs to be done in terms of financial contributions and actions. The focus needs to be on 'greening' the process of economic development, including providing clean energy to those without, improving forestry practices to slow deforestation, and putting in place sustainable urban transportation systems.

If the CDM and other market-based initiatives can successfully fund mitigation activities in developing countries, they will help ensure that limited ODA funds can be used most effectively where the private sector is not likely to be nearly as active – for example, in the field of adaptation. But we should be careful not to be too simplistic or formulaic in our prescriptions – there is a clear need for ODA to support capacity-building activities related to the CDM – for example, in helping developing countries set up national designated authorities or in helping them to develop national sustainable development criteria. The WB's Carbon Fund and the initiatives of governments such as the Netherlands (although not directly tied to ODA) have been extremely useful in helping to ensure that the CDM is a major player in the international carbon market. That said, as the carbon market matures, particularly after 2012, one hopes that the WB and national governments will play a less prominent role in developing a certified emissions reductions (CER) market, leaving it to the private sector to be the major player in that market.

Nor should we immediately dismiss prospects for private-sector participation in adaptation-related activities. Private-public partnerships, such as joint ventures between insurance and investment firms with the cooperation of multilateral development banks (MDBs) could go a long way towards funding adaptation-related activities. In addition, CDM carbon-sink investments, for example, if properly designed, can provide sustainable mitigation and adaptation benefits. Traditional climate policy tends to isolate adaptation and mitigation and assumes that one chooses from a portfolio of independent adaptation and mitigation options. It is argued that adaptation benefits are felt locally in time and space, whereas mitigation benefits (as opposed to the direct benefits of energy provision) are felt later in time and on a global scale. Even if large methodological hurdles can be overcome to allow costs and benefits to be reliably estimated on vastly different temporal and spatial scales, mitigation and adaptation measures are only substitutable at the global level and relevant only to some non-existent global decision-maker. However, such analysis provides no practical insight at the project or national/regional scale, where adaptation and mitigation decisions will actually be made. The potential for project-level integration of adaptation and mitigation is also downplayed, and likely reflects the residual northern domination of the climate debate. Instead, it might be more beneficial and effective if we examined the potential for adaptation and mitigation synergies, particularly to the extent that such activities support ecosystem-oriented poverty alleviation priorities, as counselled by the World Summit on Sustainable Development in its Plan of Implementation in 2002.[8]

Poverty is both a driver and an outcome of critical sustainable development–climate linkages such as energy deprivation, desertification, and deforestation. The ecosystem focus to poverty alleviation moves us beyond the rather platitudinous observation that the poor are endowed with the least adaptive capacity and hence are most vulnerable to climate change, towards a practical intervention policy. The WB's initiatives on the Community Development Carbon Fund,[9] as well as the Biocarbon Fund,[10] represent innovative investments that provide a twinning of adaptation and mitigation opportunities. Recognizing the challenge of delivering GHG reductions in a competitive CDM market environment, these funds are explicitly established to help small-scale projects from the local community become competitive in the global market. In focusing on adaptation opportunities, while also emphasizing GHG-reduction credit opportunities, the Bank is helping to highlight the potential role of the private sector in natural resource management activities.

The key implication is that a coherent climate policy as it relates to developing countries must become much more closely aligned with and, indeed, one aspect of a sustainable development pathway committed to poverty alleviation. Climate change mitigation is a large co-benefit of this approach. The reader is cautioned that the intersection of adaptation-mitigation benefits is not proposed as a panacea for climate policy; it is, however, proposed as a logical and equitable prerequisite to engaging the South in an eventual comprehensive post-Kyoto mitigation regime.

In relation to ODA, it must be emphasized that the extent to which the market can help bear the costs of climate change, including adaptation, is the extent to which we are dependent on ODA to deliver on an issue that is but one of many, and vastly less important than most developing countries' immediate priorities for development and poverty eradication.

4. Conclusion: Bringing It All Back home

With respect to developing countries, then, it is critical that attention be paid to domestic implementation mechanisms and priorities. In particular, institutionalization of climate change issues in domestic government agencies would effectively create 'champions' for mitigation and adaptation within governments of developing countries. This engagement is a crucial step, which would build a constituency for action and help give domestic and foreign businesses and NGOs reliable points of contact to engage governments on climate change. It also means much more effective co-ordination between aid agencies and international financial institutions (IFIs) and enhanced coherence, in turn, with the FDI flows to developing countries. And finally, above all, for OECD countries it means showing leadership at home.

OECD countries must demonstrate that they are taking significant actions at home to mitigate climate change and do so without compromising their economic objectives. Until developing countries can see that this is in fact the case, the prospects for bringing them aboard will always be limited. But what, at the end of the day, is the proper role of development aid and financial-flow considerations in the post-2012 negotiations? First of all, we would strongly advise that parties gain a much more realistic understanding of how appropriate development takes root in developing countries. For example, in the area of technology transfer, it needs to be recognized that most of these technologies are not in fact a public good, but the result of private-sector investments. Even if OECD countries strongly increase their ODA contributions, what will be made available for climate change is likely to be limited. This calls for innovative solutions, the diversity of which we have only just begun to explore. For example, the work of Lewis Milford of the Clean Energy Group in exploring the potential precedent of innovative approaches in the distribution of AIDS pharmaceutical products is a valuable contribution to this discussion.[11]

My prognostication for the future? Not entirely well-founded, but it remains (guardedly) optimistic. I would fully expect that by 2025 we would be back to where we started in a sense – a multilateral system of internationally binding targets, but the important difference would be twofold. The emission targets would be met *and* we would have a much broader community of major emitters engaged in those activities. In other words, today countries need to take a deep breath, look seriously at what they can do and by when, and with that information confidently go forward in joining an internationally binding regime that will literally determine the mode of societies' development over this century and beyond.

Notes

© 2007, International Institute for Sustainable Development. This paper is a slightly modified version of a policy paper that was previously posted at http://www.iisd.org/. It is an output of the 'Mapping Global Environmental Governance Reform' project of the International Institute for Sustainable Development (IISD). The initiative was conceived and funded by the Ministry of Foreign Affairs, Government of Denmark. The author extends his thanks to colleagues Aaron Cosbey and Adil Najam for their preliminary comments and suggestions.

1 It will be the subject of another article by this author in the future.
2 See Intergovernmental Panel on Climate Change, '16 Years of Scientific Assessment in Support of the Climate Convention (December 2004) online, http://www.ipcc.ch/about/anniversarybrochure.pdf.
3 WG III is the working group responsible for discussing the range of mitigation activities available to address climate change.
4 *United Nations Framework Convention on Climate Change*, U.N. Doc. A/AC.237/18 (Part II)/Add.1, reprinted in (1992) 31 I.L.M. 849 [hereinafter UNFCCC]; *Kyoto Protocol to the United Nations Framework Convention on Climate Change*, 10 December 1997, U.N. Doc. FCCC/CP/1997/L.7/add. 1, reprinted in (1998) 37 I.L.M. 22, [hereinafter the *Kyoto Protocol*].
5 In personal discussions with both Americans and Australians (neither country ratified the *Kyoto Protocol*), I have been told that the likelihood of either of those countries establishing national/regional systems would

have been much more difficult without the external pressure of the Kyoto regime.
6 This section is based on John Drexhage, 'The Role of Development Assistance and Investment Flows,' in T. Sugiyama, ed., *Governing Climate: The Struggle for a Global Framework Beyond Kyoto* (Winnipeg: International Institute for Sustainable Development, 2005) at 109.
7 *Energy Charter Treaty,* in *Energy Charter Treaty and Related Documents: A Legal Framework for International Energy Cooperation* (September 1994) online, http://www.encharter.org/fileadmin/user_upload/document/EN.p df.
8 'Plan of Implementation World Summit on Sustainable Development,' in *Report of the World Summit on Sustainable Development*, Johannesburg, South Africa (26 August–4 September 2002) UN Doc. A/CONF/199/20, online, http://www.un.org/jsummit/html/documents/summit_docs.html.
9 World Bank (Carbon Finance Unit), 'Community Development Carbon Fund,' online, http://carbonfinance.org/Router.cfm?Page=CDCF.
10 World Bank (Carbon Finance Unit), 'BioCarbon Fund,' online, http://carbonfinance.org/Router.cfm?Page=BioCF.
11 For information on the Clean Energy Group and the work of Lewis Milford, see, online, http://www.cleanegroup.org/.

17 Challenges and Opportunities in Canadian Climate Policy

KATHRYN HARRISON

As a country with one of the highest rates of greenhouse gas emissions per capita (table 17.1) and, not entirely coincidentally, a relatively high per capita income, Canada undeniably has a special responsibility to address climate change. However, at the same time, climate change presents particularly great political challenges for Canada relative to many other advanced industrialized countries. This chapter draws on cross-national comparisons in reviewing three obstacles to Canadian climate policy: the relatively large economic – and thus political – costs of greenhouse gas reductions; historically weak demand for action on climate change from the Canadian electorate; and political institutions that, in combination with the two preceding factors, have deterred policy change. The chapter concludes by considering the recent resurgence of public attention to the environment as a window of opportunity for the adoption of more aggressive climate policies in Canada.

The Costs of Climate Policy

In many respects, the costs of addressing climate change are greater for Canada than for most other industrialized countries. As table 17.1 shows, in response to its natural resource endowments Canada has developed a relatively greenhouse gas–intensive economy that relies heavily on both production of fossil fuels and a manufacturing sector dependent on inexpensive fossil fuel derived energy. Moreover, anticipated growth in production from Alberta's oil sands will only increase the greenhouse gas–intensive nature of Canada's economy.[1]

While the challenge of addressing climate change thus will demand more of Canada than many other countries, politics tends to be biased

Table 17.1 Greenhouse gas emissions per capita and per $ GDP

Country	2004 GHG emissions (without LULUCF) tonnes/person	2004 emissions intensity relative to GDP, kg/$US
Canada	23.3	0.8
Japan	10.6	0.3
United States	24.1	0.6
EU 15	11.1	0.3

Sources: Emissions as reported by individual countries to the UN FCCC (http://unfccc.int/ghg_emissions_data/items/38954.php); Population data from US Census Bureau (http://www.census.gov/ipc/www/idbrank.html); GDP data from UN Statistics Division (http://unstats.un.org/unsd/snaama/selectionbasicFast.asp).

in favour of the status quo. The sectors that stand to lose the most from greenhouse gas reductions employ hundreds of thousands of Canadians and are highly engaged politically, while cleaner industries that might be expected to emerge under an aggressive plan to address climate change remain largely hypothetical, and thus silent, in contemporary political debates. Similarly, even though it has been argued that the benefits of action will significantly outweigh the costs,[2] most of the beneficiaries of actions that might be taken today to address climate change are not yet born, and in any case will not live in Canada.

The challenge for Canada has been compounded by uneven demands made of different parties to the *Kyoto Protocol*. Although the formal targets in the *Kyoto Protocol* for Canada, Japan, the United States, and the European Union (EU) are similar, ranging from –6 per cent to –8 per cent relative to 1990 levels, if one considers the magnitude of departure from a business-as-usual trajectory much greater disparities are evident (table 17.2). In particular, while at the time of ratification Canada anticipated that it would need to make a 29 per cent cut below business-as-usual in order to comply with its *Kyoto Protocol* target, the EU anticipated that it would need only a three to nine per cent cut below business as usual.

The EU target is less demanding, in part, because of its more stable population. While the populations of Canada and the United States both grew by 17 per cent from 1990 to 2005, that of the original 15 EU members covered by the EU 'bubble' grew by only 4.5 per cent.[3] Arguably more important in the EU case, however, are the windfall emissions reductions of two key member states, Germany and the United

Table 17.2. Kyoto Protocol targets relative to business-as-usual trajectory

	Kyoto Protocol Target, reduction relative to 1990 by 2008–2012	Anticipated Reduction relative to 'business as usual' emissions in 2010
Russia	0%	>0%
EU15	–8%	–3% to –9%
Japan	–6%	–12%
Canada	–6%	–29%
United States	–7%	–31%

Sources: Henry and Sundstrom, *supra* note 5; European Environmental Agency, *Greenhouse Gas Emission Trends and Projections in Europe* (2002), online, http://reports.eea.europa.eu/report_2002_1205_091750/en; European Environmental Agency, *Analysis and Comparison of National and EU-Wide Projections of Greenhouse Gas Emissions* (2002), online, http://reports.eea.europa.eu/topic_report_2002_1/en; Tiberghien and Schreurs, *supra* note 6; Government of Canada, *Climate Change Plan for Canada* (2002); United States of America, *Climate Action Report 2002* (2002), online, http://www.gcrio.org/CAR2002/.

Kingdom. Germany experienced a significant decline in its greenhouse gas emissions as a result of economic restructuring following reunification, while the UK has seen deep emissions reductions as offshore gas has replaced coal. These two states alone account for more than 100 per cent of the EU15's commitment under the *Kyoto Protocol*,[4] which has allowed other states within the EU to avoid significant reductions.

The result of varying Kyoto commitments has been varying levels of domestic opposition to ratification by the business community. At one extreme, Russia, which won a *Kyoto Protocol* target *above* its anticipated emissions for the commitment period, saw a strong business lobby in favour of ratification.[5] The EU and Japan, which committed to less ambitious, but nonetheless real, reductions compared to Canada and the United States, faced a mix of business support and opposition. At the other end of the spectrum, the business communities in both Canada and the United States were virtually unanimous in their opposition to ratification.[6] Moreover, when the United States confirmed that it would not seek ratification of the *Kyoto Protocol* in 2001, the economic stakes for Canada and, correspondingly, the strength of business opposition to climate policy measures only increased.

Public Opinion

A second challenge for Canadian climate policy lies in public opinion. Although Canadians have long professed support for actions to address climate change, until recently they simply have not been paying much attention to the issue. In contrast, they *have* been paying greater attention to economic issues that are potentially in tension with serious efforts to address climate change. Even during the high profile political debate over ratification of the *Kyoto Protocol* in the fall of 2002, only 8 per cent of Canadians considered the environment the 'most important problem' facing the country, compared to 33 per cent who cited health care.[7] Indeed, just a few months after the ratification debate consumed the front pages and evening news, roughly half of Canadians did not know that Canada had already ratified the *Kyoto Protocol*.[8] In the years before and since 2002, the environment seldom fared better than fifth place among Canadians' priorities (until 2006, as discussed below). Finally, cross-national polls suggest that voters in Canada have been less concerned about climate change than their counterparts in Western Europe and Japan.[9]

In that context, Canadian politicians have had ample reason to be sceptical of the public's professed willingness to pay for measures to address climate change. Not surprisingly, they have responded with a host of symbolic programs that have a high profile but relatively little impact, such as the One Tonne Challenge, and spending programs that subsidize individuals and the private sector. In contrast to regulation or taxes, subsidies typically do not provoke political opposition both because recipients invariably welcome government handouts and because the costs are widely diffused among inattentive taxpayers. The downside, however, is that subsidies are seldom a cost-effective means of achieving emissions reductions because governments end up paying many individuals and firms to do what they were going to do anyway, with the federal transit tax credit introduced in 2006 a classic case in point.[10]

Political Institutions

The third challenge for Canadian climate policy lies in our political institutions. Canada's first-past-the-post electoral system has yielded broad-based political parties that compete for the median voter, who, as discussed above, has been notably inattentive to climate change. In contrast, proportional representation electoral systems facilitate the emergence of political parties that speak for smaller subsets of voters. Proportional representation has allowed Green Parties to play an influential role in key EU member states, in some cases as a partner in coalition governments, as well as in the EU Parliament, thus providing a stronger push for more aggressive climate policies in Europe.[11]

The effects of federalism are undeniably complex. The quasi-federal system in the EU has been instrumental in fostering EU leadership on the international stage as well as stronger national commitments within the EU.[12] Similarly, led by California, U.S. state governments are increasingly filling the void left by federal inaction.[13] In contrast, Canadian federalism, especially in combination with regionally distinct economies, has been a significant obstacle to adoption of policies to abate greenhouse gas emissions.[14] A critical difference lies in the interests of powerful players. With the advantage of the windfall reductions noted above, it was relatively easy for Germany and the UK to press for strong commitments by the EU. Moreover, these two states were able to facilitate compromise by ceding their reductions to the EU as a whole, thus allowing half of EU member states to continue to increase their emissions. Similarly, in the United States, the leading 'laboratory of democracy,' California, does not have a particularly greenhouse gas–intensive economy.[15] In contrast, in Canada Alberta protects its oil industry, Ontario protects its auto industry, and Quebec defends what it sees as exclusive provincial jurisdiction.

Opportunities for Change

Viewed in this context, Canada faces an uphill battle in comparison to many other advanced industrialized countries. Moreover, there is little prospect for significant change on the first two factors. The auto manufacturing and oil industries will continue to be economically significant, and thus politically powerful, for the foreseeable future, though the possibility of policy change in the United States after the 2008 presidential election would reduce the competitiveness concerns of Canadian industry. Moreover, Canadians are unlikely to undertake wholesale reform of their political institutions, as demonstrated by voters' recent rejection of reforms of electoral systems in two provinces.

There has, however, been significant movement on the remaining factor, public opinion, in recent months. The politics of climate change in Canada shifted dramatically in 2006 as the per cent of Canadians citing the environment as the 'most important issue' facing the country increased from 4 per cent in January 2006 to 26 per cent in January 2007, at which point it was the most frequently cited issue.[16] As in two previous 'green waves,' in the late 1960s and late 1980s,[17] politicians from all political parties scrambled to impress newly attentive voters by strengthening their environmental policy commitments. That fact, combined with a minority government federally in which opposition parties perceive leverage over the governing party on climate change and the prospect of a federal election during a period of heightened salience of environmental issues, would seem to suggest a window of opportunity for policy change.

That said, Canadians have many times before been promised both regulations and spending programs that never materialized. The possibility remains that politicians and the electorate are caught in a chicken-and-egg dilemma. Sceptical of voters' professed willingness to pay for climate measures, for two decades Canadian politicians have embraced the 'sustainable development' mantra that economic prosperity and environmental protection go hand in hand: Canada can save the planet and get rich too. However, in turn, voters may not be willing to sacrifice even now that the time is ripe for action simply because they have been told time and again that they do not need to so. In that regard, it is noteworthy that at the time of Canada's ratification, 78 per cent of voters supported ratification, but 68 per cent also believed, contrary to all government studies, that compliance with the *Kyoto Protocol* would be good for Canadian business.[18] Whether the increased strength of public concern for climate change will be sufficient to prompt significant policy change in Canada thus remains to be seen.

Notes

1 Natural Resources Canada, *Canada's Energy Outlook: The Reference Case 2006* (2006), online, http://www.nrcan.gc.ca/inter/publications/peo_e.html.
2 N. Stern, *The Economics of Climate Change: The Stern Review* (Cambridge: Cambridge University Press, 2006).
3 Population data drawn from US Census Bureau, online, http://www.census.gov/ipc/www/idbrank.html.
4 Germany's and the UK's promised reductions of 21 per cent and 12.5 per cent below 1990 emissions would yield 257.5 MT and 97 MT respectively, which together exceed the EU's commitment of 341.2 MT.
5 L.A. Henry and L. McIntosh Sundstrom, 'Russia and the Kyoto Protocol:

Seeking an Alignment of Interests and Image' (2007) 7(4) Global Environmental Politics 47.
6 M.A. Schreurs and Y. Tiberghien, 'Multi-level Reinforcement: Explaining European Union Leadership in Climate Change Mitigation' (2007) 7(4) Global Environmental Politics 19; Y. Tiberghien and M.A. Schreurs, 'High Noon in Japan: Embedded Symbolism and post-2001 Kyoto Protocol Politics' (2007) 7(4) Global Environmental Politics 70; K. Harrison, 'The Road Not Taken: Climate Change Policy in Canada and the United States' (2007) 7(4) Global Environmental Politics 92.
7 IPSOS-Reid, 'The Public's Year-End Agenda' (2002) 17(6) *Issue Watch*.
8 EKOS Research Associates Inc., *Canadian Attitudes toward Climate Change: Spring 2003 Tracking Study, Final Report* (2003).
9 K. Harrison and L. McIntosh Sundstrom, 'The Comparative Politics of Climate Change' (2007) 7(4) Global Environmental Politics 1.
10 'Transit Tax Credit Proposal Expensive Way to Cut CO2,' *Edmonton Journal* (7 April 2006) B12.
11 Schreurs and Tiberghien, *supra* note 6.
12 Ibid.
13 B.G. Rabe, *Statehouse and Greenhouse: The Stealth Politics of American Climate Policy* (Washington, DC: Brookings Institution Press, 2004).
14 Harrison, *supra* note 6.
15 California also has exported some of the costs of meeting its goals. Although control of automobile emissions is a necessary component of any plan to address California's greenhouse gas emissions, it is relatively easy for California to regulate the automobile sector because the vehicles purchased by Californians are manufactured elsewhere. California also is moving aggressively to limit greenhouse gas emissions associated with electricity production, including from sources outside the state, but counts reductions by out-of-state sources for the purposes of assessing progress toward California's targets.
16 Strategic Counsel, 'A Report to the Globe and Mail and CTV. State of Canadian Public Opinion: The Greening of Canada' (26 January 2007) online, http://www.thestrategiccounsel.com/our_news/polls.asp.
17 K. Harrison, *Passing the Buck: Federalism and Canadian Environmental Policy* (Vancouver: UBC Press, 1996).
18 EKOS Research Associates Inc., *Climate Change and Kyoto* (October–December 2002).

Contributors

Scott Barrett is professor of Environmental Economics and International Political Economy at the Johns Hopkins University School of Advanced International Studies, where he also directs the International Policy Program and the new Global Health and Foreign Policy Initiative. He is the author of *Environment and Statecraft: The Strategy of Environmental Treaty-Making* (published in paperback by Oxford University Press in 2005) and numerous research and policy papers on climate change. He has also advised a number of international bodies on the subject, and was a lead author of the second assessment report by the Intergovernmental Panel on Climate Change. Formerly on the faculty of the London Business School, he has also been a Distinguished Visiting Fellow at the Yale Center for the Study of Globalization. His latest book, *Why Cooperate? The Incentive to Supply Global Public Goods*, was published by Oxford University Press in September 2007.

Steven Bernstein is an associate professor of Political Science and associate director of the Centre for International Studies, University of Toronto. His publications include *Global Liberalism and Political Order: Toward a New Grand Compromise?* co-edited with Louis W. Pauly (SUNY Press, 2007) and *The Compromise of Liberal Environmentalism* (Columbia University Press, 2001), as well as a number of journal articles and book chapters. His current research focuses on the problem of legitimacy in global governance.

Jutta Brunnée is professor of Law and holds the Metcalf Chair in Environmental Law at the University of Toronto. Her teaching and research interests are in the areas of public international law and international environmental law. Brunnée is co-editor of the *Oxford Handbook of International Environmental Law* (Oxford University Press 2007) and the author of *Acid Rain and Ozone Layer Depletion: International Law and Regulation* (1988), as well as of numerous articles on topics of international environmental law and international law. She is a member of the International Law Association's Committee on the Use of Force and of World Conservation Union's (IUCN) Environmental Law Commission and serves on the board of editors of the *American Journal of International Law*.

Meinhard Doelle is a faculty member at Dalhousie Law School, specializing in environmental law, and currently serves as the associate director of the Marine and Environmental Law Institute. From 1996 to 2001, he was the Executive Director of Clean Nova Scotia, with a mandate to pursue an environmentally healthy and sustainable society in Nova Scotia. He is Environmental Counsel to the Atlantic Canada law firm of Stewart McKelvey, where he has been involved in prosecutions, issues of common law liability, environmental assessments, environmental audits, site assessments, and other environmental law matters. He was the principle legislative drafter of the draft Nova Scotia Environment Act (1995). Doelle has written on a variety of environmental law topics, including climate change, invasive species, environmental assessments, and public participation in environmental decision making. His most recent book is entitled From *Hot Air to Action: Climate Change, Compliance and the Future of International Environmental Law* (Thomson Carswell, 2005).

John Drexhage is director of the Climate Change and Energy Program of the International Institute for Sustainable Development (IISD). His work on climate change is based on 12 years of experience on the issue, first as a domestic advisor and international negotiator on climate change and then as an expert analyst and manager for IISD. His expertise covers a broad range of areas related to climate change, and he is currently focusing on regulatory frameworks for greenhouse gas emissions, post-2012 climate change regimes, market based instruments and more fully exploring linkages between adaptation, mitigation and sustainable development. Drexhage is also a lead author with Working Group 3 of the Intergovernmental Panel on Climate Change. He is involved in energy policy and analysis and modeling as the leader of the climate and energy group at IISD, which provides regular consulting services to energy companies and has also conducted extensive analysis for government agencies at the federal level (Environment

Canada, Foreign Affairs Canada and Natural Resources Canada) and provincial level (Manitoba Government, Alberta Government and Ontario Government).

David M. Driesen is the Angela S. Cooney Professor, Syracuse University College of Law, where he teaches environmental law (domestic and international) and constitutional law. His has written extensively about emissions trading, especially in the climate change context and is a member of the editorial board of the *Carbon and Climate Change Law Review* (Lexxion, Berlin). His writing includes *Environmental Law: A Conceptual and Pragmatic Approach* (Aspen, 2007) (with Robert Adler); *The Economic Dynamics of Environmental Law* (MIT Press, 2003), as well as numerous articles.

David G. Duff is an associate professor at the University of Toronto Faculty of Law, which he joined in 1996. Prior to joining the Faculty, he was a tax associate at the Toronto office of Stikeman, Elliott. He was also employed as a researcher with the Ontario Fair Tax Commission from 1991 to 1993 and as a tax policy analyst with the Ontario Ministry of Finance in 1993–1994. He has been a visiting scholar at the Faculty of Law of Oxford University, at the University of Sydney Law Faculty, and at the Faculty of Law at McGill University. His teaching and research interests are in the areas of tax law, tax policy, environmental taxation, comparative and international taxation, statutory interpretation, and distributive justice. He has published numerous articles in the areas of taxation, torts and family law, and a textbook/casebook on *Canadian Income Tax Law*.

Andrew J. Green is an assistant professor at the Faculty of Law and the School of Public Policy and Governance, University of Toronto. His research interests focus on environmental law; international trade and administrative law, including how international trade rules constrain countries' ability to implement domestic environmental policy; instrument choice in environmental law, including instruments for fostering renewable energy; and the role of law, including administrative law, in fostering individuals' environmental values. Prior to joining the Faculty, he practiced environmental law in Toronto for six years.

Kathryn Harrison is a professor of Political Science at the University of British Columbia. Before joining the faculty of UBC, she worked as a policy analyst for both Environment Canada and the United States Congress. She is the author of *Passing the Buck: Federalism and Canadian Environmental Policy* (UBC Press, 1996), coauthor of *Risk, Science, and Politics: Regulation of Toxic Substances in Canada and the United States* (McGill-Queen's University Press, 1994), co-editor of *Managing the Environmental Union* (Queen's University School of Policy Studies, 2000), and editor of *Racing to the Bottom? Provincial Interdependence in the Canadian Federation* (UBC Press, 2000). She has published recent articles in *Global Environmental Politics*, the *Canadian Journal of Political Science*, the *Canadian Journal of Economics*, and the *Journal of Policy Analysis and Management*. Her current research focuses on environmental regulation in the context of economic globalization, the efficacy of alternative policy instruments, and comparative climate change policy.

Matthew J. Hoffmann is assistant professor of Political Science at the University of Toronto. His research interests include global environmental governance, the politics of climate change and international relations theory. His recent book *Ozone Depletion and Climate Change: Constructing a Global Response* (SUNY Press, 2005) examined how evolving international norms influence the governance of ozone depletion and climate change.

Thomas Homer-Dixon holds the George Ignatieff Chair of Peace and Conflict Studies at the Trudeau Centre for Peace and Conflict Studies at University College, University of Toronto, where he leads several research projects studying the links between environmental stress and violence in developing countries. Recently, his research has focused on threats to global security in the 21st century and on how societies adapt to complex economic, ecological, and technological change. His books include *The Upside of Down: Catastrophe, Creativity, and the Renewal of Civilization* (Knopf, Island Press, 2006), which won the 2006 National Business Book Award; *The Ingenuity Gap* (Knopf, 2000), which won the 2001 Governor General's Non-fiction Award; and *Environment, Scarcity, and Violence* (Princeton University Press, 1999), which won the Caldwell Prize of the American Political Science Association.

David B. Hunter is assistant professor of Law and director of the Environmental Law Program at American University's Washington College of Law. He teaches U.S. Environmental Law, Comparative Law and the Law of Torts. He is the former executive director of the Center for International Environmental Law, a non-governmental organization dedicated to protecting the global environment through the use of international law. He is also the president of Peregrine Environmental Consulting, and currently serves on the board of directors of the Environmental Law Alliance World Wide–US (chair), EarthRights International, the Project on Government Oversight (chair), and Greenpeace USA. Hunter is author of many articles on international environmental law, and is co-author of the leading textbook in the field: *International Environmental Law and Policy* (Foundation Press, 2001).

Kelly Levin is a doctoral candidate at Yale University's School of Forestry and Environmental Studies. The goal of her research is to enhance the policy response to the problem of climate impacts to biodiversity. She is examining the current disconnect between scientific knowledge and adaptation policies in an effort to recommend prescriptions to advance biodiversity conservation in a changing climate. In addition to her PhD work, Levin is lead researcher for an upcoming Discovery Channel series on climate change; works with the World Resources Institute to perform an annual review of the major climate change science literature; and is a writer for the Earth Negotiations Bulletin. She holds a BA in Ecology and Evolutionary Biology from Yale College and Master of Environmental Management from Yale's School of Forestry and Environmental Studies.

Bradley C. Parks is a development policy officer in the Department of Policy and International Relations at the Millennium Challenge Corporation in Washington, DC, and senior researcher at the Center for International Policy Research at the College of William and Mary.

Lavanya Rajamani is an associate professor at the Centre for Policy Research in New Delhi. She is an international lawyer specializing in environment law and policy. She was previously University lecturer in Environmental Law and Fellow and director of studies in Law at Queens' College, Cambridge. She is the author of *Differential Treatment in International Environmental Law* (Oxford University Press, 2006) and numerous articles. In her current research she is exploring ways of further integrating developing countries into international environmental regimes, in particular the climate change regime, and studying national laws and policies in select developing countries (Brazil, China and India) implementing international climate change law. She is also writing a book provisionally titled *International Environmental Law in*

Indian Courts: the Vanishing Line between Rhetoric and Law. She has been invited to serve as director of studies for the 2008 research session on *Implementation of International Environmental Law* at the Hague Academy of International Law. She works as a consultant to the UN Framework Convention on Climate Change Secretariat, is associated with the Yale Centre for Environmental Law and Policy, and serves on the editorial board of the *Review of European Community and International Environmental Law*.

Ian H. Rowlands is an associate professor in Department of Environment and Resource Studies at the University of Waterloo. He is also the associate dean (Research) in the University's Faculty of Environmental Studies. Rowlands has research and teaching interests in the areas of energy management strategies and policy, corporate environmentalism and international environmental relations. He was one of the Canadian representatives to the International Energy Agency – Demand Side Management Program's Task XIII on Demand Response. He is also leading an Ontario Centre for Energy project with Milton Hydro on conservation and demand response strategies. Before joining the faculty of the University of Waterloo in 1998, he was a researcher at the United Nations Collaborating Centre on Energy and Environment in Denmark (1996–97) and a lecturer in International Relations and Development Studies at the London School of Economics and Political Science (1991–96).

J. Timmons Roberts is professor of Sociology and interim director of Environmental Science and Policy at the College of William and Mary. He is co-author of over 40 articles and book chapters, and five books: *From Modernization to Globalization: Perspectives on Social Change and Development* (Blackwell, 2000, with Amy Hite); *Chronicles from the Environmental Justice Frontline* (Cambridge, 2001, with Melissa Toffolon-Weiss); and *Trouble in Paradise: Globalization and Environmental Crises in Latin America* (Routledge, 2003, with Nikki Thanos); *The Globalization and Development Reader* (Blackwell, 2007, with Amy Hite), and *A Climate of Injustice: Global Inequality, North-South Politics, and Climate Policy* (MIT Press, 2007, with Bradley Parks. Forthcoming is *Greening Aid: Understanding Environmental Foreign Assistance to Developing Countries*, with an interdisciplinary research team from William and Mary (Oxford University Press, 2008).

Mark S. Winfield is an assistant professor with York University's Faculty of Environmental Studies. Prior to joining the FES Winfield was program and policy director with the Pembina Institute, and director of research with the Canadian Institute for Environmental Law and Policy. Winfield has published reports, book chapters and papers on a wide range of environmental policy issues.